Soil Mechanics

Soil Mechanics
Calculations, Principles, and Methods

Victor N. Kaliakin
University of Delaware, Newark, DE

Butterworth-Heinemann
An imprint of Elsevier
elsevier.com

Butterworth-Heinemann is an imprint of Elsevier
The Boulevard, Langford Lane, Kidlington, Oxford OX5 1GB, United Kingdom
50 Hampshire Street, 5th Floor, Cambridge, MA 02139, United States

Library of Congress Cataloging-in-Publication Data
A catalog record for this book is available from the Library of Congress

British Library Cataloguing-in-Publication Data
A catalogue record for this book is available from the British Library

ISBN: 978-0-12-804491-9

For information on all Butterworth-Heinemann publications
visit our website at https://www.elsevier.com/books-and-journals

Working together
to grow libraries in
developing countries

www.elsevier.com • www.bookaid.org

Publisher: Joe Hayton
Acquisition Editor: Ken McCombs
Editorial Project Manager: Peter Jardim
Production Project Manager: Mohanapriyan Rajendran
Designer: Maria Ines Cruz

Typeset by TNQ Books and Journals

In Memory of Professors

Kandiah ("Arul") Arulanandan
and
Edward A. Nowatzki

Contents

Preface

The first course in soil mechanics typically proves to be challenging for undergraduate students. This is due to the fact that soils are three-phase particulate materials, and thus must be treated differently than other engineering materials that undergraduates are introduced to as part of their curriculum. The situation is further complicated by the need to account for the presence of pore fluid, both under hydrostatic and transient conditions, as well as the subject of shear strength.

One of the biggest difficulties in teaching soil mechanics is the lack of lecture time in which to present a sufficient number of example problems, with varying degrees of difficulty, that illustrate the concepts associated with the subject. This book has been written to address the aforementioned shortcoming. It presents worked example problems that will facilitate a student's understanding of topics presented in lecture. This book is not meant to replace existing soil mechanics textbooks but to serve as a supplementary resource.

Victor N. Kaliakin

Acknowledgments

Professor Namunu (Jay) Meegoda from the New Jersey Institute of Technology first suggested the idea for the present book and encouraged me to undertake the task of writing it. The example problems presented in the book have been developed over several years of teaching soil mechanics. Some of the more challenging problems are patterned after similar ones that were provided by my former University of Delaware colleague, Dr. Dov Leshchinsky. Finally, special thanks goes to my current colleague, Dr. Kalehiwot Nega Manahiloh for critically reviewing select chapters of the book and for providing some ideas for example problems.

Cheers,
Victor N. Kaliakin

Chapter 1

Example Problems Involving Phase Relations for Soils

1.0 GENERAL COMMENTS

Soils are prime examples of complex engineering materials, whereas in elementary physics, solid, liquid, and gaseous states are distinguished. Soils are not simple bodies that can be placed in one of these three groups. Soils are generally composed of *solid*, *liquid*, and *gas*, with the solid part being a porous medium made up of numerous particles. Soils are thus *particulate* materials.

The behavior of soils is largely determined by the *relative* amounts of the aforementioned constituents. To quantify these relative amounts requires knowledge of the "mass–volume" or "weight–volume" relations. These relations quantify a soil's *aggregate properties*.

1.1 GENERAL DEFINITIONS

The volume of the various constituents of a soil is quantified by following quantities:

V = total volume of a soil. In some books V_t denotes the total volume.
V_v = volume of the voids (pores).
V_s = volume of the solid phase.
V_a = volume of the gas in the voids.
V_w = volume of the liquid in the voids.

Thus, for all soils

$$V = V_v + V_s = (V_a + V_w) + V_s \qquad (1.1)$$

The mass of the various constituents of a soil is quantified by following quantities:

M = total mass of a soil. In some books M_t denotes the total mass.
M_a = mass of the gas in the voids (pores) = 0.
M_w = mass of the liquid in the voids.
M_s = mass of the solid phase.

Soil Mechanics. http://dx.doi.org/10.1016/B978-0-12-804491-9.00001-X

1

The weight of the various constituents of a soil is quantified by following quantities:

W = total weight of a soil. In some books W_t denotes the total weight.
W_a = weight of the gas in the voids (pores) = 0.
W_w = weight of the liquid in the voids.
W_s = weight of the solid phase.

Thus, for all soils

$$W = W_w + W_s \tag{1.2}$$

Remark: If $V_v = V_w (\Rightarrow V_a = 0)$ and $W_w \neq 0$, the soil is said to be *saturated*; otherwise it is *unsaturated*.

A very convenient, although somewhat idealized, way in which to visualize the mass–volume and weight–volume relations is through the use of phase diagrams. A phase diagram depicts the three phases of a soil as being segregated. For example, Figure 1.1 shows a phase diagram that relates the volume and mass of the three phases.

Figure 1.2 shows a similar phase diagram that relates the volume and weight of the three phases.

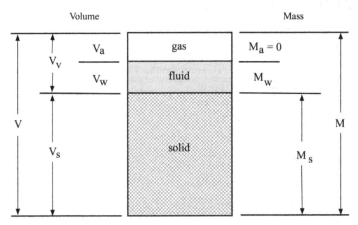

FIGURE 1.1 Phase diagram showing the relationship between volume and mass of gas, fluid, and solid phases in a soil.

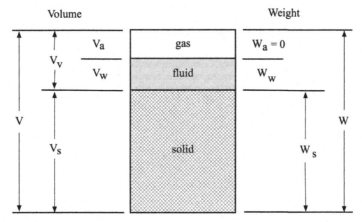

FIGURE 1.2   Phase diagram showing the relationship between volume and weight of gas, fluid, and solid phases in a soil.

1.2 MASS DENSITIES

The following mass densities are used to quantify the relative amounts of a soil's constituents:

- Soil (moist) mass density:

$$\rho = \frac{M}{V} \tag{1.3}$$

- Solid mass density:

$$\rho_s = \frac{M_s}{V_s} \tag{1.4}$$

- Dry mass density:

$$\rho_d = \frac{M_s}{V} \tag{1.5}$$

- Mass density of water:

$$\rho_w = \frac{M_w}{V_w} \tag{1.6}$$

At 4°C, $\rho_w = \rho_0 = 1000 \text{ kg/m}^3 = 1 \text{ g/cm}^3 = 1 \text{ Mg/m}^3 = 1.941 \text{ slug/ft}^3$. For ordinary engineering applications at other temperatures, $\rho_w \cong \rho_0$.

1.3 UNIT WEIGHTS

The following unit weights are used to quantify the relative amounts of a soil's constituents:

- Soil (moist) unit weight:

$$\gamma = \frac{W}{V} = \frac{Mg}{V} = \rho g \tag{1.7}$$

- Solid unit weight:

$$\gamma_s = \frac{W_s}{V_s} = \frac{M_s g}{V_s} = \rho_s g \tag{1.8}$$

- Dry unit weight:

$$\gamma_d = \frac{W_s}{V} = \frac{M_s g}{V} = \rho_d g \tag{1.9}$$

- Unit weight of water:

$$\gamma_w = \frac{W_w}{V_w} \tag{1.10}$$

At 4°C, $\gamma_w = \gamma_0 = 9810 \text{ N/m}^3 = 9.81 \text{ kN/m}^3 = 62.4 \text{ lb/ft}^3$. For ordinary engineering applications at other temperatures, $\gamma_w \approx \gamma_0$. In the equations, $g = 9.81 \text{ m/s}^2 = 32.2 \text{ ft/s}^2$ is the gravitational acceleration.

1.4 DEFINITION OF FUNDAMENTAL QUANTITIES

The specific gravity of solids is defined as follows:

$$G_s = \frac{\gamma_s}{\gamma_0} \approx \frac{\gamma_s}{\gamma_w} = \frac{W_s}{V_s \gamma_w} \tag{1.11}$$

Remark: G_s normalizes the solid unit weight of a material.

The volume of voids is defined by two quantities, namely the porosity n, and the void ratio e, where,

$$n = \left(\frac{V_v}{V}\right) * 100\% \tag{1.12}$$

and

$$e = \frac{V_v}{V_s} \tag{1.13}$$

The *relative* weight and volume of the pore fluid is quantified by the moisture content (w) and the degree of saturation (S), where,

$$w = \left(\frac{W_w}{W_s}\right) * 100\% \tag{1.14}$$

and

$$S = \left(\frac{V_w}{V_v}\right) * 100\% \tag{1.15}$$

For a *saturated* soil, $V_w = V_v$ and $S = 100\%$.

1.5 RELATIONS DERIVED FROM FUNDAMENTAL QUANTITIES

The basic quantities G_s, n, e, w, and S can be suitably combined to form relations that are particularly useful for particular types of problems. These relations *do not*, however, constitute any new definitions of quantities used to describe the phase relations for soils. Some specific examples of such relations are given in the following section.

1.5.1 Case 1.1: Relation Between Void Ratio and Porosity

Rewriting the void ratio definition in terms of the volume of voids (V_v) and then dividing through by the total volume (V) gives the following relation:

$$e = \frac{V_v}{V_s} = \frac{V_v}{V - V_v} = \frac{V_v/V}{1 - (V_v/V)} = \frac{n}{1 - n} \tag{1.16}$$

where n is understood to be a decimal number.

1.5.2 Case 1.2: Relation Between Porosity and Void Ratio

Rewriting the porosity definition by expanding the total volume (V) and then dividing through by the volume of solids (V_s) gives the following relation:

$$n = \left(\frac{V_v}{V}\right) * 100\% = \left(\frac{V_v}{V_s + V_v}\right) * 100\% = \left(\frac{V_v/V_s}{1 + V_v/V_s}\right) * 100\%$$
$$= \left(\frac{e}{1 + e}\right) * 100\% \tag{1.17}$$

This result could likewise have been obtained by solving the equation derived in Case 1.1 for porosity in terms of void ratio.

1.5.3 Case 1.3: Relation Between Moisture Content, Specific Gravity of Solids, Void Ratio, and Degree of Saturation

The weight of the solid phase is written in terms of G_s as follows:

$$G_s = \frac{\gamma_s}{\gamma_w} = \frac{W_s}{V_s \gamma_w} \Rightarrow W_s = G_s V_s \gamma_w \qquad (1.18)$$

Next, the weight of the pore fluid is written in terms of γ_w.

$$\gamma_w = \frac{W_w}{V_w} \Rightarrow W_w = V_w \gamma_w \qquad (1.19)$$

Substituting Eqs. (1.18) and (1.19) into the definition of the moisture content (Eq. 1.14) gives,

$$w = \left(\frac{W_w}{W_s}\right) * 100\% = \left(\frac{V_w \gamma_w}{G_s V_s \gamma_w}\right) * 100\% = \left(\frac{V_w}{G_s V_s}\right) * 100\% \qquad (1.20)$$

The volume of pore fluid is next written in terms of the degree of saturation; i.e.,

$$S = \left(\frac{V_w}{V_v}\right) * 100\% \Rightarrow V_w = \left(\frac{S}{100\%}\right) V_v = \left(\frac{S}{100\%}\right) e V_s \qquad (1.21)$$

where the definition of the void ratio has been used. Substituting Eq. (1.21) into Eq. (1.20) gives the desired relation; i.e.,

$$w = \frac{Se}{G_s} \quad \text{or} \quad Se = G_s w \qquad (1.22)$$

where w and S are understood to be decimal numbers.

The aforementioned expression shows that the moisture content (w) is thus a function of *three* quantities, namely e, S, and G_s. The upper bound on w corresponds to the case of full saturation (i.e., $S = 100\%$), when $w \equiv w_{sat} = e/G_s$. The lower bound on w is zero, which corresponds to a completely dry soil for which $S = 0\%$.

1.5.4 Case 1.4: Relation Between Dry Unit Weight, Specific Gravity of Solids, and Void Ratio

Beginning with the definition of the dry unit weight given by Eq. (1.9), substituting for W_s in terms of G_s gives,

$$\gamma_d = \frac{W_s}{V} = \frac{G_s V_s \gamma_w}{V_s + V_v} \qquad (1.23)$$

Dividing through the resulting expression by V_s gives the desired relation; i.e.,

$$\gamma_d = \frac{G_s \gamma_w}{1 + e} \qquad (1.24)$$

1.5.5 Case 1.5: Relation Between Moist Unit Weight, Specific Gravity of Solids, Moisture Content, and Void Ratio

Beginning with the definition of the moist unit weight given by Eq. (1.7), and representing the weight of the pore fluid in terms of the w and W_s gives

$$\gamma = \frac{W}{V} = \frac{W_s + W_w}{V_s + V_v} = \frac{W_s(1+w)}{V_s + V_v} \tag{1.25}$$

where w is understood to be a decimal number. Substituting for W_s in terms of G_s (i.e., $W_s = G_s V_s \gamma_w$) and diving through the resulting expression by V_s gives the desired relation; i.e.,

$$\gamma = \frac{G_s \gamma_w (1+w)}{1+e} \tag{1.26}$$

1.5.6 Case 1.6: Relation Between Moist Unit Weight, Dry Unit Weight, and Moisture Content

In light of Eq. (1.24), the relation for γ derived in of Case 1.5 becomes

$$\gamma = \gamma_d(1+w) \quad \text{or} \quad \gamma_d = \frac{\gamma}{(1+w)} \tag{1.27}$$

where w is understood to be a decimal number.

1.5.7 Case 1.7: Relation Between Moist Unit Weight, Specific Gravity of Solids, Degree of Saturation, and Void Ratio

Replacing the moisture content in Eq. (1.26) with the relation derived in Case 1.3 (i.e., $w = Se/G_s$) gives

$$\gamma = \frac{\gamma_w(G_s + Se)}{1+e} \tag{1.28}$$

1.5.8 Case 1.8: Unit Weight of Submerged Soil and Its Relation to Moist Unit Weight

Consider a saturated soil that is submerged in water. According to Archimedes' principle, the buoyancy force acting on a body is equal to the weight of the fluid displaced by the body.

Since the soil is saturated, $S = 100\%$ and $V_w = V_v$. The buoyant unit weight is thus

$$\gamma_b = \frac{(W_s - V_s\gamma_w) + (W_w - V_v\gamma_w)}{V_s + V_v} \tag{1.29}$$

Writing W_s in terms of G_s and W_w in terms γ_w gives

$$\gamma_b = \frac{(G_s V_s \gamma_w - V_s \gamma_w) + (V_v \gamma_w - V_v \gamma_w)}{V_s + V_v} = \frac{\gamma_w V_s (G_s - 1)}{V_s + V_v} \qquad (1.30)$$

Dividing through the equation by V_s gives the final expression for the buoyant unit weight; i.e.,

$$\gamma_b = \frac{\gamma_w(G_s - 1)}{1 + e} \qquad (1.31)$$

For a saturated soil the expression for moist unit weight given by Eq. (1.28) reduces to

$$\gamma = \gamma_{sat} = \frac{\gamma_w(G_s + e)}{1 + e} \qquad (1.32)$$

Manipulating this expression gives the relationship between the saturated and buoyant unit weights; i.e.,

$$\gamma_{sat} = \frac{\gamma_w(G_s + e)}{1 + e} = \frac{\gamma_w(G_s - 1)}{1 + e} + \frac{\gamma_w(1 + e)}{1 + e} = \gamma_b + \gamma_w \qquad (1.33)$$

or

$$\gamma_b = \gamma_{sat} - \gamma_w \qquad (1.34)$$

EXAMPLE PROBLEM 1.1

General Remarks

Knowing the definitions of the basic quantities e, n, w, S, and G_s, it is relatively straightforward to derive more specific relations than those presented in Cases 1.1–1.8.

Problem Statement

Derive an expression for void ratio (e) in terms of the total weight (W), total volume (V), the unit weight of water (γ_w), the degree of saturation (S), and the specific gravity of solids (G_s).

Solution

Recall the relation for moist unit weight derived in Case 1.7 (Eq. 1.28); i.e.,

$$\gamma = \frac{W}{V} = \frac{\gamma_w(G_s + Se)}{1 + e} \qquad (1.1.1)$$

Solving for the void ratio leads to the following results:

$$e + 1 = \left(\frac{V\gamma_w}{W}\right)(G_s + Se) \Rightarrow e\left(1 - \frac{V}{W}\gamma_w S\right) = \frac{V}{W}\gamma_w G_s - 1 \qquad (1.1.2)$$

Multiplying both sides of the equation by W and solving for the void ratio gives the desired relation

$$e = \frac{V\gamma_w G_s - W}{W - V\gamma_w S} \tag{1.1.3}$$

EXAMPLE PROBLEM 1.2

General Remarks

Knowing the definitions of the basic quantities e, n, w, S, and G_s, it is relatively straightforward to derive more specific relations than those presented in Cases 1.1−1.8. In this problem alternate expressions for the degree of saturation are derived.

Problem Statement

a) Derive an expression for the degree of saturation (S) in terms of the moisture content (w), the specific gravity of solids (G_s), the moist unit weight (γ), and γ_w.

b) Derive an expression for S in terms of w, G_s, and porosity (n).

Solution

a) Beginning with the definition of the specific gravity of solids

$$G_s = \frac{W_s}{V_s \gamma_w} \Rightarrow W_s = G_s V_s \gamma_w \tag{1.2.1}$$

From the definition of the moisture content (Eq. 1.14),

$$w = \left(\frac{W_w}{W_s}\right) \Rightarrow W_w = w W_s \tag{1.2.2}$$

Substituting Eq. (1.2.1) into Eq. (1.2.2) gives

$$W_w = w(G_s V_s \gamma_w) \tag{1.2.3}$$

From the definition of the unit weight of water,

$$\gamma_w = \frac{W_w}{V_w} \Rightarrow V_w = \frac{W_w}{\gamma_w} \tag{1.2.4}$$

Substituting Eq. (1.2.3) for W_w into Eq. (1.2.4) gives

$$V_w = \frac{w(G_s V_s \gamma_w)}{\gamma_w} = w G_s V_s \tag{1.2.5}$$

From the definition of the moist unit weight,

$$\gamma = \frac{W_w + W_s}{V_v + V_s} \Rightarrow V_v = \frac{W_w + W_s}{\gamma} - V_s \qquad (1.2.6)$$

Substituting Eqs. (1.2.1) and (1.2.3) for W_s and W_w, respectively, gives

$$V_v = \frac{w(G_s V_s \gamma_w) + G_s V_s \gamma_w}{\gamma} - V_s = G_s V_s \frac{\gamma_w}{\gamma}(1 + w) - V_s \qquad (1.2.7)$$

Finally, recalling the definition of the degree of saturation and substituting Eqs. (1.2.5) and (1.2.7) gives the desired expression; i.e.,

$$S = \left(\frac{V_w}{V_v}\right) = \frac{w G_s V_s}{G_s V_s \frac{\gamma_w}{\gamma}(1 + w) - V_s} = \frac{w G_s}{G_s \frac{\gamma_w}{\gamma}(1 + w) - 1}$$

or

$$S = \frac{w}{\frac{\gamma_w}{\gamma}(1 + w) - \frac{1}{G_s}} \qquad (1.2.8)$$

b) Returning to the definition of the degree of saturation; i.e., $S = V_w/V_v$, the definition of the unit weight of water is used to give

$$S = \frac{W_w/\gamma_w}{V_v} = \frac{w W_s}{\gamma_w V_v} \qquad (1.2.9)$$

where the definition of the moisture content has been used to rewrite W_w in terms of W_s. Next the definition of the specific gravity of solids is used to replace weight of the solid phase, and the resulting expression is divided through by the total volume V, giving

$$S = \frac{w(G_s V_s \gamma_w)}{\gamma_w V_v} = \frac{w G_s(V - V_v)}{V_v} = \frac{w G_s(1 - V_v/V)}{V_v/V}$$

or

$$S = \frac{w G_s(1 - n)}{n} \qquad (1.2.10)$$

where w, S, and the porosity (n) are written as decimal numbers.

EXAMPLE PROBLEM 1.3

General Remarks

In this problem an alternate expression for the moist unit weight is derived.

Problem Statement

Derive an expression the moist unit weight (γ) in terms of the dry unit weight (γ_d), the degree of saturation (S), the porosity (n), and γ_w.

Solution

Beginning with the expression derived in Case 1.6 (Eq. 1.27) and using the definition of the moisture content and dry unit weight gives

$$\gamma = \gamma_d(1+w) = \gamma_d + \gamma_d\left(\frac{W_w}{W_s}\right) = \gamma_d + \left(\frac{W_s}{V}\right)\left(\frac{\gamma_w V_w}{W_s}\right) = \gamma_d + \frac{\gamma_w V_w}{V}$$

$$(1.3.1)$$

From the definition of the degree of saturation

$$S = \frac{V_w}{V_v} \Rightarrow V_w = SV_v$$

$$(1.3.2)$$

Substituting Eq. (1.3.2) into Eq. (1.3.1) gives

$$\gamma = \gamma_d + \frac{\gamma_w SV_v}{V}$$

$$(1.3.3)$$

Recalling the definition of the porosity ($n = V_v/V$) and substituting it into Eq. (1.3.3) gives the final expression; i.e.,

$$\gamma = \gamma_d + Sn\gamma_w$$

$$(1.3.4)$$

If the soil is *saturated*, $S = 100\%$. Eq. (1.3.4) thus reduces to

$$\gamma \equiv \gamma_{sat} = \gamma_d + n\gamma_w$$

$$(1.3.5)$$

EXAMPLE PROBLEM 1.4

General Remarks

In this problem an alternate expression for the buoyant unit weight is derived.

Problem Statement

Derive an expression the buoyant unit weight (γ_b) in terms of the dry unit weight (γ_d), the porosity (n), and the unit weight of water (γ_w).

Solution

The relation between the buoyant and saturated unit weights was derived in Case 1.8 (Eq. 1.34); i.e.,

$$\gamma_b = \gamma_{sat} - \gamma_w$$

$$(1.4.1)$$

Beginning with the basic definition of the saturated unit weight, and using the definition of the unit weight of water gives

$$\gamma_{sat} = \frac{W_w + W_s}{V} = \frac{W_w}{V} + \gamma_d = \frac{\gamma_w V_w}{V} + \gamma_d \qquad (1.4.2)$$

But for a saturated soil, $V_w = V_v$. Thus, Eq. (1.4.2) becomes

$$\gamma_{sat} = \frac{\gamma_w V_v}{V} + \gamma_d = n\gamma_w + \gamma_d \qquad (1.4.3)$$

Using this expression for the saturated unit weight, Eq. (1.4.1) is written as follows:

$$\gamma_b = \gamma_{sat} - \gamma_w = (n\gamma_w + \gamma_d) - \gamma_w$$

or

$$\gamma_b = \gamma_d + \gamma_w(n - 1) \qquad (1.4.4)$$

EXAMPLE PROBLEM 1.5

General Remarks

In this problem an expression for the moisture content associated with a saturated soil is derived.

Problem Statement

Derive an expression for the moisture content (w_{sat}) associated with a saturated soil in terms of the saturated unit weight (γ_{sat}), the porosity (n), and the unit weight of water (γ_w).

Solution

Since the desired expression is to involve the saturated unit weight, begin with the expression derived in Case 1.8; i.e.,

$$\gamma_{sat} = \frac{\gamma_w(G_s + e)}{1 + e} \qquad (1.5.1)$$

Specializing the relation derived in Case 1.3 for a saturated soil ($S = 100\%$; $w = w_{sat}$) gives

$$G_s = \frac{e}{w_{sat}} \qquad (1.5.2)$$

Substituting Eq. (1.5.2) for G_s, the saturated unit weight becomes

$$\gamma_{sat} = \frac{\gamma_w\left(\dfrac{e}{w_{sat}} + e\right)}{1 + e} = \frac{e}{1 + e}\left(\frac{1}{w_{sat}} + 1\right)\gamma_w \qquad (1.5.3)$$

In Case 1.1 the void ratio and porosity were related in the following manner:

$$e = \frac{n}{1-n} \tag{1.5.4}$$

From Eq. (1.5.4) it follows that

$$1 + e = 1 + \frac{n}{1-n} = \frac{(1-n)+n}{1-n} = \frac{1}{1-n} \tag{1.5.5}$$

Combining Eqs. (1.5.4) and (1.5.5) gives

$$\frac{e}{1+e} = \frac{\frac{n}{1-n}}{\frac{1}{1-n}} = n \tag{1.5.6}$$

Substituting Eq. (1.5.6) into Eq. (1.5.3) gives

$$\gamma_{sat} = \frac{e}{1+e}\left(\frac{1}{w_{sat}}+1\right)\gamma_w = n\left(\frac{1}{w_{sat}}+1\right)\gamma_w \Rightarrow \frac{1}{w_{sat}} = \left(\frac{\gamma_{sat}}{n\gamma_w}-1\right) \tag{1.5.7}$$

Inverting this result gives the desired expression for w_{sat}

$$w_{sat} = \frac{n\gamma_w}{\gamma_{sat}-n\gamma_w} \tag{1.5.8}$$

If the relations derived in Cases 1.1, 1.3, and 1.8 are not readily available at the time that the calculations are performed, the aforementioned result can always be determined from the fundamental quantities defined in Section 1.4. For example, begin with the definition of the saturated unit weight

$$\gamma_{sat} = \frac{W_w + W_s}{V_w + V_s} = \frac{W_w + W_s}{V_v + V_s} = \frac{W_w + G_s V_s \gamma_w}{V_v + V_s} \tag{1.5.9}$$

where the definition of the specific gravity of solids has been used and $V_v = V_w$ because the soil is saturated.

From the definition of the moisture content, and noting again that $V_v = V_w$, gives

$$w_{sat} = \frac{W_w}{W_s} = \frac{\gamma_w V_w}{G_s V_s \gamma_w} = \frac{V_v}{G_s V_s} = \frac{e}{G_s} \Rightarrow G_s = \frac{e}{w_{sat}} \tag{1.5.10}$$

Substituting Eq. (1.5.10) into Eq. (1.5.9) gives

$$\gamma_{sat} = \frac{W_w + G_s V_s \gamma_w}{V_v + V_s} = \frac{\gamma_w V_v + \left(\dfrac{e}{w_{sat}}\right)V_s \gamma_w}{V_v + V_s} = \frac{\gamma_w\left(V_v/V_s + \dfrac{e}{w_{sat}}\right)}{V_v/V_s + 1}$$

$$= \frac{\gamma_w e\left(1 + \dfrac{1}{w_{sat}}\right)}{1+e} \tag{1.5.11}$$

where the definition of the unit weight of water has been used. Since this result is *identical* to Eq. (1.5.7), it once again leads to Eq. (1.5.8).

EXAMPLE PROBLEM 1.6

General Remarks

This example problem illustrates the manner in which the quantities used to describe the phase relations are computed; it involves both densities and unit weights.

Problem Statement

A sample of gray silty clay has a mass of 126 kg. Laboratory tests results give a moist density (ρ) of 2.05 g/cm^3, a specific gravity of solids (G_s) of 2.71, and a moisture content (w) of 15.7%.

First determine all entries in the phase diagram. Then determine the void ratio (e), the porosity (n), the degree of saturation (S), the dry density (ρ_d), the dry unit weight (γ_d), and the moist unit weight (γ).

Solution

Since the moist density ($\rho = 2.05$ g/cm$^3 = 2050$ kg/m^3) and the total mass (M) are known, the total volume is thus

$$V = \frac{M}{\rho} = \frac{126.0 \text{ kg}}{2050 \text{ kg/m}^3} = 0.0615 \text{ m}^3 \qquad (1.6.1)$$

From the definition of the moisture content (Eq. 1.14),

$$w = \left(\frac{W_w}{W_s}\right) * 100\% = \left(\frac{M_w}{M_s}\right) * 100\% \Rightarrow M_w = wM_s \qquad (1.6.2)$$

where w is understood to be a decimal number. The total mass is thus written as $M = M_s + M_w = (1 + w)M_s$. Solving for the mass of the solids gives

$$M_s = \frac{M}{1+w} = \frac{126 \text{ kg}}{(1+0.157)} = 108.9 \text{ kg} \qquad (1.6.3)$$

Thus,

$$M_w = M - M_s = 126 - 108.90 = 17.10 \text{ kg} \qquad (1.6.4)$$

The volumes of the constituents are next computed. From the definition of the specific gravity of solids,

$$V_s = \frac{M_s}{G_s \rho_w} = \frac{108.90 \text{ kg}}{(2.71)(1000 \text{ kg/m}^3)} = 0.0402 \text{ m}^3 \qquad (1.6.5)$$

From the definition of the mass density of water,

$$V_w = \frac{M_w}{\rho_w} = \frac{17.10 \text{ kg}}{(1000 \text{ kg/m}^3)} = 0.0171 \text{ m}^3 \qquad (1.6.6)$$

The volume of the gaseous phase is then

$$V_a = V - V_s - V_w = 0.0615 - 0.0402 - 0.0171 = 0.0042 \text{ m}^3 \qquad (1.6.7)$$

Finally, the volume of the voids is computed as follows

$$V_v = V_a + V_w = 0.0402 + 0.0171 = 0.0213 \text{ m}^3 \qquad (1.6.8)$$

Figure Ex. 1.6 shows the phase diagram associated with this soil. The void ratio is next computed as follows

$$e = \frac{V_v}{V_s} = \frac{0.0213}{0.0402} = \mathbf{0.530} \qquad (1.6.9)$$

or, alternatively, as

$$e = \frac{V - V_s}{V_s} = \frac{V}{V_s} - 1 = \frac{0.0615}{0.0402} - 1 = \mathbf{0.530} \qquad (1.6.10)$$

The porosity follows from

$$n = \left(\frac{V_v}{V}\right) * 100\% = \left(\frac{V_a + V_w}{V}\right) * 100\%$$

$$\qquad (1.6.11)$$

$$= \left(\frac{0.0042 + 0.0171}{0.0615}\right) * 100\% = \mathbf{34.6\%}$$

or, alternatively (recall Case 1.2), from

$$n = \left(\frac{e}{1 + e}\right) * 100\% = \left(\frac{0.530}{1 + 0.530}\right) * 100\% = \mathbf{34.6\%} \qquad (1.6.12)$$

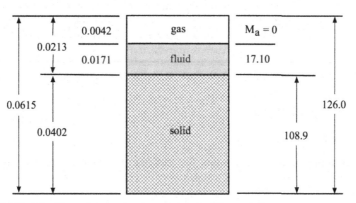

FIGURE EX. 1.6 Phase diagram showing relationship between volume and mass for the gray silty clay.

The degree of saturation is next computed as

$$S = \left(\frac{V_w}{V_v}\right) * 100\% = \left(\frac{0.0171}{0.0042 + 0.0171}\right) * 100\% = \mathbf{80.3\%} \qquad (1.6.13)$$

The dry density is thus

$$\rho_d = \frac{M_s}{V} = \frac{108.90 \text{ kg}}{0.0615 \text{ m}^3} = 1772 \text{ kg/m}^3 = \mathbf{1.772 \text{ g/cm}^3} = \mathbf{1.772 \text{ Mg/m}^3} \qquad (1.6.14)$$

The dry density can likewise be computed from the moisture content; i.e.,

$$\rho_d = \frac{G_s \rho_w}{1+e} = \frac{(2.71)(1000 \text{ kg/m}^3)}{1+0.530} = 1772 \text{ kg/m}^3 \qquad (1.6.15)$$
$$= \mathbf{1.772 \text{ g/cm}^3} = \mathbf{1.772 \text{ Mg/m}^3}$$

which verifies the previous result. The dry unit weight is thus

$$\gamma_d = \rho_d g = \left(1772 \text{ kg/m}^3\right)(9.81 \text{ m/s}) = 17,383 \text{ N/m}^3 = \mathbf{17.38 \text{ kN/m}^3} \qquad (1.6.16)$$

Finally, the moist unit weight can be computed directly from the dry density; i.e.,

$$\gamma = \gamma_d(1+w) = \left(17.38 \text{ kN/m}^3\right)(1+0.157) = \mathbf{20.11 \text{ kN/m}^3} \qquad (1.6.17)$$

The moist unit weight can likewise be computed as follows:

$$\gamma = \rho g = \left(2050 \text{ kg/m}^3\right)\left(9.81 \text{ m/s}^2\right) = 20,111 \text{ N/m}^3 = \mathbf{20.11 \text{ kN/m}^3} \qquad (1.6.18)$$

which verifies the previous result.

Remark: This problem has shown that the values for many of the quantities that are used to describe the phase relations can be computed in different ways. Consequently, this fact typically serves as a good check on the results obtained.

EXAMPLE PROBLEM 1.7

General Remarks

This example problem illustrates the manner in which quantities that are used to describe the phase relations are computed for a *cubical* soil specimen.

Problem Statement

A cubical sample of San Francisco Bay mud (a soft marine silty clay) has been prepared for testing in a true triaxial device. The dimensions of the sample are 76 mm by 76 mm by 76 mm. The sample has a moisture content (w) of 68.5% and a moist density (ρ) of 1.44 g/cm^3. The average specific gravity of solids (G_s) is 2.55.

Determine (a) the void ratio (e) and degree of saturation (S), (b) the moist density (γ) and moisture content (w) if the soil becomes saturated at the same total volume, and (c) the density (ρ) of the soil if all pore fluid is dried off with no change in total volume.

Solution

A general solution for the problem is first developed. Beginning with the definition of the moisture content (Eq. 1.14); i.e.,

$$w = \left(\frac{W_w}{W_s}\right) * 100\% = \left(\frac{M_w}{M_s}\right) * 100\% \tag{1.7.1}$$

the moist density is written as follows:

$$\rho = \frac{M}{V} = \frac{M_s + M_w}{V} = \frac{M_s(1+w)}{V} \tag{1.7.2}$$

Thus,

$$M_s = \frac{\rho V}{1+w} \tag{1.7.3}$$

and $M_w = wM_s$, where w is understood to be a decimal number. Since the void ratio is sought in this problem, the volumes of the respective phases are also required. As such, recalling the general definition for the specific gravity of solids, the volume of solids is computed from M_s as follows:

$$G_s = \frac{\gamma_s}{\gamma_w} = \frac{M_s}{V_s \rho_w} \Rightarrow V_s = \frac{M_s}{G_s \rho_w} \tag{1.7.4}$$

Since the volume of the voids is $V_v = V - V_s$, the void ratio is computed as follows:

$$e = \frac{V_v}{V_s} = \frac{V - V_s}{V_s} \tag{1.7.5}$$

The volume occupied by the pore fluid is obtained from the mass of the fluid (here assumed to be water) according to

$$\rho_w = \frac{M_w}{V_w} \Rightarrow V_w = \frac{M_w}{\rho_w} \tag{1.7.6}$$

Finally, if needed, the volume occupied by air is $V_a = V_v - V_w$. Attention is next focused on the present problem.

a) The total volume of the cubical sample is

$$V = (76 \text{ mm})^3 (\text{m}/1000 \text{ mm})^3 = 4.390 \times 10^{-4} \text{ m}^3 \qquad (1.7.7)$$

The moist density is converted to units of kilograms and meters

$$\rho = (1.44 \text{ g/cm}^3)(\text{kg}/1000 \text{ g})(100 \text{ cm/m})^3 = 1.440 \times 10^3 \text{ kg/m}^3 \quad (1.7.8)$$

The mass of the solid phase is thus

$$M_s = \frac{\rho V}{1+w} = \frac{(1.440 \times 10^3 \text{ kg/m}^3)(4.390 \times 10^{-4} \text{ m}^3)}{1+0.685} = 3.752 \times 10^{-1} \text{ kg}$$
$$(1.7.9)$$

The mass of the fluid phase is next computed as follows

$$M_w = wM_s = (0.685)(3.752 \times 10^{-1} \text{ kg}) = 2.570 \times 10^{-1} \text{ kg} \qquad (1.7.10)$$

For completeness, the total mass of the soil sample is found to be

$$M = M_s + M_w = (3.752 \times 10^{-1} \text{ kg}) + (2.570 \times 10^{-1} \text{ kg}) = 6.322 \times 10^{-1} \text{ kg}$$
$$(1.7.11)$$

The necessary volumes are next computed

$$V_s = \frac{M_s}{G_s \rho_w} = \frac{(3.752 \times 10^{-1} \text{ kg})}{(2.55)(1000 \text{ kg/m}^3)} = 1.471 \times 10^{-4} \text{ m}^3 \qquad (1.7.12)$$

$$V_w = \frac{M_w}{\rho_w} = \frac{2.570 \times 10^{-1} \text{ kg}}{1000 \text{ kg/m}^3} = 2.570 \times 10^{-4} \text{ m}^3 \qquad (1.7.13)$$

$$V_v = V - V_s = (4.390 \times 10^{-4} \text{ m}^3) - (1.471 \times 10^{-4} \text{ m}^3) = 2.919 \times 10^{-4} \text{ m}^3$$
$$(1.7.14)$$

$$V_a = V_v - V_w = (2.919 \times 10^{-4} \text{ m}^3) - (2.570 \times 10^{-4} \text{ m}^3) = 3.490 \times 10^{-5} \text{ m}^3$$
$$(1.7.15)$$

The void ratio is thus

$$e = \frac{V_v}{V_s} = \frac{2.919 \times 10^{-4} \text{ m}^3}{1.471 \times 10^{-4} \text{ m}^3} = \mathbf{1.984} \qquad (1.7.16)$$

Finally, the degree of saturation is computed as

$$S = \left(\frac{V_w}{V_v}\right) * 100\% = \left(\frac{2.570 \times 10^{-4} \text{ m}^3}{2.919 \times 10^{-4} \text{ m}^3}\right) * 100\% = \mathbf{88.0\%} \qquad (1.7.17)$$

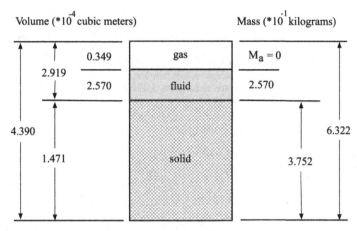

FIGURE EX. 1.7A   Phase diagram showing relationship between volume and mass for a sample of San Francisco Bay mud.

The degree of saturation could also have been computed from the relation developed in Case 1.3; i.e.,

$$S = \left(\frac{G_s w}{e}\right) * 100\% = \left[\frac{(2.55)(0.685)}{1.984}\right] * 100\% = \mathbf{88.0\%} \qquad (1.7.18)$$

which confirms the previous result. Figure Ex. 1.7a shows the phase diagram associated with this soil.

b) If the soil becomes saturated at the *same* total volume, then the volume occupied by the fluid phase is equal to the volume of the voids (Figure Ex. 1.7b).

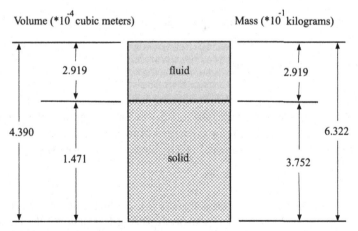

FIGURE EX. 1.7B   Phase diagram showing relationship between volume and mass for a saturated sample of San Francisco Bay mud.

Thus,

$$V_w = V_v = 2.919 \times 10^{-4} \text{ m}^3 \tag{1.7.19}$$

The mass of water contained in the voids is thus

$$M_w = \rho_w V_w = (1000 \text{ kg/m}^3)(2.919 \times 10^{-4} \text{ m}^3) = 2.919 \times 10^{-1} \text{ kg} \tag{1.7.20}$$

The resulting moisture content is

$$w = \left(\frac{M_w}{M_s}\right) * 100\% = \left(\frac{2.919 \times 10^{-1} \text{ kg}}{3.752 \times 10^{-1} \text{ kg}}\right) * 100\% = \mathbf{77.8\%} \tag{1.7.21}$$

The associated moist density is thus

$$\rho = \frac{M_s + M_w}{V} = \frac{(3.752 \times 10^{-1} \text{ kg}) + (2.919 \times 10^{-1} \text{ kg})}{4.390 \times 10^{-4} \text{ m}^3} \tag{1.7.22}$$
$$= \mathbf{1520 \text{ kg/m}^3 = 1.520 \text{ Mg/m}^3}$$

The moist density can likewise be computed from the moisture content; i.e.,

$$\rho = \frac{M_s(1 + w)}{V} = \frac{(3.752 \times 10^{-1} \text{ kg})(1 + 0.778)}{4.390 \times 10^{-4} \text{ m}^3} \tag{1.7.23}$$
$$= \mathbf{1520 \text{ kg/m}^3 = 1.520 \text{ Mg/m}^3}$$

which verifies the previous result.
c) If all pore fluid is dried off with no change in total volume, the moist density will then be equal to the dry density; i.e.,

$$\rho = \rho_d = \frac{M_s}{V} = \frac{3.752 \times 10^{-1} \text{ kg}}{4.390 \times 10^{-4} \text{ m}^3} = \mathbf{854.6 \text{ kg/m}^3 = 0.855 \text{ Mg/m}^3} \tag{1.7.24}$$

Figure Ex. 1.7c shows the phase diagram associated with this material state.

EXAMPLE PROBLEM 1.8

General Remarks

This example problem illustrates the manner in which quantities that are used to describe the phase relations are computed from data that are given in a form very similar to actual experimental results obtained in a laboratory.

Problem Statement

For an undisturbed soil, the total volume is 5.1 ft^3, the moist weight is 601 lb, the dry weight is 523 lb, and the porosity is 37.5%. Calculate (a) the moisture content (w), (b) the dry unit weight (γ_d), (c) the moist unit weight (γ), (d) the

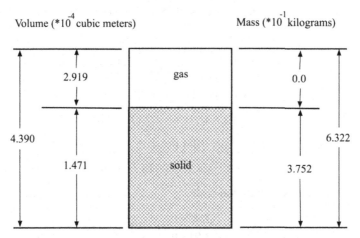

FIGURE EX. 1.7C   Phase diagram showing relationship between volume and mass for a sample of San Francisco Bay mud with all pore fluid dried off.

degree of saturation (S), (e) the void ratio (e), and (f) the specific gravity of solids (G_s).

Solution

a) The weight of the water is simply $W_w = W - W_s = 601 - 523 = 78.0$ lb. The moisture content is thus

$$w = \left(\frac{W_w}{W_s}\right) * 100\% = \left(\frac{78.0}{523}\right) * 100\% = \textbf{14.9\%} \qquad (1.8.1)$$

b) The dry unit weight is

$$\gamma_d = \frac{W_s}{V} = \frac{523 \text{ lb}}{5.1 \text{ ft}^3} = \textbf{102.5 lb/ft}^3 \qquad (1.8.2)$$

c) Similarly, the moist unit weight is

$$\gamma = \frac{W}{V} = \frac{601 \text{ lb}}{5.1 \text{ ft}^3} = \textbf{117.8 lb/ft}^3 \qquad (1.8.3)$$

d) The degree of saturation requires both the volume of water (V_w) and the volume of voids (V_v). The latter is most simply computed from the porosity as follows:

$$n = \left(\frac{V_v}{V}\right) * 100\% \Rightarrow V_v = \left(\frac{n}{100\%}\right)V = (0.375)(5.1 \text{ ft}^3) = 1.913 \text{ ft}^3$$

$$(1.8.4)$$

The volume of water is computed from W_w and the unit weight of water (γ_w); i.e.,

$$\gamma_w = \frac{W_w}{V_w} \Rightarrow V_w = \frac{W_w}{\gamma_w} = \frac{78 \text{ lb}}{62.4 \text{ lb/ft}^3} = 1.250 \text{ ft}^3 \qquad (1.8.5)$$

The volume of the gaseous phase is thus

$$V_a = V_v - V_w = 1.913 - 1.250 = 0.663 \text{ ft}^3 \qquad (1.8.6)$$

Finally, the volume of the solid phase is

$$V_s = V - V_v = 5.1 - 1.913 = 3.187 \text{ ft}^3 \qquad (1.8.7)$$

The degree of saturation is thus

$$S = \left(\frac{V_w}{V_v}\right) * 100\% = \left(\frac{1.250 \text{ ft}^3}{1.913 \text{ ft}^3}\right) * 100\% = \mathbf{65.4\%} \qquad (1.8.8)$$

e) The void ratio can be computed either from

$$e = \frac{n}{1-n} = \frac{0.375}{1-0.375} = \mathbf{0.600} \qquad (1.8.9)$$

or from

$$e = \frac{V_v}{V_s} = \frac{1.913 \text{ ft}^3}{(5.1 - 1.913 \text{ ft}^3)} = \mathbf{0.600} \qquad (1.8.10)$$

f) Finally, the specific gravity of solids is computed from its definition; i.e.,

$$G_s = \frac{\gamma_s}{\gamma_w} = \frac{W_s}{V_s \gamma_w} = \frac{523 \text{ lb}}{(3.187 \text{ ft}^3)(62.4 \text{ lb/ft}^3)} = \mathbf{2.63} \qquad (1.8.11)$$

The specific gravity of solids can likewise be determined from the relation between e, w, S, and G_s derived in Case 1.3; i.e.,

$$G_s = \frac{Se}{w} = \frac{(0.654)(0.600)}{(0.149)} = \mathbf{2.63} \qquad (1.8.12)$$

Figure Ex. 1.8 shows the phase diagram associated with this soil.

EXAMPLE PROBLEM 1.9

General Remarks

This example problem illustrates the manner in which volume changes are computed for saturated soils.

Problem Statement

A volume of one million cubic meters of slurry (saturated soil with high water content) is compressed for a period of 2.5 years. Initially, the slurry has a

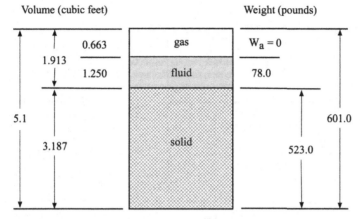

FIGURE EX. 1.8 Phase diagram showing relationship between volume and weight for an undisturbed soil.

moisture content of 119%. 2.5 years later, the moisture content drops to 52%. The specific gravity of solids (G_s) for the slurry is 2.72.

Determine the change in volume that has taken place during the 2.5 years of compression.

Solution

A key fact to note is that during compression, the weight (W_s) and volume of the solid phase (V_s) and G_s remain unchanged. In addition, the slurry remains saturated throughout the compression process.

From the general definition of moisture content,

$$w = \left(\frac{W_w}{W_s}\right) * 100\% \Rightarrow W_w = wW_s \tag{1.9.1}$$

From the definition of the specific gravity of solids,

$$G_s = \frac{W_s}{V_s\gamma_w} \Rightarrow W_s = G_sV_s\gamma_w \tag{1.9.2}$$

Combining the two equations gives

$$W_w = wW_s = wG_sV_s\gamma_w \tag{1.9.3}$$

Substituting the aforementioned result into the definition of the unit weight of water gives

$$\gamma_w = \frac{W_w}{V_w} \Rightarrow V_w = \frac{W_w}{\gamma_w} = \frac{wG_sV_s\gamma_w}{\gamma_w} = wG_sV_s \tag{1.9.4}$$

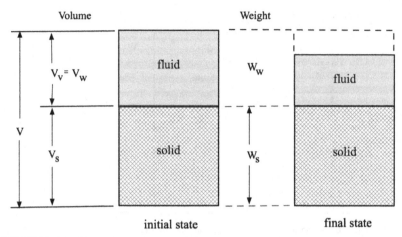

FIGURE EX. 1.9 Phase diagram showing relationship between volume and weight for a saturated slurry.

The general expression for the total volume of the saturated slurry is

$$V = V_w + V_s = wG_sV_s + V_s = (wG_s + 1)V_s \qquad (1.9.5)$$

Initially, $V_0 = 1.0 \times 10^6$ m^3 and $w_0 = 119\%$. Thus,

$$V_0 = (w_0G_s + 1)V_s \Rightarrow V_s = \frac{V_0}{w_0G_s + 1} = \frac{1.0 \times 10^6 \text{ m}^3}{(1.19)(2.72) + 1} = 236,027 \text{ m}^3$$

$$(1.9.6)$$

After 2.5 years,

$$V_f = (w_fG_s + 1)V_s = [(0.52)(2.72) + 1](236,027 \text{ m}^3) = 569,864 \text{ m}^3$$

$$(1.9.7)$$

The desired change in volume is thus

$$\Delta V = V_0 - V_f = 1,000,000 - 569,864 = \mathbf{430,136 \text{ m}^3} \qquad (1.9.8)$$

Figure Ex. 1.9 shows the phase diagram associated with this soil.

EXAMPLE PROBLEM 1.10

General Remarks

This example problem deals with a saturated soil below the groundwater table. It illustrates how to compute the buoyant unit weight.

Problem Statement

A sample of soil was taken from below the groundwater table. The moisture content was determined to be 52% and the specific gravity of solids (G_s) was found to be 2.69. Determine (a) the moist unit weight, (b) the dry unit weight, (c) the buoyant (submerged) unit weight, (d) the void ratio, and (e) the porosity.

Solution

Since the soil sample was taken from below the groundwater table, it is *saturated* (i.e., $S = 100\%$). The void ratio is computed from the equation $e = wG_s/S$ (recall Case 1.3 and Eq. (1.22)). Since the soil is saturated, this reduces to

$$e = G_s w = (2.69)(0.52) = \mathbf{1.399} \tag{1.10.1}$$

The porosity is computed using Eq. (1.17); i.e.,

$$n = \left(\frac{e}{1+e}\right) * 100\% = \left(\frac{1.399}{1+1.399}\right) * 100\% = \mathbf{58.3\%} \tag{1.10.2}$$

a) The moist unit weight is computed from Eq. (1.26); i.e.,

$$\gamma = \frac{G_s \gamma_w (1+w)}{1+e} = \frac{(2.69)(9.81 \text{ kN/m}^3)(1+0.52)}{1+1.399} = \mathbf{16.72 \text{ kN/m}^3} \tag{1.10.3}$$

b) The dry unit weight follows directly from Eq. (1.27); i.e.,

$$\gamma_d = \frac{\gamma}{1+w} = \frac{16.72 \text{ kN/m}^3}{1+0.52} = \mathbf{11.00 \text{ kN/m}^3} \tag{1.10.4}$$

c) Finally, the buoyant unit weight is computed from Eq. (1.34); i.e.,

$$\gamma_b = \gamma - \gamma_w = (16.72 - 9.81) = \mathbf{6.91 \text{ kN/m}^3} \tag{1.10.5}$$

where it is noted that for this problem γ is the saturated unit weight.

EXAMPLE PROBLEM 1.11

General Remarks

This example problem illustrates the manner in which quantities that are used to describe the phase relations are computed from data that are given in a form very similar to actual experimental results obtained in a laboratory.

Problem Statement

A sample of moist soil was found to have the following characteristics:

- Total volume: 0.01456 m^3
- Total mass: 25.74 kg

- Mass after oven drying: 22.10 kg
- Specific gravity of solids: 2.69

Determine (a) the moist density, (b) the moist unit weight, (c) the moisture content, (d) the void ratio, (e) the porosity, and (f) the degree of saturation of the moist soil.

Solution

a) The moist density is simply (recall Eq. 1.3)

$$\rho = \frac{M}{V} = \frac{25.74 \text{ kg}}{0.01456 \text{ m}^3} = \textbf{1768 kg/m}^3 \tag{1.11.1}$$

b) The moist unit weight follows directly from Eq. (1.7); i.e.,

$$\gamma = \rho g = \left(1768 \text{ kg/m}^3\right)\left(9.981 \text{ m/s}^2\right) = 17,343 \text{ N/m}^3 = \textbf{17.34 kN/m}^3 \tag{1.11.2}$$

c) The moisture content is next computed from the given masses. First,

$$M_w = M - M_s = 25.74 - 22.10 = 3.64 \text{ kg} \tag{1.11.3}$$

Thus,

$$w = \left(\frac{M_w}{M_s}\right) * 100\% = \left(\frac{3.64}{22.10}\right) * 100\% = \textbf{16.5\%} \tag{1.11.4}$$

The determination of the remaining quantities requires the computation of the volume of solids and voids. The former quantity is computed from the specific gravity of solids as follows

$$G_s = \frac{W_s}{V_s \gamma_w} \Rightarrow V_s = \frac{M_s g}{G_s \gamma_w} = \frac{(22.10 \text{ kg})(9.81 \text{ m/s}^2)(\text{kN}/1000 \text{ N})}{(2.69)(9.81 \text{ kN/m}^3)} \tag{1.11.5}$$
$$= 8.216 \times 10^{-3} \text{ m}^3$$

The volume of the voids is thus

$$V_v = V - V_s = \left(1.456 \times 10^{-2}\right) - \left(8.216 \times 10^{-3}\right) = 6.344 \times 10^{-3} \text{ m}^3 \tag{1.11.6}$$

d) The void ratio and porosity are thus

$$e = \frac{V_v}{V_s} = \frac{6.344 \times 10^{-3} \text{ m}^3}{8.216 \times 10^{-3} \text{ m}^3} = \textbf{0.772} \tag{1.11.7}$$

e) The porosity can be computed from known volumes according to

$$n = \left(\frac{V_v}{V}\right) * 100\% = \left(\frac{6.344 \times 10^{-3}\ m^3}{0.01456}\right) * 100\% = \mathbf{43.5\%} \qquad (1.11.8)$$

It can likewise be computed directly from the void ratio (recall Case 1.2); i.e.,

$$n = \left(\frac{e}{1+e}\right) * 100\% = \left(\frac{0.772}{1+0.772}\right) * 100\% = \mathbf{43.5\%} \qquad (1.11.9)$$

f) The volume of water is determined from the mass of pore fluid and the density of water

$$\rho_w = \frac{M_w}{V_w} \Rightarrow V_w = \frac{M_w}{\rho_w} = \frac{3.64\ kg}{1000\ kg/m^3} = 3.640 \times 10^{-3}\ m^3 \qquad (1.11.10)$$

The degree of saturation is thus

$$S = \left(\frac{V_w}{V_v}\right) * 100\% = \left(\frac{3.640 \times 10^{-3}\ m^3}{6.340 \times 10^{-3}\ m^3}\right) * 100\% = \mathbf{57.4\%} \qquad (1.11.11)$$

As a check, use the relation $Se = wG_s$ (recall Case 1.3), which gives

$$S = \left(\frac{wG_s}{e}\right) * 100\% = \frac{(0.165)(2.69)}{0.772} * 100\% = \mathbf{57.4\%} \qquad (1.11.12)$$

and confirms the previous result.

To complete the determination of quantities appearing in the phase diagram, the volume of the gaseous phase (air) is computed as follows:

$$V_a = V_v - V_w = \left(6.344 \times 10^{-3}\ m^3\right) - \left(3.640 \times 10^{-3}\ m^3\right) = 2.704 \times 10^{-3}\ m^3 \qquad (1.11.13)$$

Figure Ex. 1.11 shows the phase diagram associated with this soil.

EXAMPLE PROBLEM 1.12

General Remarks

This example problem illustrates the manner in which quantities that are used to describe the phase relations are computed from data that are given in a form very similar to experimental readings in the laboratory.

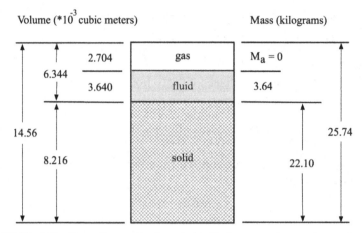

Volume (*10^{-3} cubic meters)   Mass (kilograms)

FIGURE EX. 1.11 Phase diagram showing relationship between volume and mass for a soil tested in the laboratory.

Problem Statement

The mass of a wet sample of soil and its container is 0.33 kg. The dry mass of the soil and its container is 0.29 kg. The mass of the container is 0.06 kg, and its volume is 0.00015 m³. The soil completely fills the container. Finally, the specific gravity of solids is found to be 2.71. Determine the following quantities:

(a) The moist unit weight, (b) the dry unit weight, (c) the void ratio, (d) the moisture content, (e) the degree of saturation, (f) the saturated unit weight, and (g) the buoyant unit weight. (h) Finally, what is the maximum dry unit weight to which this soil can be compacted *without changing its moisture content*?

Solution

a) The moist unit weight is computed as follows (recall Eq. (1.7)):

$$\gamma = \frac{W}{V} = \frac{(0.33 - 0.06 \text{ kg})(9.81 \text{ m/s}^3)}{0.00015 \text{ m}^3} = 17,658 \text{ N/m}^3 = \mathbf{17.66 \text{ kN/m}^3}$$

$$(1.12.1)$$

b) The weight of the solid phase is computed from the given information; i.e.,

$$W_s = (0.29 - 0.06 \text{ kg})(9.81 \text{ m/s}^2)(\text{kN}/1000 \text{ N}) = 2.256 \times 10^{-3} \text{ kN}$$

$$(1.12.2)$$

The dry unit weight is

$$\gamma_d = \frac{W_s}{V} = \frac{2.256 \times 10^{-3} \text{ kN}}{0.00015 \text{ m}^3} = \textbf{15.04 kN/m}^3 \qquad (1.12.3)$$

c) The volume of the solid phase is computed from the specific gravity of solids and the weight of the solids; i.e.,

$$V_s = \frac{W_s}{G_s \gamma_w} = \frac{2.256 \times 10^{-3} \text{ kN}}{(2.71)(9.81 \text{ kN/m}^3)} = 8.487 \times 10^{-5} \text{ m}^3 \qquad (1.12.4)$$

The volume of the voids is thus

$$V_v = V - V_s = \left(1.500 \times 10^{-4} - 8.487 \times 10^{-5}\right) = 6.513 \times 10^{-5} \text{ m}^3 \qquad (1.12.5)$$

Finally, the void ratio is computed as follows:

$$e = \frac{V_v}{V_s} = \frac{6.513}{8.487} = \textbf{0.767} \qquad (1.12.6)$$

d) The moisture content can be computed from the given information as follows:

$$w = \left(\frac{W_w}{W_s}\right) * 100\% = \left(\frac{M_w}{M_s}\right) * 100\% = \left(\frac{0.33 - 0.29}{0.29 - 0.06}\right) * 100\% = \textbf{17.4\%} \qquad (1.12.7)$$

e) The degree of saturation requires the knowledge of the volume of pore fluid. This is computed from the known weight of the fluid and the unit weight of water; i.e.,

$$W_w = (0.33 - 0.29 \text{ kg})(9.81 \text{ m/s}^2)(\text{kN}/1000 \text{ N}) = 3.924 \times 10^{-4} \text{ kN} \qquad (1.12.8)$$

$$V_w = \frac{W_w}{\gamma_w} = \frac{3.924 \times 10^{-4} \text{ kN}}{9.81 \text{ kN/m}^3} = 4.000 \times 10^{-5} \text{ m}^3 \qquad (1.12.9)$$

Thus,

$$S = \left(\frac{V_w}{V_v}\right) = \left(\frac{4.000 \times 10^{-5}}{6.513 \times 10^{-5}}\right) * 100\% = \textbf{61.4\%} \qquad (1.12.10)$$

The degree of saturation can likewise be computed from the relation (recall Case 1.3)

$$S = \left(\frac{G_s w}{e}\right) * 100\% = \left[\frac{(2.71)(0.1739)}{0.767}\right] * 100\% = \textbf{61.4\%} \qquad (1.12.11)$$

which serves a check on the results obtained.

f) When the soil is saturated,

$$V_v = V_w = 6.513 \times 10^{-5} \text{ m}^3 \qquad (1.12.12)$$

Thus,

$$W_w = \gamma_w V_w = (9.81 \text{ kN/m}^3)(6.513 \times 10^{-5} \text{ m}^3) = 6.389 \times 10^{-4} \text{ kN} \qquad (1.12.13)$$

The saturated unit weight is thus

$$\gamma_{sat} = \frac{W_s + W_w}{V} = \frac{(2.256 \times 10^{-3} \text{ kN}) + (6.389 \times 10^{-4} \text{ kN})}{0.00015 \text{ m}^3} = \mathbf{19.30 \ kN/m^3} \qquad (1.12.14)$$

The saturated unit weight can likewise be computed from the relation (recall Case 1.6)

$$\gamma_{sat} = \frac{\gamma_w(G_s + e)}{1 + e} = \frac{(9.81 \text{ kN/m}^3)(2.71 + 0.767)}{1 + 0.767} = \mathbf{19.30 \ kN/m^3} \qquad (1.12.15)$$

which serves a check on the results obtained.

g) The buoyant unit weight follows directly from the saturated unit weight according to (recall Case 1.8)

$$\gamma_b = \gamma_{sat} - \gamma_w = 19.30 - 9.81 = \mathbf{9.49 \ kN/m^3} \qquad (1.12.16)$$

h) Since the moisture content remains unchanged, and since W_s does not change (no solids are removed or added to the soil), it follows that W_w must likewise remain unchanged. As a consequence of these observations, it follows that V_s and V_w must remain unchanged. Thus, the compaction alluded to in this part of the problem can only displace the air present in the voids. The resulting dry density will thus be

$$\gamma_d = \frac{W_s}{V} = \frac{W_s}{V_s + V_w} = \frac{2.256 \times 10^{-3} \text{ kN}}{[(8.487 \times 10^{-5}) + (4.000 \times 10^{-5})] \text{m}^3} = \mathbf{18.07 \ kN/m^3} \qquad (1.12.17)$$

The dry density can likewise be computed from the relation (recall Case 1.6)

$$\gamma_d = \frac{\gamma}{1 + w} \qquad (1.12.18)$$

where

$$\gamma = \frac{W_s + W_w}{V_s + V_w} = \frac{[(2.256 \times 10^{-3}) + (3.924 \times 10^{-4})] \text{kN}}{[(8.487 \times 10^{-5}) + (4.000 \times 10^{-5})] \text{m}^3} = 21.21 \text{ kN/m}^3 \qquad (1.12.19)$$

Thus,

$$\gamma_d = \frac{\gamma}{1+w} = \frac{21.21 \text{ kN/m}^3}{1+0.1759} = \textbf{18.04 kN/m}^3 \qquad (1.12.20)$$

which serves a check on the results obtained (within round-off error).

EXAMPLE PROBLEM 1.13

General Remarks

In this problem gives insight into the issues associated with the inundation (flooding) of a soil.

Problem Statement

A dry sand layer with specific gravity of solids (G_s) equal to 2.66 was compacted to a dry unit weight of 16.7 kN/m^3. It was subsequently inundated with water. Determine the moisture content (w) and the moist unit weight (γ) of the inundated sand.

Solution

The initial void ratio is computed from the dry unit weight as follows (recall Case 1.4):

$$\gamma_d = \frac{G_s \gamma_w}{1+e} \Rightarrow e = \frac{G_s \gamma_w}{\gamma_d} - 1 = \frac{(2.66)(9.81 \text{ kN/m}^3)}{16.7 \text{ kN/m}^3} - 1 = 0.563 \quad (1.13.1)$$

After inundation, the volume and weight of the solid phase remain *unchanged*. The moisture content is most easily computed using the relation derived in Case 1.3; i.e., $w = Se/G_s$. This, however, requires that *assumptions* be made on the degree of saturation (S) and on the value of the void ratio.

With the void ratio known and moisture content known, the moist unit weight is computed from the expression derived in Case 1.5; i.e.,

$$\gamma = \frac{G_s \gamma_w (1+w)}{1+e} \qquad (1.13.2)$$

It is reasonable to assume that the sand has become saturated ($S = 100\%$), particularly if it has been inundated for a relatively long period of time. If the void ratio is also assumed to remain unchanged after inundation, the desired moisture content is thus

$$w = \frac{Se}{G_s} = \frac{(1.00)(0.563)}{2.66} * 100\% = \textbf{21.2\%} \qquad (1.13.3)$$

The moist unit weight is then

$$\gamma = \frac{G_s\gamma_w(1+w)}{1+e} = \frac{(2.66)(9.81 \text{ kN/m}^3)(1+0.212)}{1+0.563} = \mathbf{20.2 \text{ kN/m}^3}$$

(1.13.4)

EXAMPLE PROBLEM 1.14

General Remarks

This example problem further illustrates the manner in which variables that are used to quantify the phase relations are computed.

Problem Statement

The porosity (n) of a poorly graded sand is 37% and its specific gravity of solids (G_s) is 2.67. Determine the void ratio (e) and the dry unit weight (γ_d). If the sand is 30% saturated, determine the moisture content (w) and the moist unit weight (γ). Finally, determine moisture content and the moist unit weight if the sand is *saturated*.

Solution

The void ratio is determined from the known porosity as (recall Case 1.1)

$$e = \frac{n}{1-n} = \frac{0.37}{1-0.37} = \mathbf{0.587}$$

(1.14.1)

Since an explicit volume of the soil is not given, assume that $V_s = 1.0 \text{ m}^3$. From the definition of the void ratio, it follows that $V_v = e \cdot V_s = e$ (in units of m^3). The weight of the solid phase is then computed from the definition of the specific gravity of solids as

$$G_s = \frac{W_s}{V_s\gamma_w} \Rightarrow W_s = G_sV_s\gamma_w = G_s(1.0)\gamma_w$$

(1.14.2)

The dry unit weight is thus (recall Case 1.4)

$$\gamma_d = \frac{G_s\gamma_w}{1+e} = \frac{(2.67)(9.81 \text{ kN/m}^3)}{1+0.587} = \mathbf{16.51 \text{ kN/m}^3}$$

(1.14.3)

In general,

$$S = \frac{V_w}{V_v} \Rightarrow V_w = SV_v = Se$$

(1.14.4)

Relating the weight of the pore fluid to its volume through the unit weight of water gives the following general expression (recall Case 1.3):

$$w = \left(\frac{W_w}{W_s}\right) * 100\% = \left(\frac{\gamma_w V_w}{G_s \gamma_w}\right) * 100\% = \left(\frac{Se}{G_s}\right) * 100\% \qquad (1.14.5)$$

The general expression for the moist unit weight is thus (recall Case 1.7)

$$\gamma = \frac{W}{V} = \frac{W_s + W_w}{1 + e} = \frac{\gamma_w(G_s + Se)}{1 + e} \qquad (1.14.6)$$

If $S = 30\%$, then

$$w = \left[\frac{(0.30)(0.587)}{2.67}\right] * 100\% = \mathbf{6.60\%} \qquad (1.14.7)$$

and

$$\gamma = \frac{\gamma_w(G_s + Se)}{1 + e} = \frac{(9.81\ \text{kN/m}^3)[2.67 + (0.30)(0.587)]}{1 + 0.587} = \mathbf{17.59\ kN/m^3}$$
$$(1.14.8)$$

For a *saturated* soil, $S = 100\%$, thus

$$w = \left[\frac{(1.0)(0.587)}{2.67}\right] * 100\% = \mathbf{22.0\%} \qquad (1.14.9)$$

and

$$\gamma = \gamma_{sat} = \frac{\gamma_w(G_s + e)}{1 + e} = \frac{(9.81\ \text{kN/m}^3)[2.67 + (0.587)]}{1 + 0.587} = \mathbf{20.13\ kN/m^3}$$
$$(1.14.10)$$

EXAMPLE PROBLEM 1.15

General Remarks

This example problem investigates a soil with a nonhomogeneous solid phase.

Problem Statement

Consider two soils. The solid phase of the first soil is composed of pure quartz ($G_s = 2.66$). The solid phase of the second soil is composed of a mixture of 64% quartz, 28% mica ($G_s = 2.75$), and 8% iron oxide ($G_s = 5.40$).

How much difference in the unit weights and specific gravities of solids is there between the two soils? Assume both soils are saturated and have void ratios of 0.632.

Solution

- Soil 1:

  Since an explicit total volume for the soil has not been specified, assume $V_s = 1.0$, implying that $V_v = V_w = e = 0.632$. Thus,

  $$W_s = G_s V_s \gamma_w = (2.66)(1.0)\gamma_w \qquad (1.15.1)$$

  Since $S = 100\%$, and recalling that $W_w = \gamma_w \cdot V_w = 0.632 \cdot \gamma_w$, it follows that

  $$\gamma = \left(\frac{W_s + W_w}{1 + e}\right) = \frac{2.66\gamma_w + 0.632\gamma_w}{1 + 0.632} = 2.017\gamma_w \qquad (1.15.2)$$

  If $\gamma_w = 9.81$ kN/m^3, then $\gamma = \mathbf{19.8\ kN/m^3}$; if $\gamma_w = 62.4$ lb/ft^3, then $\gamma = \mathbf{125.9\ lb/ft^3}$.

  In summary, for *Soil 1*: $G_s = 2.66$ and $\gamma = 19.8$ kN/m^3 = 125.9 lb/ft^3.

- Soil 2:

  Using the definition of the specific gravity of solids, since the solids are apportioned by volume, it follows that

  $$V_{quartz} = 0.64 V_s = \frac{W_{s\_quartz}}{G_{s\_quartz}\gamma_w} \qquad (1.15.3)$$

  Thus,

  $$W_{s\_quartz} = 0.64(1.0)(2.66)\gamma_w \qquad (1.15.4)$$

  Similarly,

  $$W_{s\_mica} = 0.28(1.0)(2.75)\gamma_w \qquad (1.15.5)$$

and

  $$W_{s\_FeO} = 0.08(1.0)(5.40)\gamma_w \qquad (1.15.6)$$

  The total weight of the solids is thus

  $$W_s = [(0.64)(2.66) + (0.28)(2.75) + (0.08)(5.40)]\gamma_w = 2.904\gamma_w \qquad (1.15.7)$$

  The specific gravity of Soil 2 is thus

  $$G_s = \frac{W_s}{V_s \gamma_w} = \frac{2.904\gamma_w}{(1.0)\gamma_w} = \mathbf{2.90} \qquad (1.15.8)$$

  It is important to point out that this value represents a weighted average of specific gravities of solids for the composite soil, and not the specific gravity of solids for one of the constituent soils.

The associated unit weight is

$$\gamma = \left(\frac{W_s + W_w}{1+e}\right) = \frac{2.904\gamma_w + 0.632\gamma_w}{1+0.632} = 2.167\gamma_w \qquad (1.15.9)$$

If $\gamma_w = 9.81$ kN/m³, then $\gamma = \mathbf{21.3 \ kN/m^3}$; if $\gamma_w = 62.4$ lb/ft³, then $\gamma = \mathbf{135.2 \ lb/ft^3}$.

In summary, for *Soil 2*: $G_s = 2.90$ and $\gamma = 21.3$ kN/m³ $= 135.2$ lb/ft³. The difference in unit weight between *Soil 1* and *Soil 2* is thus

$$21.3 - 19.8 = \mathbf{1.50 \ kN/m^3} \qquad (1.15.10)$$

or

$$135.2 - 125.9 = \mathbf{9.30 \ lb/ft^3} \qquad (1.15.11)$$

The difference in specific gravity of solids between *Soil 1* and *Soil 2* is thus

$$2.90 - 2.66 = \mathbf{0.24} \qquad (1.15.12)$$

EXAMPLE PROBLEM 1.16

General Remarks

This example problem also investigates a soil with a nonhomogeneous solid phase.

Problem Statement

Consider the following two soils:

- *Soil 1*: solid phase is composed of pure silica ($G_s = 2.66$).
- *Soil 2*: solid phase is composed of a mixture of 35% silica, 45% feldspar ($G_s = 2.75$), and 20% ferro magnesium ($G_s = 3.45$). These percentages refer to volume fractions.

Assume both soils have a moisture content of 22% and a void ratio of 0.787.

a) What is the degree of saturation, the moist unit weight, and the dry unit weight for Soil 1?

b) What is the specific gravity of solids, the degree of saturation, the moist unit weight, and the dry unit weight for Soil 2?

Solution

a) Since the solid phase in Soil 1 is homogeneous, the determination of the desired quantities is straightforward. The degree of saturation is determined as follows (recall Case 1.3):

$$S = \frac{G_s w}{e} = \frac{(2.66)(0.22)}{0.787} = 0.744 = \mathbf{74.4\%} \tag{1.16.1}$$

The moist unit weight is next computed (recall Case 1.6)

$$\gamma = \frac{\gamma_w(G_s + Se)}{1 + e} = \frac{(9.81 \text{ kN/m}^3)[2.66 + (0.744)(0.787)]}{1 + 0.787} = \mathbf{17.82 \text{ kN/m}^3} \tag{1.16.2}$$

Finally, the dry unit weight is computed from the moist unit weight as follows (recall Case 1.6):

$$\gamma_d = \frac{\gamma}{1 + w} = \frac{17.82 \text{ kN/m}^3}{1 + 0.22} = \mathbf{14.60 \text{ kN/m}^3} \tag{1.16.3}$$

b) The composite nature of the solid phase for Soil 2 complicates the determination of the specific gravity of solids. From the definition of G_s it follows that in general,

$$G_s = \frac{W_s}{V_s \gamma_w} \Rightarrow V_s = \frac{W_s}{G_s \gamma_w} \tag{1.16.4}$$

For simplicity, assume $V_s = 1.0 \text{ m}^3$. Thus, from the given volume fractions,

$$V_{s\_silica} = 0.35 V_s = 0.35(1) = \frac{W_{s\_silica}}{(2.66)\gamma_w} \Rightarrow W_{s\_silica} = 0.35(2.66)\gamma_w \tag{1.16.5}$$

$$V_{s\_feldspar} = 0.45 V_s = 0.45(1) = \frac{W_{s\_feldspar}}{(2.75)\gamma_w} \Rightarrow W_{s\_feldspar} = 0.45(2.75)\gamma_w \tag{1.16.6}$$

$$V_{s\_ferro} = 0.20 V_s = 0.20(1) = \frac{W_{s\_ferro}}{(3.45)\gamma_w} \Rightarrow W_{s\_ferro} = 0.20(3.45)\gamma_w \tag{1.16.7}$$

The total weight of the solid phase for Soil 2 is thus

$$W_s = W_{s\_silica} + W_{s\_feldspar} + W_{s\_ferro} = [0.35(2.66) + 0.45(2.75) + 0.20(3.45)]\gamma_w \tag{1.16.8}$$

The specific gravity of solids for Soil 2 is thus

$$G_s = \frac{W_s}{V_s \gamma_w} = \frac{[0.35(2.66) + 0.45(2.75) + 0.20(3.45)]\gamma_w}{(1.0)\gamma_w} = \textbf{2.86} \quad (1.16.9)$$

It is important to note that this value represents a weighted average of specific gravities of solids for the composite soil, and not the specific gravity of solids for one of the constituent soils. The degree of saturation and moist and dry unit weights are then computed in the same manner as for Soil 1; i.e.,

$$S = \frac{G_s w}{e} = \frac{(2.86)(0.22)}{0.787} = 0.799 = \textbf{79.9\%} \quad (1.16.10)$$

$$\gamma = \frac{\gamma_w(G_s + Se)}{1 + e} = \frac{(9.81 \text{ kN/m}^3)[2.86 + (0.799)(0.787)]}{1 + 0.787} = \textbf{19.15 kN/m}^3$$
$$(1.16.11)$$

$$\gamma_d = \frac{\gamma}{1 + w} = \frac{19.15 \text{ kN/m}^3}{1 + 0.22} = \textbf{15.69 kN/m}^3$$

EXAMPLE PROBLEM 1.17

General Remarks

This example problem involves the constraint of constant volume.

Problem Statement

A soil has a unit weight of 109 lb/ft^3 and a moisture content of 6%. How much water, in cubic feet, should be added to each cubic yard of soil to raise its moisture content to 13%? Assume that the void ratio remains constant (i.e., the added water simply displaces air in the voids.

Solution

Since the weight and volume of the solids remains unchanged, and since the void ratio is constrained to remain constant, it follows that $V = V_v + V_s$ remains constant. Beginning with the expression for the moist unit weight, and using the definition of the moisture content gives

$$\gamma = \frac{W}{V} = \frac{W_s + W_w}{V} = \frac{W_s(1 + w)}{V} \Rightarrow W_s = \frac{\gamma V}{1 + w} \quad (1.17.1)$$

and

$$W_w = w W_s = \frac{w \gamma V}{1 + w} \quad (1.17.2)$$

Initially, $\gamma = 109$ lb/ft^3, $w = 6\%$, and $V = 1$ yd$^3 = 27$ ft^3. Thus,

$$W_s = \frac{(109 \text{ lb/ft}^3)(27 \text{ ft}^3)}{1 + 0.06} = 2776.4 \text{ lb} \qquad (1.17.3)$$

and

$$W_w = wW_s = (0.06)(2776.4) = 166.6 \text{ lb} \qquad (1.17.4)$$

After water is added to the soil, W_s is unchanged, and

$$W_{w_{new}} = w_{new}W_s = (0.13)(2776.4 \text{ lb}) = 360.9 \text{ lb} \qquad (1.17.5)$$

The change in weight of the water is thus

$$\Delta W_w = W_{w_{new}} - W_w = 360.9 - 166.6 = 194.3 \text{ lb} \qquad (1.17.6)$$

Finally, the volume of water to be added to attain a moisture content of 13% is computed as follows:

$$\Delta V_w = \frac{\Delta W_w}{\gamma_w} = \frac{194.3 \text{ lb}}{62.4 \text{ lb/ft}^3} = \mathbf{3.11 \text{ ft}^3} \qquad (1.17.7)$$

EXAMPLE PROBLEM 1.18

General Remarks

This example problem also involves the constraint of constant volume.

Problem Statement

A soil has a moist unit weight of 128 lb/ft^3 and a moisture content of 12%. What will be the moisture content if the soil dries out to a moist unit weight of 123 lb/ft^3 and the void ratio remains unchanged? Assume $G_s = 2.68$.

Solution

Since an explicit volume of the soil is not given, assume that $V_s = 1$. From the definition of the void ratio, it follows that $V_v = e \cdot V_s = e$. From the definition of the moist unit weight (also recall Case 1.5)

$$\gamma = \frac{W}{V} = \frac{W_s + W_w}{V_s + V_v} = \frac{W_s(1 + w)}{V_s + V_v} = \frac{G_s\gamma_w(1 + w)}{1 + e} \qquad (1.18.1)$$

where the definition of the specific gravity of solids has been used. Solving for the void ratio gives

$$e = \frac{G_s\gamma_w(1 + w)}{\gamma} - 1 \qquad (1.18.2)$$

Denote the initial moisture content and moist unit weight by w_{init} and γ_{init}, respectively, and the corresponding values after drying out by w_{final} and γ_{final}, respectively. Since the void ratio remains unchanged during drying, it follows that

$$\frac{G_s \gamma_w (1 + w_{init})}{\gamma_{init}} - 1 = \frac{G_s \gamma_w (1 + w_{final})}{\gamma_{final}} - 1 \qquad (1.18.3)$$

or

$$\frac{(1 + w_{init})}{\gamma_{init}} = \frac{(1 + w_{final})}{\gamma_{final}} \qquad (1.18.4)$$

Solving for the moisture content after drying gives

$$w_{final} = \frac{\gamma_{final}}{\gamma_{init}}(1 + w_{init}) - 1 = \frac{(123 \text{ lb/ft}^3)}{(128 \text{ lb/ft}^3)}(1 + 0.12) - 1 = 0.076 = \mathbf{7.60\%}$$

$$(1.18.5)$$

EXAMPLE PROBLEM 1.19

General Remarks

This example problem also involves the constraint of constant volume.

Problem Statement

A sample of clayey soil has a moisture content (w) of 15.6%, a specific gravity of solids (G_s) of 2.72, and a degree of saturation (S) equal to 72%. If the soil soaks up water during a rain event and the degree of saturation increases to 92.5%, what is the new moisture content? The change of volume during this soaking is negligible.

Solution

The key to this problem is the fact that both the volume of the solid phase (V_s) and the total volume (V) remain constant during soaking. Since $V_v + V_s = V$, it follows that V_v will likewise remain constant. Stated alternately, the void ratio $e = V_v/V_s$ remains constant.

It is thus desirable to obtain an expression for V_v in terms of the information given in the problem. From the definition of the degree of saturation,

$$S = \frac{V_w}{V_v} \Rightarrow V_v = \frac{V_w}{S} = \frac{(W_w/\gamma_w)}{S} \qquad (1.19.1)$$

But from the definition of the moisture content, $W_w = w \cdot W_s$, so

$$V_v = \frac{w W_s}{\gamma_w S} \qquad (1.19.2)$$

Finally, from the definition of the specific gravity of solids, $W_s = G_s \cdot V_s \cdot \gamma_w$. Thus, (recall Case 1.3)

$$V_v = \frac{w(G_s V_s \gamma_w)}{\gamma_w S} = \frac{w G_s V_s}{S} \Rightarrow e = \frac{w G_s}{S} \qquad (1.19.3)$$

Denoting the initial and final states with "i" and "f" subscripts, respectively, gives

$$e_i = e_f \Rightarrow \frac{w_i G_s}{S_i} = \frac{w_f G_s}{S_f} \qquad (1.19.4)$$

So,

$$w_f = \frac{w_i S_f}{S_i} = \frac{(0.156)(0.925)}{0.720} = 0.200 = \mathbf{20.0\%} \qquad (1.19.5)$$

If additional information is desired about the three soil phases, assume $V_s = 1.0 \text{ m}^3$. Thus, $V_v = e$, which remains constant during soaking. For the initial state,

$$V_v = e_i = \frac{w_i G_s}{S_i} = \frac{(0.156)(2.72)}{0.72} = 0.589 \text{ m}^3 \qquad (1.19.6)$$

The initial volume of pore fluid is computed from the degree of saturation as follows

$$S_i = \frac{V_{w_i}}{V_v} \Rightarrow V_{w_i} = S_i V_v = (0.72)(0.589 \text{ m}^3) = 0.424 \text{ m}^3 \qquad (1.19.7)$$

The initial volume of air in the pores is thus

$$V_{a_i} = V_v - V_{w_i} = 0.589 - 0.424 = 0.165 \text{ m}^3 \qquad (1.19.8)$$

The initial weight of the pore fluid is next computed from the unit weight of water; i.e.,

$$\gamma_w = \frac{W_{w_i}}{V_{w_i}} \Rightarrow W_{w_i} = \gamma_w V_{w_i} = (9.81 \text{ kN/m}^3)(0.424 \text{ m}^3) = 4.16 \text{ kN} \qquad (1.19.9)$$

The weight of the solid phase is thus

$$w_i = \frac{W_{w_i}}{W_s} \Rightarrow W_s = \frac{W_{w_i}}{w_i} = \frac{4.16 \text{ kN}}{0.156} = 26.68 \text{ kN} \qquad (1.19.10)$$

The total initial weight of the soil is thus

$$W_i = W_{w_i} + W_s = 4.16 + 26.68 = 30.84 \text{ kN} \qquad (1.19.11)$$

Figure Ex. 1.19a shows the phase diagram associated with the initial state of this soil.

In a similar manner, the final volume of the pore fluid is

$$V_{w_f} = S_f V_v = (0.925)(0.589 \text{ m}^3) = 0.545 \text{ m}^3 \qquad (1.19.12)$$

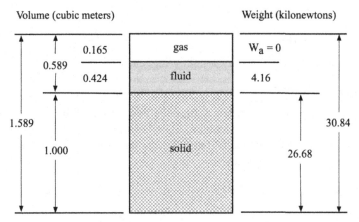

FIGURE EX. 1.19A   Phase diagram showing relationship between volume and weight for the initial state of a soil.

The final weight of the pore fluid is

$$W_{w_f} = \gamma_w V_{w_f} = (9.81 \text{ kN/m}^3)(0.545 \text{ m}^3) = 5.35 \text{ kN} \qquad (1.19.13)$$

The final volume of air in the pores is

$$V_{a_f} = V_v - V_{w_f} = 0.589 - 0.545 = 0.044 \text{ m}^3 \qquad (1.19.14)$$

Finally, the total weight of the soil in its final state is thus

$$W_f = W_{w_f} + W_s = 5.35 + 26.68 = 32.03 \text{ kN} \qquad (1.19.15)$$

Figure Ex. 1.19b shows the phase diagram associated with the final state of this soil.

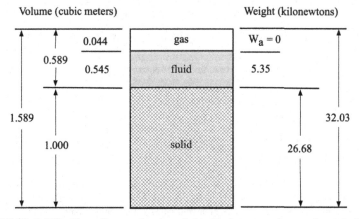

FIGURE EX. 1.19B   Phase diagram showing relationship between volume and weight for the final state of a soil.

EXAMPLE PROBLEM 1.20

General Remarks

This example problem involves a highly organic soil.

Problem Statement

A highly organic (peat[1]) soil has a saturated unit weight of 10.8 kN/m^3 and a specific gravity of solids (G_s) equal to 2.36. Determine the void ratio (e), and the moisture content (w). Finally determine the unit weight if the soil dries out *without* a change in void ratio.

Solution

Beginning with the saturated unit weight,

$$\gamma_{sat} = \frac{W}{V} = \frac{W_w + W_s}{V_v + V_s} \quad (1.20.1)$$

where, due to the fact that the soil is saturated, $V_v = V_w$. From the definition of the specific gravity of solids, $W_s = G_s \cdot V_s \cdot \gamma_w$. Similarly, from the definition of the unit weight of water, $W_w = V_w \cdot \gamma_w = V_v \cdot \gamma_w$. Substituting for W_s and W_w, the saturated unit weight is thus rewritten as

$$\gamma_{sat} = \frac{\gamma_w V_v + G_s \gamma_w V_s}{V_v + V_s} \quad (1.20.2)$$

Dividing through this equation by V_s gives (recall Case 1.7)

$$\gamma_{sat} = \frac{\gamma_w e + G_s \gamma_w}{1 + e} = \frac{\gamma_w (G_s + e)}{1 + e} \quad (1.20.3)$$

Solving for the void ratio completes the first part of the problem

$$e = \frac{\gamma_w G_s - \gamma_{sat}}{\gamma_{sat} - \gamma_w} = \frac{(9.81 \text{ kN/m}^3)(2.36) - 10.8 \text{ kN/m}^3}{10.8 \text{ kN/m}^3 - 9.81 \text{ kN/m}^3} = \textbf{12.48} \quad (1.20.4)$$

The moisture content is next computed as follows (recall Case 1.3):

$$w = \frac{W_w}{W_s} = \frac{\gamma_w V_w}{G_s V_s \gamma_w} = \frac{V_w}{G_s V_s} = \frac{e}{G_s} \quad (1.20.5)$$

Substituting for the known values gives

$$w = \left(\frac{e}{G_s}\right) * 100\% = \left(\frac{12.48}{2.36}\right) * 100\% = \textbf{528.7\%} \quad (1.20.6)$$

---

1. Peat is fibrous, partially decomposed organic matter or a soil containing large amounts of fibrous organic matter. Peats are loose (very high void ratio) and extremely compressible.

Finally, the dry unit weight of the soil is computed as follows

$$\gamma_d = \frac{W_s}{V} = \frac{G_s V_s \gamma_w}{V_s + V_v} = \frac{G_s \gamma_w}{1 + e} = \frac{(2.36)(9.81 \text{ kN/m}^3)}{1 + 12.48} = \textbf{1.72 kN/m}^3 \quad (1.20.7)$$

EXAMPLE PROBLEM 1.21

General Remarks

This example problem involves the saturation of a soil while maintaining its mass constant.

Problem Statement

Given the total mass (M), volume (V), moisture content (w), and specific gravity of solids (G_s) for a sample. How much would the volume of the sample need to be changed to achieve 100% saturation if the total mass remains unchanged?

Solution

Since the sample is to be saturated *without* changing its mass, it follows that all gas (air) must be removed from the pores. We must therefore determine V_a. From the definition of the moisture content (recall Eq. (1.14)),

$$w = \left(\frac{W_w}{W_s} \right) \Rightarrow W_w = w W_s \quad (1.21.1)$$

The total weight of the sample is then written as

$$W = Mg = W_w + W_s = (1 + w) W_s \Rightarrow W_s = \frac{Mg}{1 + w} \quad (1.21.2)$$

and

$$W_w = \frac{w M g}{1 + w} \quad (1.21.3)$$

It is next necessary to determine the volume of each constituent

$$V_s = \frac{W_s}{G_s \gamma_w} = \frac{Mg}{(1 + w)(G_s \gamma_w)} \quad (1.21.4)$$

$$V_w = \frac{W_w}{\gamma_w} = \frac{w M g}{(1 + w) \gamma_w} \quad (1.21.5)$$

The volume of the air is thus

$$V_a = V - V_w - V_s = V - \frac{w M g}{(1 + w) \gamma_w} - \frac{Mg}{(1 + w)(G_s \gamma_w)} \quad (1.21.6)$$

or

$$V_a = V - \frac{Mg}{(1 + w) \gamma_w} \left(w + \frac{1}{G_s} \right) \quad (1.21.7)$$

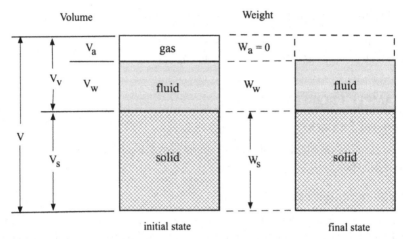

FIGURE EX. 1.21 Phase diagrams showing relationship between volume and weight for the initial and final states of the soil.

For example, if $M = 160$ g, $V = 80$ cm^3, $w = 20\%$, and $G_s = 2.70$, then

$$V_a = (80 \text{ cm}^3) - \frac{(0.160 \text{ kg})(9.81 \text{ m/s}^2)\left(\dfrac{\text{kN}}{1000 \text{ N}}\right)}{(1 + 0.20)(9.81 \text{ kN/m}^3)}$$

$$\times \left(0.20 + \frac{1}{2.70}\right)\left(\frac{100 \text{ cm}}{\text{m}}\right)^3 = \textbf{3.951 cm}^3 \qquad (1.21.8)$$

Thus, to saturate the soil sample, its initial volume of 80 cm^3 must be reduced by 3.951 cm^3. Figure Ex. 1.21 schematically illustrates the phase diagrams for the initial and final states of the soil.

EXAMPLE PROBLEM 1.22

General Remarks

Phase relationships are commonly used to quantify the results of laboratory tests on soils. This example problem illustrates the use of such relationships to check the degree of saturation in a consolidation test.

Problem Statement

A one-dimensional consolidation test was performed on a clay sample that was 6.4 cm in diameter and 2.54 cm thick. Following the test, questions were raised if the sample was indeed saturated throughout the test. Recall that in a one-dimensional consolidation test the only deformation that takes place in the sample is in the direction of load application; i.e., *vertically*. The rigid metal specimen ring prevents all lateral deformation.

The initial void ratio (e_0) of the sample was 3.356, the initial moisture content (w_0) was 105.7%, and the clay was found to have a specific gravity of solids equal to 2.70. The dry weight of the sample was 0.497 N. At the end of the test the final void ratio (e_f) was 2.638 and the final moisture content (w_f) was 95.1%. Compute (a) the dry unit weight (γ_d), (b) the moist unit weight (γ), and (c) the degree of saturation (S) of the sample at the end of the test. Was the sample saturated at the end of the test?

Solution

a) The initial total volume of the soil sample is

$$V_0 = \frac{\pi}{4}(6.4 \text{ cm})^2(2.54 \text{ cm}) = 81.71 \text{ cm}^3 \left(\frac{m}{100 \text{ cm}}\right)^3 = 8.171 \times 10^{-5} \text{ m}^3$$

(1.22.1)

At the end of the test the change in void ratio is

$$\Delta e = e_0 - e_f = 3.356 - 2.638 = 0.718$$

(1.22.2)

From the definition of the volumetric strain for infinitesimal kinematics,

$$\frac{\Delta V}{V_0} = \left(\frac{\Delta e}{1+e_0}\right) \Rightarrow \Delta V = \left(\frac{\Delta e}{1+e_0}\right)V_0 = \left(\frac{0.718}{1+3.356}\right)(8.171 \times 10^{-5} \text{ m}^3)$$
$$= 1.347 \times 10^{-5} \text{ m}^3$$

(1.22.3)

The final volume of the sample is thus

$$V_f = V_0 - \Delta V = (8.171 \times 10^{-5} \text{ m}^3) - (1.347 \times 10^{-5} \text{ m}^3) = 6.824 \times 10^{-5} \text{ m}^3$$

(1.22.4)

Since the weight of the solid phase remains unchanged, the dry unit weight at the end of the test will thus be

$$\gamma_d = \frac{W_s}{V_f} = \frac{(0.497 \text{ N})(1 \text{ kN}/1000 \text{ N})}{6.824 \times 10^{-5} \text{ m}^3} = 7.28 \text{ kN/m}^3$$

(1.22.5)

The dry unit weight can likewise be computed as follows:

$$\gamma_d = \frac{G_s\gamma_w}{1+e_f} = \frac{(2.70)(9.81 \text{ kN/m}^3)}{1+2.638} = 7.28 \text{ kN/m}^3$$

(1.22.6)

b) In either case, the moist unit weight at the end of the test is next computed as follows

$$\gamma = \gamma_d(1+w_f) = (7.28 \text{ kN/m}^3)(1+0.951) = 14.21 \text{ kN/m}^3$$

(1.22.7)

c) The degree of saturation at the end of the test is

$$S_f = \frac{G_s w_f}{e_f} = \frac{(2.70)(0.951)}{2.638} = \mathbf{97.3\%} \qquad (1.22.8)$$

The sample was thus *not* fully saturated at the end of the one-dimensional consolidation test. This could have resulted if the sample was allowed to partially dry out.

The aforementioned results could also have been obtained in an alternate fashion. Calculating first the degree of saturation (S_f), the moist unit weight at the end of the test is computed as follows

$$\gamma = \frac{\gamma_w(G_s + S_f e_f)}{1 + e_f} = \frac{(9.81 \text{ kN/m}^3)[2.70 + (0.973)(2.638)]}{1 + 2.638} = \mathbf{14.2 \text{ kN/m}^3}$$

$$(1.22.9)$$

Finally, the dry density is again computed using the value of the moisture content at the end of the test; i.e.,

$$\gamma_d = \frac{\gamma}{1 + w_f} = \frac{14.21 \text{ kN/m}^3}{1 + 0.951} = \mathbf{7.28 \text{ kN/m}^3} \qquad (1.22.10)$$

To further investigate the state of the sample at the end of the consolidation test, the associated weights and volumes can easily be computed. For example, since the weight of solids (W_s) is equal to 0.497 N, the volume of the solid phase is computed from the known value of G_s; i.e.,

$$G_s = \frac{W_s}{V_s \gamma_w} \Rightarrow V_s = \frac{W_s}{G_s \gamma_w} = \frac{(0.497 \text{ N})(1 \text{ kN/1000 N})}{(2.70)(9.81 \text{ kN/m}^3)} = 1.876 \times 10^{-5} \text{ m}^3$$

$$(1.22.11)$$

The volume of the voids is thus

$$V_v = V_f - V_s = 6.824 \times 10^{-5} - 1.876 \times 10^{-5} = 4.948 \times 10^{-5} \text{ m}^3$$

$$(1.22.12)$$

As a check, the void ratio at the end of the test is computed

$$e_f = \frac{V_v}{V_s} = \frac{4.948 \times 10^{-5} \text{ m}^3}{1.876 \times 10^{-5} \text{ m}^3} = 2.638 \qquad (1.22.13)$$

and agrees with the given value. The weight of the fluid phase is computed from the final moisture content as follows:

$$W_w = w_f W_s = (0.951)[(0.497 \text{ N})(1 \text{ kN/1000 N})] = 4.726 \times 10^{-4} \text{ kN}$$

$$(1.22.14)$$

The volume of the fluid phase is computed from the unit weight of water; i.e.,

$$\gamma_w = \frac{W_w}{V_w} \Rightarrow V_w = \frac{W_w}{\gamma_w} = \frac{4.726 \times 10^{-4} \text{ kN}}{9.81 \text{ kN/m}^3} = 4.818 \times 10^{-5} \text{ m}^2 \quad (1.22.15)$$

The volume of the air in the voids is then

$$V_a = V_v - V_w = 4.948 \times 10^{-5} - 4.818 \times 10^{-5} = 1.300 \times 10^{-6} \text{ m}^3$$
$$(1.22.16)$$

As a final check, the degree of saturation is computed

$$S = \left(\frac{V_w}{V_v}\right) * 100\% = \left(\frac{4.818 \times 10^{-5} \text{ m}^2}{4.948 \times 10^{-5} \text{ m}^3}\right) * 100\% = 97.4\% \quad (1.22.17)$$

which agrees with the previous result.

EXAMPLE PROBLEM 1.23

General Remarks

This example problem illustrates the manner in which two different pore fluids are handled in determining phase relations for an unsaturated soil.

Problem Statement

A soil in the Gulf Coast region was found to have been contaminated by crude oil. The voids of this soil consist of 20% (by volume) air, 35% seawater (specific gravity $= 1.025$), and 45% crude oil (specific gravity $= 0.876$). Given that the specific gravity of solids (G_s) is equal to 2.71 and that the moisture content (w) is 17%, determine (a) the void ratio (e) and porosity (n), (b) the moist unit weight, now defined by

$$\gamma = \frac{W_{oil} + W_w + W_s}{V}$$

where W_{oil} is the weight of the crude oil, and W_w is the weight of the seawater, (c) the dry unit weight, and (d) the degree of saturation, now defined by

$$S = \left(\frac{V_{oil} + V_w}{V_v}\right) * 100\%$$

where V_{oil} is the volume of the crude oil, and V_w is the volume of the seawater.

Solution

Assuming $V_s = 1$ implies that $V_v = e$. Thus,

$$V_a = 0.20V_v = 0.20e; \quad V_{oil} = 0.45V_v = 0.45e; \quad V_w = 0.35V_v = 0.35e$$
$$(1.23.1)$$

a) The unit weight of seawater is

$$\frac{W_w}{V_w} = 1.025\gamma_w \quad (1.23.2)$$

From the definition of the moisture content and specific gravity of solids,

$$w = \frac{W_w}{W_s} \Rightarrow W_w = wW_s = wG_sV_s\gamma_w \qquad (1.23.3)$$

where $V_s = 1$. Substituting for W_w and V_w into the expression for the unit weight of seawater given in the problem statement and solving for the void ratio gives

$$\frac{wG_s\gamma_w}{0.35e} = 1.025\gamma_w \Rightarrow e = \frac{wG_s}{(1.025)(0.35)} = \frac{(0.17)(2.71)}{(1.025)(0.35)} = \mathbf{1.284} \quad (1.23.4)$$

The porosity follows immediately from

$$n = \left(\frac{e}{1+e}\right) * 100\% = \left(\frac{1.284}{1+1.284}\right) * 100\% = \mathbf{56.2\%} \qquad (1.23.5)$$

b) To compute the moist unit weight, note that the specific gravity for the oil is

$$G_{oil} = \frac{W_{oil}}{V_{oil}\gamma_w} \qquad (1.23.6)$$

Thus,

$$\gamma = \frac{W_{oil} + W_w + W_s}{V} = \frac{G_{oil}V_{oil}\gamma_w + wG_s\gamma_w + G_s\gamma_w}{1+e}$$
$$= \frac{[G_{oil}V_{oil} + G_s(1+w)]\gamma_w}{1+e} \qquad (1.23.7)$$

Substituting the given values gives the final result; i.e.,

$$\gamma = \frac{[(0.876)(0.45)(1.284) + (2.71)(1+0.17)](9.81 \text{ kN/m}^3)}{1+1.284} = \mathbf{15.79 \text{ kN/m}^3} \quad (1.23.8)$$

c) The dry unit weight is computed as follows:

$$\gamma_d = \frac{G_s\gamma_w}{1+e} = \frac{(2.71)(9.81 \text{ kN/m}^3)}{1+1.284} = \mathbf{11.64 \text{ kN/m}^3} \qquad (1.23.9)$$

d) Finally, the degree of saturation is computed as follows

$$S = \left(\frac{V_{oil} + V_w}{V_v}\right) * 100\% = \frac{[(0.45+0.35)e]}{e} * 100\% = \mathbf{80.0\%} \quad (1.23.10)$$

EXAMPLE PROBLEM 1.24

General Remarks

This problem illustrates that the values of most quantities used to describe the phase relations fall within fairly narrow ranges.

Problem Statement

The soil at a given site has been found to have an average moisture content (w) of 30% and a degree of saturation (S) equal to 88%. The specific gravity of solids (G_s) for the soil is, however, unknown. Assuming a typical range of G_s values, determine (a) The void ratio (e), (b) the dry unit weight (γ_d), and (c) the moist unit weight (γ) of the soil.

Solution

a) The void ratio is computed using the expression derived in Case 1.3, i.e.,

$$e = \frac{G_s w}{S} \tag{1.24.1}$$

b) The dry unit weight is next computed using the expression derived in Case 1.4, i.e.,

$$\gamma_d = \frac{G_s \gamma_w}{1+e} \tag{1.24.2}$$

c) The dry unit weight is next computed using the expression derived in Case 1.6, i.e.,

$$\gamma = \gamma_d (1 + w) \tag{1.24.3}$$

For soils, the specific gravity of solids typically ranges between 2.65 and 2.75. For example, for $G_s = 2.65$,

$$e = \frac{G_s w}{S} = \frac{(2.65)(0.30)}{0.88} = \mathbf{0.903} \tag{1.24.4}$$

$$\gamma_d = \frac{G_s \gamma_w}{1+e} = \frac{(2.65)(9.81 \text{ kN/m}^3)}{1+0.903} = \mathbf{13.66 \text{ kN/m}^3} \tag{1.24.5}$$

and

$$\gamma = \gamma_d(1 + w) = (13.66 \text{ kN/m}^3)(1 + 0.30) = \mathbf{17.76 \text{ kN/m}^3} \tag{1.24.6}$$

Since the aforementioned calculations are repeated for the various values of G_s, they are particularly well suited for execution using a spreadsheet.

TABLE EX. 1.23 Sample Spreadsheet Used to Compute Void Ratio and Unit Weights for a Given Value of Specific Gravity of Solids

G_s	e	γ_d (kN/m³)	γ (kN/m³)
2.65	0.903	13.66	17.76
2.66	0.907	13.68	17.79
2.67	0.910	13.71	17.83
2.68	0.914	13.74	17.86
2.69	0.917	13.77	17.90
2.70	0.920	13.79	17.93
2.71	0.924	13.82	17.96
2.72	0.927	13.85	18.00
2.73	0.931	13.87	18.03
2.74	0.934	13.90	18.07
2.75	0.938	13.92	18.10

Table Ex. 1.23 shows a sample spreadsheet that was set up to compute the void ratio and the dry and moist unit weights. From this table it is evident that changes in G_s have a relatively minor effect on the magnitude of the dry and moist unit weights.

Chapter 2

Example Problems Related to Soil Identification and Classification

2.0 GENERAL COMMENTS

One of the foremost aims in soil mechanics has been to find methods for discriminating between different kinds of soil in a given category. The properties on which these distinctions are based are known as *index properties*. The tests required to determine the index properties are referred to as *classification tests*. When discussing soil grain properties it is convenient to divide soil into *cohesionless* (gravel, sand, silt) and *cohesive* (clay) soils.

The nature of any soil can be altered by appropriate manipulation. For example, vibrations transform a loose sand into a dense one. The behavior of a soil thus depends not only on the significant properties of the *individual* constituents of the soil mass, but also on those properties that are due to the *arrangement* of the particles within the mass. It is thus convenient to divide index properties into two classes:

- *Soil aggregate properties.* This involves analyzing a soil mass in its intact state (as much as possible). The stress history is thus taken into consideration. The most significant aggregate property of cohesionless soils is the *relative density*, whereas that of cohesive soils is the *consistency*.
- *Soil grain* (or individual) *properties.* The soil characteristics are based on the *size*, *shape*, and *distribution* of the particles. In clay soils, the mineralogical character of the smallest grains is of importance. No consideration of the stress history is included in the determination of such properties.

2.1 PARTICLE SIZES

Perhaps the most fundamental soil grain property is the size of a particle. A single linear dimension cannot uniquely define the size of a particle, other than a cube or a sphere. Thus, particle (grain) size typically refers to the *diameter* of a soil grain. The meaning of "particle size" therefore depends on the

Soil Mechanics. http://dx.doi.org/10.1016/B978-0-12-804491-9.00002-1

dimension that was recorded and how it was obtained. The sizes of soil particles vary over a very large range.

Depending on the *predominant* size of particles within a soil, it is generally referred to as being a *gravel, sand, silt,* or *clay.*

Gravels are unconsolidated rock fragments produced by mechanical weathering with occasional particles of quartz, feldspar, and other minerals.

Sand particles are produced by mechanical weathering and are made of mostly quartz and feldspar, though other mineral grains may also be present.

Silts are the microscopic soil fractions that consist of very fine quartz grains and some flake (plate)-shaped particles that are fragments of micaceous minerals; they are produced by chemical and physical weathering.

Clays are mostly flake (plate)-shaped microscopic and submicroscopic particles of mica, clay minerals, and other minerals.

- The clay minerals are products of chemical weathering of feldspars, ferromagnesians, and micas.
- The clay minerals give the plastic (putty like) property to soils.
- Clays also exhibit *cohesiveness*; i.e., the ability to stick together. As a result, such soils are thus referred to as being *cohesive*; by contrast, soils consisting of gravels, sands, and silts are referred to as *cohesionless.*
- The presence of *water* greatly affects the engineering response of clays.
- The most important clay minerals are kaolinite, illite, and montmorillonite.

Since they are larger in size, gravels and sands are referred to as being *coarse grained.* By contrast, silts and clays are referred to as being *fine grained.* A conventional dividing line between these two groups is the approximate smallest particle sizes that are visible to the naked eye, typically 0.060 or 0.075 mm.

Coarse-grained soils are thus *nonplastic* and *noncohesive* (or cohesionless). Among fine-grained soils, silts are generally *nonplastic* and *noncohesive,* while clays are both *plastic* and *cohesive.*

To describe soils by their particle size, several organizations have developed particle size classifications. Since there is no clear-cut division between the four general soil groups (i.e., gravel, sand, silt, and clay), and since particle sizes vary widely, several classifications exist, with different points of delineation. Some examples of such classifications are given below.

Remark: The range of particle sizes from 200 to 0.002 mm spans *five* (5) orders of magnitude!

Table 2.1 lists the grain size scale proposed by the Swedish chemist and agricultural scientist Albert Atterberg (1846−1916).

Table 2.2 lists the grain size scale proposed by the US Department of Agriculture (USDA).

TABLE 2.1 Grain Size Ranges According to A. Atterberg

Name of Soil	Size Range (mm)
Boulders	>200
Cobble (stone)	60–200
Coarse gravel (pebble)	20–60
Medium gravel	6–20
Fine gravel	2–6
Coarse sand	0.6–2
Medium sand	0.2–0.6
Fine sand	0.06–0.2
Coarse silt	0.02–0.06
Medium silt	0.006–0.02
Fine silt	0.002–0.006
Clay	<0.002

TABLE 2.2 Grain Size Ranges According to the US Department of Agriculture

Name of Soil	Size Range (mm)
Very coarse sand	1.00–2.00
Coarse sand	0.50–1.00
Medium sand	0.25–0.50
Fine sand	0.10–0.25
Very fine sand	0.05–0.10
Silt	0.002–0.05
Clay	<0.002

Finally, Table 2.3 lists the grain size scale proposed by the Unified Soil Classification System (USCS).

Remark: Those particles classified as being "clay" on the basis of their size may not necessarily contain clay minerals.

TABLE 2.3 Grain Size Ranges According to the Unified Soil Classification System

Name of Soil	Size Range (mm)
Boulders	>300
Cobbles	75–300
Gravel	4.75–75
Sand	0.075–4.75
Fines (silt and clay)	<0.075

Remark: Nonclay soils typically contain quartz, feldspar, or mica particles that are small enough to be less than 0.002 mm (2 μm) in size.

Remark: It is thus appropriate for particles smaller than 0.002 or 0.005 mm to be called "clay-size" particles rather than clay.

2.2 DISTRIBUTION OF GRAIN SIZES

In classifying a soil it is important to determine the size range of particles present in a soil (typically expressed as a percentage of the total dry weight). Two approaches are generally used to determine the soil particle size distribution. A *sieve analysis* is used for particle sizes greater than or equal to 0.075 mm. For particle sizes smaller than 0.075 mm, a *hydrometer analysis* is used.

Remark: Both tests significantly disturb soil samples from their in situ state.

2.2.1 Sieve Analysis

A sieve analysis involves the shaking of a soil sample through a set of sieves that has progressively smaller square openings. The results of a sieve analysis are plotted on a particle size distribution or *gradation* curve. In such a curve, the particle size (in mm) is plotted as the abscissa on a logarithmic scale. The ordinate on such curves is the percent passing a specific sieve size.

The classification of the particle size distribution of a soil is referred to as *soil gradation*. Coarse-grained soils, mainly gravels or sands, are graded as either *well graded* or *poorly graded*.

Well-graded soils contain particles having a wide range of sizes. They have a good representation of all sizes from No. 4 sieve (4.75 mm) to No. 200 sieve (0.075 mm).

Poorly graded soils do *not* have a good representation of all sizes of particles from No. 4 to No. 200 sieve. Poorly graded soils are further divided into *uniformly graded* or *gap-graded* soils.

A *uniformly graded* soil is a soil that has most of its particles at about the same size. An example of a uniformly graded soil is one in which only sand of the No. 20 sieve size (0.850 mm) is present.

A *gap-graded* soil is a soil that has an excess or deficiency of certain particle sizes or a soil that has at least one particle size *missing*. An example of a gap-graded soil is one in which sand of the No. 10 (2.00 mm) and No. 40 sizes (0.425 mm) are missing, and all the other sizes are present.

2.2.2 Quantities Computed From Gradation Curves

To facilitate the interpretation of gradation curves, let D_{10}, D_{30}, and D_{60} equal to the grain diameter (in mm) corresponding to 10%, 30%, and 60%, respectively, passing by weight (or mass).

The *coefficient of uniformity* C_u, which is a crude shape parameter, is calculated using the following equation:

$$C_u = \frac{D_{60}}{D_{10}} \qquad (2.1)$$

If $C_u = 1.0$, the soil contains primarily *one* grain size. The smaller the C_u, the *more uniform* will be the associated gradation.

The *coefficient of curvature* C_c, which is a shape parameter, is calculated using the following equation:

$$C_c = \frac{(D_{30})^2}{D_{10}D_{60}} \qquad (2.2)$$

Once the coefficient of uniformity and the coefficient of curvature have been calculated, they must be compared to published gradation criteria. For example, both C_u and C_c are used in the Unified Soil Classification System (USCS).[1] In particular, for a gravel to be classified as being *well graded*

$$C_u > 4 \quad \text{and} \quad 1 < C_c < 3 \qquad (2.3)$$

If both of these conditions are not met, the gravel is classified as being *poorly graded*. Section 2.5 gives additional details pertaining to the USCS.

If both of these criteria are met, the gravel is classified as *well graded*. If, however, both of these criteria are not met, the gravel is classified as being *poorly graded* (GP).

---

1. Section 2.5 discusses the Unified Soil Classification System in greater detail.

For a sand to be classified as *well graded*, the following criteria must be met:

$$C_u > 6 \quad \text{and} \quad 1 < C_c < 3 \qquad (2.4)$$

If both of these criteria are met, the sand is classified as *well graded*. If both of these criteria are not met, the sand is classified as being *poorly graded*.

2.2.3 Importance of Soil Gradation

Soil gradation is very important to geotechnical engineering. It is an indicator of other engineering properties such as *compressibility, shear strength*, and *hydraulic conductivity.*

In a design, the gradation of the in situ or on site soil often controls the design and ground water drainage of the site. A poorly graded soil will have better drainage than a well-graded soil because there are more void spaces in a poorly graded soil.

When a fill material is being selected for a project such as a highway embankment or earthen dam, the soil gradation is considered. A well-graded soil is able to be compacted more than a poorly graded soil. These types of projects may also have gradation requirements that must be met before the soil to be used is accepted.

When options for ground remediation techniques are being selected, the soil gradation is often a controlling factor.

2.2.4 Hydrometer Analysis

A hydrometer analysis attempts to determine the size distribution of fine-grained soils. It is based on the principle of sedimentation of soil particles settling in water. Hydrometer analyses are typically performed on particles passing the No. 200 sieve (0.074 mm). The data from a hydrometer analysis allow for the determination of the suspension density.

In determining the distribution of particle sizes, a deflocculating agent is first applied to the soil, which has been mixed with water. On mixing, the particles are separated out and form a solution. When a soil sample is dispersed in water, the particles will settle at different velocities, depending on their size, shape, weight, and on the viscosity of water.

All particles are *assumed* to be spherical (Figure 2.1).

For a spherical particle of radius r, the upward buoyancy force is equal to the volume of fluid displaced multiplied by the unit weight of water; i.e.,

$$F_b = \frac{4}{3}\pi r^3 \gamma_w \qquad (2.5)$$

The downward force is due to the weight of the particle; i.e.,

$$F_d = \frac{4}{3}\pi r^3 \gamma_s \qquad (2.6)$$

where, as before, γ_s is the unit weight of the solid phase.

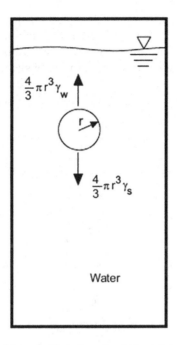

FIGURE 2.1 Schematic illustration of a single particle in suspension.

Stokes found that the net vertical force was related to the velocity of the particle according to the following relation:

$$\frac{4}{3}\pi r^3 \gamma_s - \frac{4}{3}\pi r^3 \gamma_w = 6\pi\eta r v \qquad (2.7)$$

where η is the viscosity of water and v is the velocity of the sphere as it settles in the water. Solving for the velocity and relating this to the particle diameter D gives

$$v = \frac{2}{9}\frac{r^2}{\eta}(\gamma_s - \gamma_w) = \frac{2}{9}\frac{(D/2)^2}{\eta}(\gamma_s - \gamma_w) = \frac{(\gamma_s - \gamma_w)D^2}{18\eta} \qquad (2.8)$$

Solving for this diameter gives

$$D = \sqrt{\frac{18\eta v}{(\gamma_s - \gamma_w)}}v = \sqrt{\frac{18\eta}{(\gamma_s - \gamma_w)}}\sqrt{\frac{L}{t}} \qquad (2.9)$$

Since $G_s = \gamma_s/\gamma_w \Rightarrow \gamma_s = G_s\gamma_w$, the above expression can also be written as

$$D = \sqrt{\frac{18\eta}{(G_s - 1)\gamma_w}}\sqrt{\frac{L}{t}} \qquad (2.10)$$

At a depth L (commonly called the effective length) at time t, all particles have a diameter smaller than D, that is, all larger particles would have settled beyond the zone of measurement. Hydrometers are thus designed to allow the amount of soil that is still in suspension to be determined.

In typical calculations associated with a hydrometer analysis, L is measured in centimeters (cm), the units of η are (g s)/cm^2, $\gamma_w = 1.0$ g/cm^3, t is measured in *minutes*, and it is advantageous to represent D in units of *millimeters*. The above expression for D is thus modified so as to account for the aforementioned units, giving

$$D = \sqrt{\frac{18\eta}{(G_s - 1)\gamma_w}} \sqrt{\frac{L}{t}\left(\frac{\text{min}}{60 \text{ s}}\right)} \left(\frac{10 \text{ mm}}{\text{cm}}\right) = \sqrt{\frac{30\eta}{(G_s - 1)}} \sqrt{\frac{L}{t}} \qquad (2.11)$$

The above expression is often written as

$$D = K\sqrt{\frac{L}{t}} \qquad (2.12)$$

where D has units of *millimeters*, L has units of *centimeters*, t is expressed in *minutes*, and

$$K = \sqrt{\frac{30\eta}{(G_s - 1)}} \qquad (2.13)$$

Since the viscosity η is a function of temperature, it follows that K will likewise be a function of temperature.

The standard conditions for hydrometer analyses (ASTM 152-H type hydrometer) are $G_s = 2.65$ at 20°C inside a 1000 mℓ fluid (usually distilled water) where no dispersing agent is added. If the hydrometer analysis is performed under conditions that differ from these calibration conditions, corrections need to be made. The four commonly made corrections are thus

1. *Temperature correction* (F_T): when the temperature differs from 20°C, the following correction must be made:[2]

$$F_T = -4.85 + 0.25T \qquad (2.14)$$

where the test temperature T is between 15 and 28°C.
2. *Specific gravity correction* (a): when the specific gravity of the solids is not 2.65, the following correction must be made:

$$a = \frac{1.65G_s}{2.65(G_s - 1)} \qquad (2.15)$$

3. *Zero correction* (F_z): if a dispersing (or deflocculating) agent is added to the soil−distilled water suspension, the zero reading from the hydrometer

---

2. Das, B.M., 2013. Soil Mechanics Laboratory Manual, eighth edition. Oxford University Press, New York.

will be changed. Thus, a correction to the deviation of reading from the zero mark of the hydrometer is required.

4. *Meniscus correction* (F_m): accounts for the reading error introduced when readings are taken off the upper level of the meniscus formed, at the stem of the hydrometer, by the soil–water suspension.

The effective length (L) must be related to the hydrometer reading (R), which ranges between 0 and 60 g/L. The magnitude of L is often computed from the following expression:

$$L = L_1 + \frac{1}{2}\left(L_2 - \frac{V_b}{A_c}\right) \tag{2.16}$$

where L_1 is the distance along the stem of the hydrometer from the top of its bulb to the mark for a specific hydrometer reading, L_2 is the length of the hydrometer bulb (typically equal to 14 cm), V_b is the volume of the hydrometer bulb, and A_c is the cross-sectional area of the sedimentation cylinder (typically equal to 27.8 cm^2).

Typically, L_1 is equal to 10.5 cm for a reading of $R = 0$, and 2.3 cm for a value of $R = 50$. Assuming a linear relationship between L_1 and R, it follows that for any value of R,

$$L_1 = 10.5 - \frac{(10.5 - 2.3)}{50}R = 10.5 - 0.164R \tag{2.17}$$

where L_1 and R will have the same units, and the hydrometer reading R is corrected for the meniscus. In this case, the general expression for L becomes

$$L = 10.5 - 0.164R + \frac{1}{2}\left(14.0 - \frac{V_b}{A_c}\right) \tag{2.18}$$

where L and R will have units of cm, A_c will have units of cm^2, and V_b will have units of cm^3.

2.3 PLASTICITY OF SOIL

As noted in Section 2.1, when mixed with a limited amount of water, clays exhibit *plastic* (putty like) behavior characterized by deformation *without* complete rebound on removal of load. It is necessary to somehow quantify the plasticity of predominantly clayey soils; this will tell if a soil contains clay and if so, how much.

As also noted in Section 2.1, particles classified as being "clay" on the basis of their size may not necessarily contain clay minerals. Nonclay soils typically contain quartz, feldspar, or mica particles that are small enough to be less than 0.002 mm (2 μm) in size. It is thus appropriate for particles smaller than 0.002 or 0.005 mm to be called "clay-size" particles rather than clay.

TABLE 2.4 Dimensions of Typical Clay Platelets[a]

Clay Mineral	Ratio of Dimensions	Range of Lengths and Breadths ($\times 10^{-10}$ m)	Range of Thicknesses ($\times 10^{-10}$ m)
Kaolinite	10:10:1	1000–2000	100–1000
Illite	20:20:1	1000–5000	50–5000
Montmorillonite	100:100:1	1000–5000	10–50

[a]Hough, B.K.,1957. Basic Soils Engineering. The Ronald Press Company, New York, NY.

Unlike gravels and sands that consist exclusively of quartz and feldspar, clays are composed of silicate members that can have different mineralogy. The predominant clay minerals are *kaolinite, illite*, and *montmorillonite*.

Clay particles are plate-like (flakey); i.e., they have a sheet-like shape. Table 2.4 lists some typical dimensions for the above clay minerals.

The faces of clay particles are negatively charged, and their edges are positively charged (Figure 2.2A). As a result, clay particles attract water molecules (dipoles), creating layers of tightly bound *adsorbed water* (Figure 2.2B).

In light of the above discussion, it is evident that gradation alone (e.g., sieve analysis) is *insufficient* to classify fine-grained soils. This is because the behavior of clays is affected by (1) *Mineralogy* (it controls the particle surface characteristics), and (2) *Specific surface* (i.e., surface area per unit mass of a dry particle), which determines the particle shapes. Table 2.5 presents some approximate specific surface values.

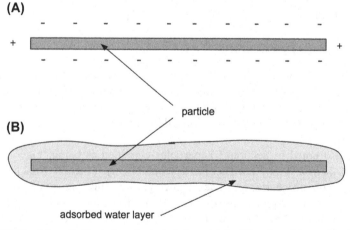

FIGURE 2.2 Schematic illustration of a typical clay particle: (A) surface charges, (B) adsorbed water layer.

TABLE 2.5 Approximate Values of Specific Surface for Common Clay Minerals[a]

Soil Particle	Specific Surface (m^2/g)
Clean sand	0.0002
Kaolinite	10–20
Illite	65–100
Montmorillonite	Up to 840

[a]*Grim, R.E., 1959. Physico-chemical properties of soils: clay minerals. Journal of the Soil Mechanics and Foundations Division, ASCE 85 (SM2), 1–17.*

Both mineralogy and specific surface are needed to properly classify clays. However, the measurement of both of these quantities is *not* practical for standard engineering applications. Instead, the nature of clays needs to be described by determining the degree of plasticity, which is attributed to the adsorbed water layers that surround a clay particle (Figure 2.2B).

2.4 ATTERBERG LIMITS

The Atterberg limits are a basic measure of the nature or consistency of a fine-grained soil. These limits correspond to specific values of moisture content (w). They were created by Albert Atterberg (1910) and later refined by Arthur Casagrande (1927). The Atterberg limits can be used to distinguish between a silt and a clay. In addition, and they can distinguish between different types of silts and clays.

Depending on the moisture content of the soil, it may appear in four states: *solid*, *semisolid*, *plastic*, and *liquid* (Figure 2.3). In each state the consistency and behavior of a soil is different and thus so are its engineering properties. Thus, the boundary between each state can be defined based on a change in the soil's behavior.

2.4.1 Basic Definitions

The Atterberg limits consist of the following key values of moisture content:

- The *Liquid Limit* (LL) is the moisture content at which a fine-grained soil no longer flows like a liquid.
- The *Plastic Limit* (PL) is the moisture content at which a fine-grained soil can no longer be remolded without cracking.
- The *Shrinkage Limit* (SL) is the moisture content at which a fine-grained soil no longer changes volume upon drying—any loss of moisture is compensated by the entry of air into the pores.

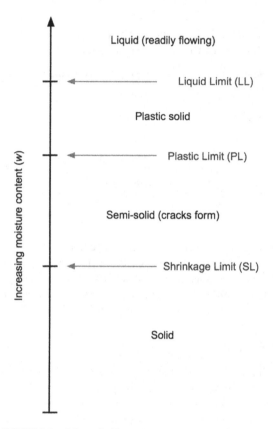

FIGURE 2.3 Schematic illustration of material states in a soil.

Remark: Whereas the LL and PL limits are *arbitrary* limits, the SL is a *definite* limit for a given soil.

Remark: The SL is useful for determining the swelling and shrinkage capacity of soils. In general, "soils that swell a lot will also shrink a lot."

Remark: Since the Atterberg limit tests are performed on remolded samples, the previous stress history of the soil is completely removed.

The Atterberg limits are useful in that they allow for soil behavior to be inferred. For example, soil having similar LL and plasticity index (PI) will typically have similar strength/water content relationships. Thus, if such a relationship is known for one soil, it can be inferred for a soil with similar Atterberg limits.

2.4.2 Derived Limits

The values of these limits are computed from the Atterberg limits (i.e., from LL, PL, and SL). There is also a close relationship between the limits and properties of a soil such as *compressibility, permeability*, and *strength*. This is thought to be very useful because as the determination of Atterberg limits is relatively simple, it is more difficult to determine compressibility, permeability, and strength. The Atterberg limits are thus not only used to identify the soil's classification, but also allow for the use of empirical correlations for some other important engineering properties.

2.4.2.1 Plasticity Index

The PI is defined as the range of moisture contents over which the soil deforms plastically. The PI is thus defined to be the difference between the LL and the PL; i.e.,

$$PI = LL - PL \qquad (2.19)$$

The PI thus is a measure of the plasticity of a soil. As such, the PI determines the amount and type of clay present in a soil. In general,

- Soils with a high PI tend to be clay,
- Those with a lower PI tend to be silt, and
- Those with a PI near zero tend to have little or no silt or clay (fines) present.

2.4.2.2 Liquidity Index

The liquidity index (LI) is used for scaling the natural water content of a soil sample to the limits. It is defined as follows:

$$LI = \frac{w - PL}{LL - PL} = \frac{w - PL}{PI} \qquad (2.20)$$

where w is the natural water content. The LI serves as a measure of soil strength. In particular,

- If $LI < 0$, the soil is in a *semisolid* state characterized by high strength and brittle response characterized by sudden fracture of the soil.
- If $0 < LI < 1$, the soil is in a *plastic* state characterized by intermediate strength; it deforms like a plastic material.
- If $LI > 1$, the soil is in a *liquid* state characterized by low strength; it deforms like a viscous fluid.

Sensitive clays are ones that, when remolded, can be transformed into a viscous form that will flow like a fluid. For such soils the in situ moisture content may be greater than the LL, implying that LI > 1.

2.4.2.3 Activity

If a small amount of sand is added to a clay, the LL and PL for the soil will decrease. The LL and PL are thus functions of not only the type of clay present, but also the amount of clay present.

Skempton[3] observed that the PI of a given soil increased approximately linearly with the percent of clay-size fraction. He defined the *activity* of a clay as

$$A = \frac{\text{PI}}{\text{percent of clay} - \text{size fraction}} \tag{2.21}$$

where the percent of clay-size fraction equals the percent by weight finer than 0.002 mm.

From the activity it is often possible to predict the dominant clay mineral present in a soil sample, and thus *indirectly* the specific surface. Table 2.6 gives ranges of activity values associated with the three most common clay minerals.

- For $0.75 \leq A \leq 1.25$, the clay is considered to be "normal".
- For $A < 0.75$, the clay is considered to be "inactive".
- For $A > 1.25$, the clay is considered to be "active".

Remark: Soils with high activity are very reactive chemically.

Remark: High activity signifies large volume increase in a soil that is wetted and large shrinkage when it is dried.

Remark: Graphically, the activity is the slope of a straight line drawn on a figure with PI as the ordinate and percentage of clay-size fraction as the abscissa. Such lines will not always pass through the origin.

---

3. Skempton, A.W., 1953. The Colloidal Activity of Clays. In: Proceedings of the Third International Conference on Soil Mechanics and Foundation Engineering (I), 57–61.

TABLE 2.6 Typical Values of Liquid Limit, Plastic Limit, and Activity

Clay Mineral	LL	PL	Activity (A)
Kaolinite	35–100	20–40	0.3–0.5
Illite	60–12	35–60	0.5–1.3
Montmorillonite	100–900	50–100	1.5–7.0

2.5 SOIL CLASSIFICATION

To describe soils more effectively based on particle size distributions, plasticity, etc., several soil classifications schemes have been developed. In this section the Unified Soil Classification System (USCS) is briefly reviewed. Additional details pertaining to this soil classification system are found in most soil mechanics textbooks.

The USCS uses the following prefix symbols to describe the various soil groups:

G = gravel or gravelly soil.
S = sand or sandy soil.
M = inorganic silt.
C = inorganic clay.
O = organic[4] silt or clay.
Pt = peat, muck, and other highly organic soils.

The above symbols are combined with the following descriptor symbols:

W = well graded.
P = poorly graded.
L = low plasticity (LL < 50%).
H = high plasticity (LL ≥ 50%).

Thus, SP denotes a poorly graded sand; CL denotes a clayey soil with low plasticity, etc. Such combinations of symbols constitute the following group symbols:
- For *coarse-grained* soils (i.e., those with more than 50% retained on the No. 200 sieve):
    The following symbols are used to classify *gravels* ("G" prefix):
    GW = well-graded gravel.

---

4. Organic matter is matter that comes from a once-living organism; it is capable of decay, or is the product of decay, or is composed of organic compounds. In a decaying material W_s and V_s may decrease with time.

GP = poorly graded gravel.

GM = silty gravel (i.e., nonclay fines).

GC = clayey gravel.

In addition, the description of gravels with 5% to 12% fines requires the following dual symbols:

GW-GM = well-graded gravel with silt.

GW-GC = well-graded gravel with clay.

GP-GM = poorly graded sand with silt.

GP-GC = poorly graded sand with clay.

The following symbols are used to classify *sand* ("S" prefix):

SW = well-graded sand.

SP = poorly graded sand.

SM = silty sand (i.e., nonclay fines).

SC = clayey sand.

In addition, the description of sand with 5% to 12% fines requires the following dual symbols:

SW-SM = well-graded sand with silt.

SW-SC = well-graded sand with clay.

SP-SM = poorly graded sand with silt.

SP-SC = poorly graded sand with clay.

- For *fine-grained* soils (i.e., those with 50% or more passing No. 200 sieve):

  For silts and clays with LL < 50%:

ML = inorganic silt.

CL = inorganic clay.

OL = organic silt and organic clay.

For silts and clays with LL $\geq$ 50%:

MH = inorganic silt.

CH = inorganic clay.

OH = organic silt and organic clay.

In addition, fines may be classified as CL-ML.

Fine-grained soils (i.e., silts and clays) are classified according to their Atterberg limits and whether or not they contain organic matter. Figure 2.4 shows the plasticity chart developed by Casagrande[5], based on the results of tests performed on soils from throughout the world.

In Casagrande's plasticity chart a value of LL = 50% delineates low plasticity silts and clays from high plasticity ones. In Figure 2.4 this dividing value is represented by the so-called "B-line".

5. Casagrande, A., 1948. Classification and Identification of Soils. Transactions, ASCE 113, 901–930.

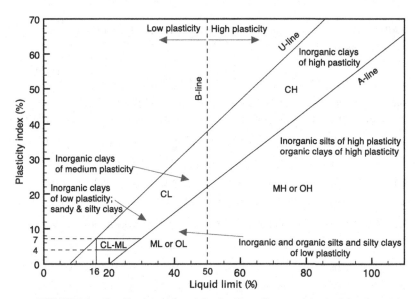

FIGURE 2.4   Classification of fine-grained soils on Casagrande's plasticity chart.

The "A-line" in Figure 2.4 is defined by

$$PI = 0.73(LL - 20) \qquad (2.22)$$

where LL and PI are expressed in percent. It represents the boundary between inorganic and organic soils.

The "U-line" in Figure 2.4 is defined by

$$PI = 0.9(LL - 8) \qquad (2.23)$$

where LL and PI are again expressed in percent. It represents the upper limit of the correlation between PI and LL for any currently known fine-grained soil.

EXAMPLE PROBLEM 2.1

General Remarks

This example problem illustrates the manner in which data from sieve analyses are manipulated so as to construct a particle-size distribution curve.

Problem Statement

A sieve analysis was performed on a clean gravelly sand. The first three columns of Table Ex. 2.1A summarize the data from this analysis. Given these data,

TABLE EX. 2.1A Results of Sieve Analyses Performed on a Clean Gravelly Sand

Sieve Number	Sieve Opening (mm)	Mass of Soil Retained (g)	Cumulative Mass of Soil Retained on Sieve (g)	Percent Finer Than a Given Sieve Size
4	4.75	0	0	100.0
7	2.80	492	0 + 492 = 492	77.7
18	1.00	898	492 + 898 = 1390	37.0
40	0.425	295	1390 + 295 = 1685	23.7
50	0.355	213	1685 + 213 = 1898	14.0
80	0.180	130	1898 + 130 = 2028	8.2
170	0.090	160	2028 + 160 = 2188	0.9
Pan	–	20	2188 + 20 = 2208	0.0
Total	–	2208	–	–

a) Plot the particle-size distribution curve, and,

b) Determine the coefficient of uniformity (C_u) and the coefficient of gradation (C_c).

Solution

a) The fourth column shows how the cumulative mass of soil retained on a given sieve is computed. In the fifth column is the percent finer than a given sieve size, which is computed as follows:

$$\%finer = \left(\frac{M_{total} - M_{cumulative}}{M_{total}}\right) * 100\% \qquad (2.1.1)$$

where $M_{cumulative}$ and M_{total} are the cumulative mass of soil retained on sieve and the total mass, respectively.

Remark: The above calculations are particularly well suited for being performed using a spreadsheet.

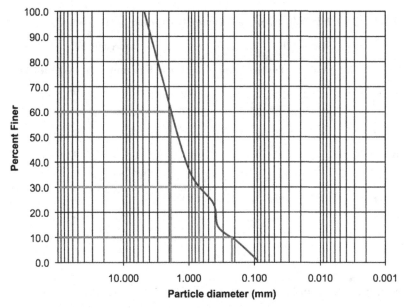

FIGURE EX. 2.1   Particle size distribution for a clean gravelly sand.

Figure Ex. 2.1 shows the particle size distribution for the clean gravelly sand.

From Figure Ex. 2.1 the diameters corresponding to 60% finer (D_{60}), 30% finer (D_{30}), and 10% finer (D_{10}) are found to be approximately 1.90, 0.70, and 0.22 mm, respectively. The coefficient of uniformity is thus

$$C_u = \frac{D_{60}}{D_{10}} = \frac{1.90}{0.22} = \textbf{8.6} \qquad (2.1.2)$$

Finally, coefficient of gradation is

$$C_c = \frac{(D_{30})^2}{D_{10}D_{60}} = \frac{(0.70)^2}{(0.22)(1.90)} = \textbf{1.17} \qquad (2.1.3)$$

EXAMPLE PROBLEM 2.2

General Remarks

This example problem further illustrates the manner in which particle-size distribution curves are drawn, and the coefficients that are computed from them.

TABLE EX. 2.2 Results of Sieve Analyses Performed on Two Soils

Glacial Till		Inorganic Silt	
Particle Size (mm)	Percent Finer	Particle Size (mm)	Percent Finer
0.295	97	0.074	96
0.147	94	0.050	89
0.074	69	0.030	72
0.055	48	0.015	47
0.035	22	0.095	34
0.025	6	0.0045	18
0.015	1	0.0015	8

Problem Statement

Sieve analyses were performed on two soils. Table Ex. 2.2 summarizes the results of these analyses. For each soil,

a) Plot the particle-size distribution curve, and,
b) Determine the coefficient of uniformity (C_u) and the coefficient of gradation (C_c).

Solution

Figure Ex. 2.2A shows the particle size distribution for the glacial till.

From Figure Ex. 2.2A the diameters corresponding to 60% finer (D_{60}), 30% finer (D_{30}), and 10% finer (D_{10}) are found to be 0.300, 0.040, and 0.007 mm, respectively. The coefficient of uniformity is thus

$$C_u = \frac{D_{60}}{D_{10}} = \frac{0.300}{0.007} = \mathbf{42.9} \qquad (2.2.1)$$

Finally, coefficient of gradation is

$$C_c = \frac{(D_{30})^2}{D_{10}D_{60}} = \frac{(0.040)^2}{(0.007)(0.300)} = \mathbf{0.76} \qquad (2.2.2)$$

Figure Ex. 2.2B shows the particle size distribution for the inorganic silt.

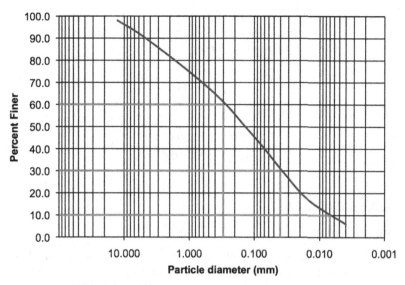

FIGURE EX. 2.2A   Particle size distribution for a glacial till.

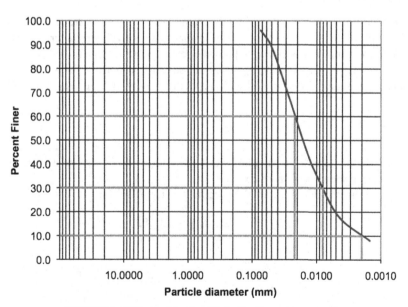

FIGURE EX. 2.2B   Particle size distribution for an inorganic silt.

From Figure Ex. 2.2B the diameters corresponding to 60% finer (D_{60}), 30% finer (D_{30}), and 10% finer (D_{10}) are found to be 0.022, 0.008, and 0.002 mm, respectively. The coefficient of uniformity is thus

$$C_u = \frac{D_{60}}{D_{10}} = \frac{0.022}{0.002} = \textbf{11.0} \qquad (2.2.3)$$

Finally, coefficient of gradation is

$$C_c = \frac{(D_{30})^2}{D_{10}D_{60}} = \frac{(0.008)^2}{(0.002)(0.022)} = \textbf{1.46} \qquad (2.2.4)$$

EXAMPLE PROBLEM 2.3

General Remarks

This example problem investigates the relationship between the initial suspension density (ϕ_0), the specific gravity of solids (G_s) and the initial concentration in a hydrometer analysis.

Problem Statement

Since the total weight (W) and total volume (V) of a sample are known, and since all of the soil is in suspension at the outset of a hydrometer test, the initial concentration is likewise known. Determine (a) a general relationship between ϕ_0, W, and V, and (b) the initial suspension density for a liter suspension containing 49 g of soil with a specific gravity of solids equal to 2.73.

Solution

a) Since the sample is saturated, its volume and weight are related in the manner shown in Figure Ex. 2.3.

From the definition of the specific gravity of solids,

$$G_s = \frac{W_s}{V_s \gamma_w} \Rightarrow V_s = \frac{W_s}{G_s \gamma_w} \qquad (2.3.1)$$

Thus,

$$V_w = V - V_s = V - \frac{W_s}{G_s \gamma_w} \qquad (2.3.2)$$

From the definition of the unit weight of water,

$$\gamma_w = \frac{W_w}{V_w} \Rightarrow W_w = \gamma_w V_w = \gamma_w \left(V - \frac{W_s}{G_s \gamma_w} \right) = V \gamma_w - \frac{W_s}{G_s} \qquad (2.3.3)$$

Volume   Weight

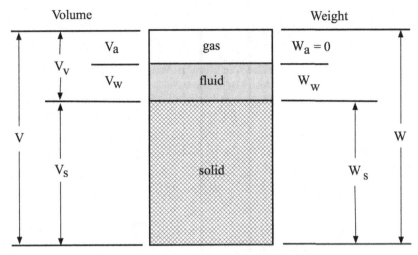

FIGURE EX. 2.3 Phase relationship associated with a hydrometer test.

The initial suspension density is thus

$$\phi_0 = \frac{W_s + W_w}{V} = \frac{W_s + \left(V\gamma_w - \dfrac{W_s}{G_s}\right)}{V} = \frac{W_s}{V}\left(1 - \frac{1}{G_s}\right) + \gamma_w \qquad (2.3.4)$$

Recalling the definition of the dry unit weight given in Eq. (1.9); i.e.,

$$\gamma_d = \frac{W_s}{V} \qquad (2.3.5)$$

The initial suspension density can likewise be written as

$$\phi_0 = \gamma_d\left(1 - \frac{1}{G_s}\right) + \gamma_w \qquad (2.3.6)$$

b) The initial suspension density for a liter suspension containing 49 g of soil with a specific gravity of solids equal to 2.73 is thus

$$\phi_0 = \frac{(49\ \text{g})}{(1000\ \text{cm}^3)}\left(1 - \frac{1}{2.73}\right) + \left(1.0\text{g/cm}^3\right) = \mathbf{1.031\text{g/cm}^3} \qquad (2.3.7)$$

EXAMPLE PROBLEM 2.4

General Remarks

This example problem illustrates the sample calculations that are performed for a hydrometer analysis that was run using an ASTM 152-H hydrometer (Figure Ex. 2.4A).

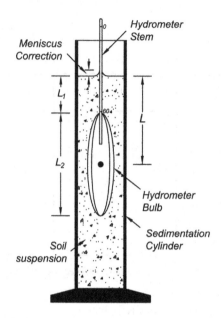

FIGURE EX. 2.4A Hydrometer in a soil suspension.

Problem Statement

Given the soil data shown in Table Ex. 2.4A, the hydrometer calibration information in Table Ex. 2.4B, and the results of a hydrometer test shown in Table Ex. 2.4C, draw the grain size distribution curve for the soil.

TABLE EX. 2.4A Soil Data Associated With a Hydrometer Analysis	
Mass Retained on No. 200 Sieve	0.0 g
Mass in suspension (M_s)	48.0 g
Specific gravity (G_s)	2.70
Meniscus correction (F_m)	0.5 g/L
Dispersing agent correction (F_z)	4.0 g/L
Diameter of sedimentation cylinder	5.95 cm
Hydrometer bulb volume	67 cm³

TABLE EX. 2.4B Hydrometer Test Data

Time (min)	Hydrometer Reading (g/L)	Temperature (°C)
1	38	22.0
2	33	22.0
3	30	21.5
4	28	21.5
8	25	21.5
15	23	21.0
30	21	21.0
60	19	21.0
240	15	19.5
900	12	18.5

Solution

Some preliminary calculations are first performed. The specific gravity correction is

$$a = \frac{1.65G_s}{2.65(G_s - 1)} = \frac{1.65(2.70)}{2.65(2.70 - 1)} = 0.989 \qquad (2.4.1)$$

The cross-sectional area of the sedimentation cylinder is

$$A_c = \frac{\pi}{4}(5.95 \text{ cm})^2 = 27.8 \text{ cm}^2 \qquad (2.4.2)$$

The general expression for the effective length is now specialized for the ASTM 152-H hydrometer, as follows:

$$L_1 = 10.5 - \left(\frac{10.5 - 2.3}{50}\right)R_{CL}; \quad L_2 = 14 \text{ cm} \qquad (2.4.3)$$

Then,

$$L = L_1 + \frac{1}{2}\left(L_2 - \frac{V_b}{A_c}\right) = 10.5 - 0.164R_{CL} + \frac{1}{2}\left(14.0 - \frac{67 \text{ cm}^3}{27.8 \text{ cm}^2}\right) \qquad (2.4.4)$$

$$= 16.295 - 0.164R_{CL}$$

TABLE EX. 2.4C Results of Hydrometer Analysis Calculations

Time (min)	R (g/L)	Temperature (°C)	F_t	R_{CP} (g/L)	% Finer	R_{CL} (g/L)	L (cm)	K	D (mm)
1	38	22.0	0.485	34.49	71.05	38.50	9.98	0.01312	0.04145
2	33	22.0	0.485	29.49	60.75	33.50	10.80	0.01312	0.03049
3	30	21.5	0.364	26.36	54.31	30.50	11.29	0.01312	0.02545
4	28	21.5	0.364	24.36	50.19	28.50	11.62	0.01312	0.02236
8	25	21.5	0.364	21.36	44.01	25.50	12.11	0.01312	0.01614
15	23	21.0	0.243	19.24	39.64	23.50	12.44	0.01312	0.01195
30	21	21.0	0.243	17.24	35.52	21.50	12.77	0.01312	0.00856
60	19	21.0	0.243	15.24	31.40	19.50	13.10	0.01312	0.00613
240	15	19.5	-0.121	10.88	22.41	15.50	13.75	0.01312	0.00314
900	12	18.5	-0.364	7.64	15.73	12.50	14.24	0.01312	0.00165

Notes: Meniscus correction (F_m) = 0.5 g/L, Dispersing agent correction (F_z) = 4.0 g/L.

For the first hydrometer reading, $R = 38$ g/L at a temperature of 22.0°C. The temperature correction for this reading is thus

$$F_T = -4.85 + 0.2425T = -4.85 + 0.2425(22.0) = 0.485 \qquad (2.4.5)$$

The corrected hydrometer reading for percent finer is then

$$R_{CP} = R + F_T - F_z = 38 + 0.485 - 4.0 = 34.49\text{g/L} \qquad (2.4.6)$$

The percent finer is next computed as follows:

$$\%finer = \left(\frac{aR_{CP}}{M_s}\right) * 100\% = \left(\frac{(0.989)(34.49\text{g/L})}{48.0\text{g}}\right) * 100\% = \mathbf{71.05\%}$$

$$(2.4.7)$$

The corrected hydrometer reading for determination of effective length is next computed as follows:

$$R_{CL} = R + F_m = 38 + 0.5 = 38.5 \text{ g/L} \qquad (2.4.8)$$

The effective length corresponding to R_{CL} is then

$$L = 16.295 - 0.164(38.5\text{g/L}) = 9.981 \text{ cm} \qquad (2.4.9)$$

The value of K must next be computed. At 22.0°C the viscosity of water is equal to 9.754×10^{-6} (g s)/cm², thus

$$K = \sqrt{\frac{30\eta}{(G_s - 1)}} = \sqrt{\frac{30(9.754 \times 10^{-6}\text{g s/cm}^2)}{(2.70 - 1)}} = 1.312 \times 10^{-2} \qquad (2.4.10)$$

Finally, the particle diameter is computed as follows:

$$D = K\sqrt{\frac{L}{t}} = \left(1.312 \times 10^{-2}\right)\sqrt{\frac{9.981 \text{ cm}}{1.0 \text{ min}}} = \mathbf{0.0414 \text{ mm}} \qquad (2.4.11)$$

The percent finer value computed above, and the particle diameter, defines a point on the grain size distribution curve. Table Ex. 2.4C summarizes the results computed for all of the hydrometer readings.

Figure Ex. 2.4B shows the particle size distribution curve that was developed using data from the above hydrometer analysis.

EXAMPLE PROBLEM 2.5

General Remarks

This example problem illustrates how the liquid limit is computed using data obtained from a Casagrande device.

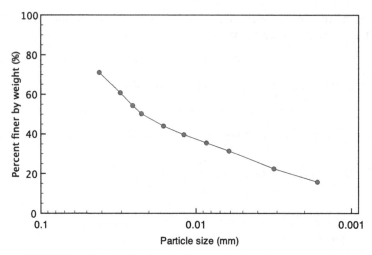

FIGURE EX. 2.4B Particle size distribution curve from hydrometer analysis.

Problem Statement

A series of Atterberg limit tests was performed on samples of a gray–brown silty clay. Table Ex. 2.5A summarizes the results of four liquid limit tests. Using this data,

a) Draw the *flow curve* for the soil.
b) Determine the *flow index* for the soil.
c) Calculate the *liquid limit*.
d) Calculate the *plasticity index* for the soil.

In the same series of Atterberg limit tests, the average plastic limit (w_{PL}) was found to be 38.5%.

TABLE EX. 2.5A Liquid Limit Data Obtained From a Casagrande Device

Trial Number	1	2	3	4
Number of blows	24	37	27	22
Weight of wet sample + container (g)	22.50	25.94	25.86	25.27
Weight of dry sample + container (g)	18.89	21.10	21.59	20.96
Weight of container (g)	14.12	14.61	15.82	15.34

Solution

Consider trial number 1. The weight of the pore fluid is first computed as follows:

$$W_w = 22.50 - 18.89 = 3.61 \text{ g} \qquad (2.5.1)$$

The weight of the solid phase is next computed

$$W_s = 18.89 - 14.12 = 4.77 \text{ g} \qquad (2.5.2)$$

The moisture content for trial number 1 is thus

$$w = \left(\frac{3.61 \text{ g}}{4.77 \text{ g}}\right) * 100\% = 75.7\% \qquad (2.5.3)$$

Similar calculations are performed for the remaining three trials. Table Ex 2.5B lists the complete results for all four trials.

a) To construct the flow curve, the moisture content is plotted versus the number of blows (in logarithmic scale). Figure Ex. 2.5 shows the resulting curve.

b) The *flow index* for the soil is equal to the absolute value of the slope of the flow curve. As such, it is equal to **0.146**.

c) The liquid limit is then the moisture content corresponding to 25 blows. In this case,

$$w_{LL} \approx \textbf{75.5\%} \qquad (2.5.4)$$

d) The *plasticity index* (I_p) is the difference between the liquid limit and the plastic limit. It represents the range of moisture content over which the

TABLE EX. 2.5B Computed Values Obtained Using Liquid Limit Data Obtained From a Casagrande Device

Trial Number	1	2	3	4
Number of blows	24	37	32	22
Weight of wet sample + container (g)	22.50	25.94	25.86	25.27
Weight of dry sample + container (g)	18.89	21.10	21.59	20.96
Weight of container (g)	14.12	14.61	15.82	15.34
Weight of pore fluid (g)	3.61	4.84	4.27	4.31
Weight of solid phase (g)	4.77	6.49	5.77	5.62
Moisture content (%)	75.7	74.6	74.0	76.7

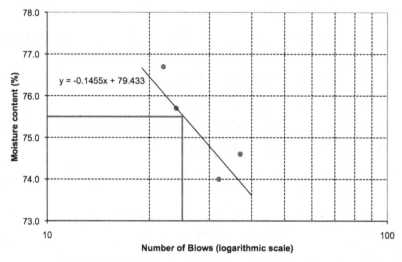

FIGURE EX. 2.5 Flow curve obtained from a Casagrande device.

soil behaves plastically (i.e., it is moldable). The value of I_p is important in classifying fine-grained soils. For the soil in question, the plasticity index is

$$I_p = w_{LL} - w_{PL} = 75.5 - 38.5 = \mathbf{37.0\%} \qquad (2.5.5)$$

EXAMPLE PROBLEM 2.6

General Remarks

The liquid limit can also be computed using data obtained from a cone penetrometer.

Problem Statement

Given data obtained as a result of a liquid limit test (Table Ex. 2.6),

a) Draw the *flow curve* for the soil.
b) Determine the *flow index* for the soil.
c) Calculate the *liquid limit*.
d) Calculate the *plasticity index* for the soil.

In the same series of Atterberg limit tests, the plastic limit (w_{PL}) was found to be 27.2%.

TABLE EX. 2.6 Liquid Limit Data Obtained From a Cone Penetrometer

Specimen	Moisture Content (%)	Penetration (mm)
1	37.2	16
2	39.1	27
3	44.3	36
4	46.0	54
5	49.8	69

Solution

a) To construct the flow curve, the moisture content is plotted versus the penetration. Figure Ex. 2.6 shows the resulting curve.

b) The *flow index* for the soil is equal to the absolute value of the slope of the flow curve. As such, it is equal to **0.236**.

c) The liquid limit is then the moisture content corresponding to 20 mm penetration. Thus,

$$w_{LL} \approx \mathbf{38.5\%} \tag{2.6.1}$$

d) Finally, the plasticity index is

$$I_p = w_{LL} - w_{PL} = 38.5 - 27.2 = \mathbf{11.3\%} \tag{2.6.2}$$

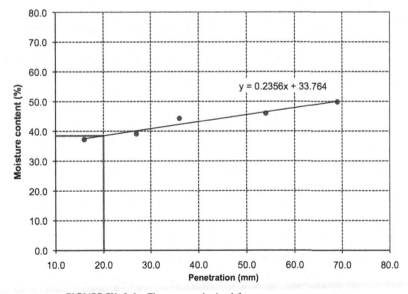

FIGURE EX. 2.6 Flow curve obtained from a cone penetrometer.

EXAMPLE PROBLEM 2.7

General Remarks

The shrinkage limit is investigated in this problem. As a saturated soil is slowly dried, capillary menisci form between the individual soil particles. As a result, the interparticle (effective) stresses increase, and the soil decreases in volume. A point is eventually reached where this volume change stops, even though the degree of saturation is still essentially 100%. The moisture content at which this occurs is defined as the *shrinkage limit* (w_{SL}), which, as noted above, is one of the Atterberg limits.

Figure Ex 2.7 shows the relation between the total weight and total volume of the soil. In this figure W_i and V_i denote the initial weight and volume of the saturated soil. The weight and volume associated with the shrinkage limit are W_{SL} and V_{dry}, respectively. Finally, W_s and V_s again denote the weight and volume of the solid phase. The values of W_i, V_i, W_s, and V_{dry} are easily measured in the laboratory.

By definition, the shrinkage limit is given by

$$w_{SL} = \left(\frac{W_{SL} - W_s}{W_s} \right) * 100\% \qquad (2.7.1)$$

Since the soil is essentially saturated, it follows that

$$W_i - W_{SL} = \gamma_w(V_i - V_{dry}) \Rightarrow W_{SL} = W_i - \gamma_w(V_i - V_{dry}) \qquad (2.7.2)$$

Substituting this result into the previous equation gives the following expression for w_{SL}:

$$
\begin{aligned}
w_{SL} &= \left[\frac{W_i - \gamma_w(V_i - V_{dry}) - W_s}{W_s} \right] * 100\% \\
&= \left[\frac{W_i - W_s - \gamma_w(V_i - V_{dry})}{W_s} \right] * 100\% = w_i - \left[\frac{\gamma_w(V_i - V_{dry})}{W_s} \right] * 100\%
\end{aligned}
$$
$$(2.7.3)$$

where w_i is expressed in percent.

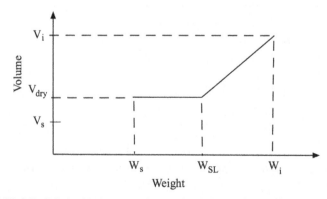

FIGURE EX. 2.7   Relationship between volume and weight associated with the shrinkage limit.

Problem Statement

To illustrate the use of the above equation, consider the following laboratory data:

- Weight of wet sample + container = 40.64 g
- Weight of dry sample + container = 30.42 g
- Weight of container = 18.36 g
- Initial volume of soil (V_i) = 15.07 cm^3
- Volume of soil after drying (V_{dry}) = 7.12 cm^3

  Compute the shrinkage limit for the above data.

Solution

The weight of the pore fluid is thus

$$W_w = 40.64 - 30.42 = 10.22 \text{ g} \qquad (2.7.4)$$

The weight of the solid is

$$W_s = 30.42 - 18.36 = 12.06 \text{ g} \qquad (2.7.5)$$

The initial moisture content is thus (recall Eq. 1.14)

$$w_i = \left(\frac{W_w}{W_s}\right) * 100\% = \left(\frac{10.22 \text{ g}}{12.06 \text{ g}}\right) * 100\% = 84.7\% \qquad (2.7.6)$$

Using the above values in Eq. (2.7.3), the shrinkage limit is thus

$$
\begin{aligned}
w_{SL} &= w_i - \left[\frac{\gamma_w(V_i - V_{dry})}{W_s}\right] * 100\% \\
&= 84.7\% - \left[\frac{(1.0 \text{g/cm}^3)(15.07 - 7.12 \text{ cm}^3)}{12.06 \text{ g}}\right] * 100\% \qquad (2.7.7) \\
&= \mathbf{18.8\%}
\end{aligned}
$$

Two quantities related to the shrinkage limit are sometimes also computed. The first quantity is the linear shrinkage ratio

$$LS = 1 - \left(\frac{V_{dry}}{V_i}\right)^{\frac{1}{3}} \qquad (2.7.8)$$

The second quantity is the shrinkage ratio,

$$SR = \frac{W_s}{V_{dry}\gamma_w} \qquad (2.7.9)$$

For the given data, Eqs. (2.7.8) and (2.7.9) give

$$LS = 1 - \left(\frac{7.12 \text{ cm}^3}{15.07 \text{ cm}^3}\right)^{\frac{1}{3}} = \mathbf{0.22} \qquad (2.7.10)$$

$$SR = \frac{W_s}{V_{dry}\gamma_w} = \frac{12.06 \text{ g}}{(7.12 \text{ cm}^3)(1.0\text{g}/\text{cm}^3)} = \mathbf{1.69} \qquad (2.7.11)$$

The next several problems give additional insight into some computations related to the Atterberg limits.

EXAMPLE PROBLEM 2.8

General Remarks

This example problem illustrates the computation of the shrinkage limit.

Problem Statement

In a test to determine the shrinkage limit of a cohesive soil, after drying, a sample had a volume of 50 cm^3 and a weight of 88.0 g. The specific gravity of solids was 2.71. Determine the shrinkage limit (w_{SL}).

Solution

Referring to Figure Ex 2.7, from the given information $W_s = 88.0$ g. From the definition of the specific gravity of solids,

$$G_s = \frac{W_s}{V_s\gamma_w} \Rightarrow V_s = \frac{W_s}{G_s\gamma_w} = \frac{88.0 \text{ g}}{(2.71)(1.0\text{g}/\text{cm}^3)} = 32.47 \text{ cm}^3 \qquad (2.8.1)$$

Since the soil is assumed to be saturated, the volume of the pore fluid is thus

$$V_w = 50.0 \text{ cm}^3 - V_s = 50.0 - 32.47 = 17.53 \text{ cm}^3 \qquad (2.8.2)$$

The weight of the pore fluid at the shrinkage limit is thus

$$W_w = V_w\gamma_w = (17.53 \text{ cm}^3)(1.0\text{g}/\text{cm}^3) = 17.53 \text{ g} \qquad (2.8.3)$$

The shrinkage limit is thus

$$w_{SL} = \left(\frac{W_w}{W_s}\right) * 100\% = \left(\frac{17.53 \text{ g}}{88.0 \text{ g}}\right) * 100\% = \mathbf{21.9\%} \qquad (2.8.4)$$

EXAMPLE PROBLEM 2.9

General Remarks

This example problem illustrates some of the calculations that are performed using Atterberg limits.

Problem Statement

The liquid limit (w_{LL}) of a clay soil is 54%, and its plasticity index (I_P) is 15%. (a) What is the plastic limit (w_{PL}) of the soil? (b) In what state of consistency is this material at a moisture content of 40%? (c) At the minimum volume reached during shrinkage, a sample of this soil has a void ratio of 0.87. If the specific gravity of solids is equal to 2.72, compute the shrinkage limit (w_{SL}).

Solution

a) The plastic limit is computed from the definition of the plasticity index as follows:

$$I_P = w_{LL} - w_{PL} \Rightarrow w_{PL} = w_{LL} - I_P = 54 - 15 = \mathbf{39\%} \qquad (2.9.1)$$

b) Since 40% > 39%, it follows that the soil is still **plastic**.

c) Since, at the minimum volume reached during shrinkage the soil is assumed to be saturated, the moisture content associated with the shrinkage limit is thus

$$w_{SL} = \left(\frac{e}{G_s}\right) * 100\% = \left(\frac{0.87}{2.72}\right) * 100\% = \mathbf{32.0\%} \qquad (2.9.2)$$

EXAMPLE PROBLEM 2.10

General Remarks

This example problem illustrates some of the calculations involving the shrinkage limit.

Problem Statement

A clay soil has a liquid limit of 60% and a shrinkage limit of 25%. If a specimen of this soil shrinks from an initial volume of 10.0 cm^3 at the liquid limit to a volume of 6.39 cm^3 at the shrinkage limit, what is the specific gravity of solids (G_s)?

Letting w_{LL} and w_{SL} denote the liquid and shrinkage limits, respectively, the moisture content for both limits is thus

$$w_{LL} = \frac{W_{w_L}}{W_s} \Rightarrow W_{w_L} = w_{LL}W_s \qquad (2.10.1)$$

$$w_{SL} = \frac{W_{w_S}}{W_s} \Rightarrow W_{w_S} = w_{SL}W_s \qquad (2.10.2)$$

Since the soil is assumed to be saturated, the change in the weight of the fluid phase is given by

$$\Delta W_w = (W_{w_L} - W_{w_S}) = \gamma_w(V_{LL} - V_{dry}) \qquad (2.10.3)$$

Substituting for the weight of the fluid gives

$$\Delta W_w = (w_{LL} - w_{SL})W_s = \gamma_w(V_{LL} - V_{dry}) \Rightarrow W_s = \frac{\gamma_w(V_{LL} - V_{dry})}{w_{LL} - w_{SL}}$$

$$(2.10.4)$$

The next quantity required to compute G_s is the volume of the solid phase (V_s). The volume of the pore fluid at the shrinkage limit is first computed

$$V_{ws} = \frac{W_{ws}}{\gamma_w} = \frac{w_{SL}W_s}{\gamma_w} = \frac{w_{SL}(V_{LL} - V_{dry})}{w_{LL} - w_{SL}} \tag{2.10.5}$$

Since the soil is saturated,

$$V_s = V_{dry} - V_{ws} = V_{dry} - \frac{w_{SL}(V_{LL} - V_{dry})}{w_{LL} - w_{SL}} = \frac{V_{dry}w_{LL} - V_{LL}w_{SL}}{w_{LL} - w_{SL}} \tag{2.10.6}$$

The specific gravity of solids is thus

$$G_s = \frac{W_s}{V_s\gamma_w} = \frac{\gamma_w(V_{LL} - V_{dry})}{(w_{LL} - w_{SL})} \frac{(w_{LL} - w_{SL})}{(V_{dry}w_{LL} - V_{LL}w_{SL})\gamma_w} = \frac{V_{LL} - V_{dry}}{V_{dry}w_{LL} - V_{LL}w_{SL}} \tag{2.10.7}$$

Substituting for all of the known quantities gives

$$G_s = \frac{(10.0 - 6.38)\text{cm}^3}{(6.38 \text{ cm}^3)(0.60) - (10.0 \text{ cm}^3)(0.25)} = \textbf{2.73} \tag{2.10.8}$$

EXAMPLE PROBLEM 2.11

General Remarks

This example problem illustrates the use of quantities derived from the Atterberg limits.

Problem Statement

A fine-grained soil has a liquid limit (LL) of 110% and a plastic limit (PL) of 56%. The clay content is 68% and the field moisture content of the soil is 60%.

a) Compute the plasticity index (PI), the liquidity index (LI), and the activity (A).
b) What is the predominant mineral in the soil?
c) What is the soil state in the field?

Solution

a) The PI is

$$\text{PI} = 110 - 56 = \textbf{54\%} \tag{2.11.1}$$

The LI is

$$LI = \frac{w - PL}{PI} = \frac{60 - 56}{54} = \mathbf{0.074} \qquad (2.11.2)$$

The activity is

$$A = \frac{PI}{\text{percent of clay} - \text{size fraction}} = \frac{54}{68} = \mathbf{0.79} \qquad (2.11.3)$$

b) Based on the activity, the predominant mineral is **illite**.
c) Since the field moisture content is greater than the PL but is less than the LL, it follows that the soil is in a **plastic** state. This is also confirmed by the fact that the LI falls in the range $0 < LI < 1$.

EXAMPLE PROBLEM 2.12

General Remarks

This example problem illustrates some of the calculations involving the shrinkage limit.

Problem Statement

A saturated sample of clay has a field moisture content (w) of 32% and a shrinkage limit (SL) of 20%. What is the ratio of the dry volume (V_{dry}) of the sample to its original (natural) volume (V) if $G_s = 2.67$?

Solution

For the field conditions, from the general definition of the moisture content given in Eq. (1.14),

$$w = \frac{W_w}{W_s} \Rightarrow W_w = wW_s \qquad (2.12.1)$$

The total volume of the saturated soil in the field is thus

$$V = V_w + V_s = \frac{W_w}{\gamma_w} + \frac{W_s}{G_s\gamma_w} \qquad (2.12.2)$$

Since the weight of the solid phase remains unchanged in the SL test, combine the two above equations so as to eliminate W_w, giving

$$V = \frac{W_w}{\gamma_w} + \frac{W_s}{G_s\gamma_w} = \frac{W_s}{\gamma_w}\left(w + \frac{1}{G_s}\right) \qquad (2.12.3)$$

At the SL,

$$w_{SL} = \frac{W_{w_{SL}}}{W_s} \Rightarrow W_{w_{SL}} = w_{SL} W_s \tag{2.12.4}$$

The total volume of the soil at the SL is thus

$$V_{dry} = V_{w_{SL}} + V_s = \frac{W_{w_{SL}}}{\gamma_w} + \frac{W_s}{G_s \gamma_w} = \frac{W_s}{\gamma_w}\left(w_{SL} + \frac{1}{G_s}\right) \tag{2.12.5}$$

The desired ratio of volumes is thus

$$\frac{V_{dry}}{V} = \frac{\dfrac{W_s}{\gamma_w}\left(w_{SL} + \dfrac{1}{G_s}\right)}{\dfrac{W_s}{\gamma_w}\left(w + \dfrac{1}{G_s}\right)} = \frac{w_{SL} + \dfrac{1}{G_s}}{w + \dfrac{1}{G_s}} \tag{2.12.6}$$

Substituting the given values of w, w_{SL}, and G_s gives the desired result; i.e.,

$$\frac{V_{dry}}{V} = \frac{w_{SL} + \dfrac{1}{G_s}}{w + \dfrac{1}{G_s}} = \frac{0.20 + \dfrac{1}{2.67}}{0.32 + \dfrac{1}{2.67}} = \mathbf{0.83} \tag{2.12.7}$$

Alternate Solution

It is also possible to solve this problem beginning with the following expression for the dry unit weight (recall Case 1.4 in Chapter 1)

$$\gamma_d = \frac{W_s}{V} = \frac{G_s \gamma_w}{1 + e} \Rightarrow W_s = \left(\frac{G_s \gamma_w}{1 + e}\right)V \tag{2.12.8}$$

For the field conditions,

$$W_s = \left(\frac{G_s \gamma_w}{1 + e}\right)V \tag{2.12.9}$$

At the SL,

$$W_s = \left(\frac{G_s \gamma_w}{1 + e_{SL}}\right)V_{dry} \tag{2.12.10}$$

Since the weight of the solid phase remains unchanged in the SL test, it follows that

$$\left(\frac{G_s \gamma_w}{1 + e}\right)V = \left(\frac{G_s \gamma_w}{1 + e_{SL}}\right)V_{dry} \Rightarrow \frac{V_{dry}}{V} = \frac{1 + e_{SL}}{1 + e} \tag{2.12.11}$$

Recalling Case 1.3 in Chapter 1, for saturated soil ($S = 100\%$), $e = wG_s$. Thus,

$$\frac{V_{dry}}{V} = \frac{1 + e_{SL}}{1 + e} = \frac{1 + w_{SL}G_s}{1 + wG_s} \qquad (2.12.12)$$

Eq. (2.12.12) is seen to be equivalent to Eq. (2.12.7).

EXAMPLE PROBLEM 2.13

General Remarks

This example problem illustrates how a soil is classified according to the Unified Soil Classification System (USCS).

Problem Statement

A wet, dark brown soil that exudes an organic odor has 100% passing the No. 200 sieve. The natural liquid limit (LL) for the soil is 39%, the oven-dried liquid limit is 25%, and the plastic limit is 28%. Classify this soil according to the USCS.

Solution

Since 100% of the soil passes the No. 200 sieve, it is *fine grained*.
  Since the LL is less than 50%, the soil will have *low plasticity*.
  The plasticity index (PI) for the soil is PI $= 39-28 = 11\%$.
  Figure Ex. 2.13 shows the location of the soil on Casagrande's plasticity chart.
  The fact that the soil exudes an organic odor is consistent with the location of the soil on the plasticity chart in Figure Ex. 2.13, and indicates that the soil is organic.
  Since

$$\frac{LL_{oven-dried}}{LL_{natural}} = \frac{25}{39} = 0.64 < 0.75 \qquad (2.13.1)$$

the appropriate USCS group symbol will thus be **OL**. The associated group name will be *organic clay*.

EXAMPLE PROBLEM 2.14

General Remarks

This example problem illustrates how a soil is classified according to the Unified Soil Classification System (USCS).

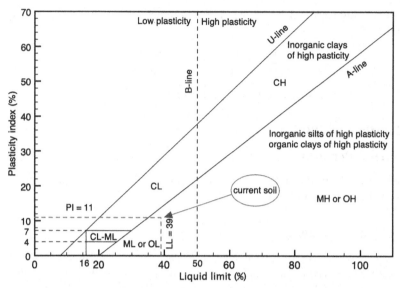

FIGURE EX. 2.13   Casagrande plasticity chart and location of current soil.

Problem Statement

A sample of gravel with sand has 74% fine to coarse subangular gravel, 24% coarse angular sand, and 2% fines. The maximum size of particles in the soil is 74 mm. The coefficient of curvature (C_c) is 2.7, and the coefficient of uniformity (C_u) is 12.6. Classify this soil according to the USCS.

Solution

Since the particle sizes in the soil do not exceed 75 mm, there are no cobbles present.

Since less than 5% fines are present, the soil will be a "clean gravel."

Since $C_u > 4$ and $C_c < 3$, the appropriate group symbol will thus be **GW**. The associated USCS group name will be a *well-graded gravel* (with sand).

EXAMPLE PROBLEM 2.15

General Remarks

This example problem illustrates how a soil is classified according to the Unified Soil Classification System (USCS).

Problem Statement

A sample of soil has 100% passing the No. 4 sieve and 75% passing the No. 200 sieve. The liquid limit (LL) for the soil was found to be 60%, while the plasticity index (PI) was 24%. Classify this soil according to the USCS.

Solution

Since more than 50% passes the No. 200 sieve, the soil is considered to be "fine grained".

The point on the A-line corresponding to LL = 60% is

$$PI_{A-line} = 0.73(60 - 20) = 29.2\% \tag{2.15.1}$$

Since the soil's PI (24%) is less than 29.2%, it will plot *below* the A-line. As such, the appropriate group symbol will thus be **MH**; i.e., a high plasticity silt.

In addition, since 25% was retained on the No. 200 sieve, and since 0% was retained on the No. 4 sieve, the soil contains 25% sand.

The associated USCS group name will be *elastic silt with sand*.

EXAMPLE PROBLEM 2.16

General Remarks

This example problem illustrates how a soil is classified according to the Unified Soil Classification System (USCS).

Problem Statement

A sample of soil has 60% passing the No. 4 sieve and 42% passing the No. 200 sieve. The liquid limit (LL) for the soil is 28%, and the plasticity index (PI) is 4.5%. Classify this soil according to the USCS.

Solution

Since less than 50% passes the No. 200 sieve, the soil is considered to be "coarse grained".

Since 60% > 50% passes the No. 4 sieve, the soil will be a sand.

The point on the A-line corresponding to LL = 28% is

$$PI_{A-line} = 0.73(28 - 20) = 5.8\% \tag{2.16.1}$$

Since the soil's PI (4.5%) is less than 5.8%, it will plot *below* the A-line. Since the percent fines (42%) exceed 12%, the soil will contain clay.

The appropriate group symbol will thus be **SM-SC**; i.e., a *silty, clayey sand*.

Since 58% was retained on the No. 200 sieve, the "coarse" fraction is 58%. In addition, since 60% passed the No. 4 sieve, the gravel fraction = 100−60 = 40%. Finally, the sand fraction = 58−40 = 18%.

The associated USCS group name will be *silty, clayey sand with gravel*.

EXAMPLE PROBLEM 2.17

General Remarks

This example problem illustrates how a soil is classified according to the Unified Soil Classification System (USCS).

Problem Statement

A sample of soil has 98% passing the No. 4 sieve and 74% passing the No. 200 sieve. The liquid limit (LL) for the soil is 58%, and the plasticity index (PI) is 30%. Classify this soil according to the USCS.

Solution

Since more than 50% passes the No. 200 sieve, the soil is considered to be "fine grained."

The point on the A-line corresponding to LL = 58% is

$$PI_{A-line} = 0.73(58 - 20) = 27.7\% \qquad (2.17.1)$$

Since the soil's PI (30%) exceeds 27.7%, it will plot *above* the A-line.

The appropriate group symbol will thus be **CH**; i.e., a *high plasticity clay*.

Since 100−74 = 26% was retained on the No. 200 sieve, the "coarse" fraction is 26%. In addition, since 98% passed the No. 4 sieve, the gravel fraction = 100−98 = 2%. Finally, the sand fraction = 26−2 = 24%.

Since the percent gravel is less than 15%, the associated USCS group name will be *sandy fat clay*.

Chapter 3

Example Problems Related to Compaction of Soils

3.0 GENERAL COMMENTS

Soil compaction is the *densification* (reduction in void ratio) of soil through the expulsion of *air* from the voids. Compaction is one of the most popular techniques for improving soils. The soil microfabric is forced into a denser configuration by the reorientation of particles, and thus the expulsion of air from the voids, as a result of some form of mechanical effort.

3.1 FUNDAMENTAL DEFINITIONS

Two of the key quantities used in solving problems involving compaction are the moisture content (w) and the dry unit weight (γ_d). Recalling the definition of the specific gravity of solids gives

$$G_s = \frac{W_s}{V_s \gamma_w} \quad \Rightarrow \quad W_s = G_s V_s \gamma_w \qquad (3.1)$$

Substituting this expression for W_s into the definition of the dry unit weight and dividing through the resulting expression by V_s gives (recall Case 1.4 in Chapter 1)

$$\gamma_d = \frac{W_s}{V} = \frac{G_s V_s \gamma_w}{V_s + V_v} = \frac{G_s \gamma_w}{1 + e} \qquad (3.2)$$

Recalling the definition of the unit weight of water ($\gamma_w = W_w/V_w$), and that of the moisture content, and substituting Eq. (3.1) gives

$$w = \frac{W_w}{W_s} = \frac{\gamma_w V_w}{G_s V_s \gamma_w} = \frac{V_w}{G_s V_s} \qquad (3.3)$$

Recalling the definition of the degree of saturation gives

$$S = \frac{V_w}{V_v} \quad \Rightarrow \quad V_w = S V_v \qquad (3.4)$$

Soil Mechanics. http://dx.doi.org/10.1016/B978-0-12-804491-9.00003-3

Substituting Eq. (3.4) into Eq. (3.3) gives the desired result (also recall Case 1.3 in Chapter 1)

$$w = \frac{V_w}{G_s V_s} = \frac{SV_v}{G_s V} = \frac{Se}{G_s} \quad \Rightarrow \quad e = \frac{G_s w}{S} \tag{3.5}$$

Substituting this expression for the void ratio into Eq. (3.2) gives

$$\gamma_d = \frac{G_s \gamma_w}{1+e} = \frac{G_s \gamma_w}{1 + \left(\dfrac{G_s w}{S}\right)} = \frac{\gamma_w}{\dfrac{1}{G_s} + \dfrac{w}{S}} \tag{3.6}$$

Remark: For a given soil (i.e., G_s), the *smaller* the void ratio (e), the *greater* will be the dry unit weight (γ_d).

Remark: The *maximum* dry unit weight (i.e., $\gamma_{d_{max}}$) is just another way of expressing the *minimum* void ratio (i.e., e_{min}) or minimum porosity (i.e., n_{min}).

The dry unit weight associated with *saturation* represents an upper limit in that all air has been forced from the voids. Since now $S = 1.0$, Eq. (3.6) reduces to the so-called "zero air voids" (*ZAV*) dry unit weight, i.e.,

$$\gamma_{d_{ZAV}} = \frac{G_s \gamma_w}{1 + G_s w} = \frac{\gamma_w}{\dfrac{1}{G_s} + w} \tag{3.7}$$

The definition of the dry unit weight of a soil is rewritten as follows (recall Case 1.4 of Chapter 1):

$$\gamma_d = \frac{W_s}{V} = \frac{G_s V_s \gamma_w}{V_s + V_v} = \frac{G_s \gamma_w}{1+e} \tag{3.8}$$

Solving for the void ratio gives

$$e = \frac{G_s \gamma_w}{\gamma_d} - 1 \tag{3.9}$$

It follows that for a given soil (i.e., G_s), the void ratio is *inversely* proportional to the dry unit weight. Thus, $\gamma_{d_{max}}$ corresponds to e_{min} and $\gamma_{d_{min}}$ corresponds to e_{max}, i.e.,

$$\gamma_{d_{min}} = \frac{G_s \gamma_w}{1 + e_{max}} \tag{3.10}$$

and

$$\gamma_{d_{max}} = \frac{G_s \gamma_w}{1 + e_{min}} \tag{3.11}$$

For cohesionless soils, particularly free-draining gravels and sands, it is useful to use e_{min} and e_{max} to quantify how dense or loose a soil is relative to the certain laboratory determined index maximum and minimum void ratio values. This is done by defining the following *relative* or *index density* (D_r):

$$D_r = \left(\frac{e_{max} - e}{e_{max} - e_{min}}\right) * 100\% \tag{3.12}$$

The relative density can also be written in terms of the dry unit weight. Solving Eqs. (3.10) and (3.11) for e_{min} and e_{max} gives

$$e_{max} = \frac{G_s \gamma_w}{\gamma_{d_{min}}} - 1 \tag{3.13}$$

and

$$e_{min} = \frac{G_s \gamma_w}{\gamma_{d_{max}}} - 1 \tag{3.14}$$

Substituting Eqs. (3.13) and (3.14) into Eq. (3.12) gives

$$D_r = \left[\frac{\left(\dfrac{G_s \gamma_w}{\gamma_{d_{min}}} - 1\right) - \left(\dfrac{G_s \gamma_w}{\gamma_d} - 1\right)}{\left(\dfrac{G_s \gamma_w}{\gamma_{d_{min}}} - 1\right) - \left(\dfrac{G_s \gamma_w}{\gamma_{d_{max}}} - 1\right)}\right] * 100\% = \frac{(\gamma_d - \gamma_{d_{min}})}{(\gamma_{d_{max}} - \gamma_{d_{min}})}\left(\frac{\gamma_{d_{max}}}{\gamma_d}\right)$$

$$* 100\% = I_d\left(\frac{\gamma_{d_{max}}}{\gamma_d}\right) \tag{3.15}$$

where

$$I_d = \left(\frac{\gamma_d - \gamma_{d_{min}}}{\gamma_{d_{max}} - \gamma_{d_{min}}}\right) * 100\% \tag{3.16}$$

is the so-called "density index". Since $\gamma_d = \rho_d g$ [recall Eq. (1.5)], etc., it follows that

$$D_r = \left(\frac{\rho_d - \rho_{d_{min}}}{\rho_{d_{max}} - \rho_{d_{min}}}\right)\left(\frac{\rho_{d_{max}}}{\rho_d}\right) * 100\% = I_d\left(\frac{\rho_{d_{max}}}{\rho_d}\right) \tag{3.17}$$

where the "density index" is now given by

$$I_d = \left(\frac{\rho_d - \rho_{d_{min}}}{\rho_{d_{max}} - \rho_{d_{min}}}\right) * 100\% \tag{3.18}$$

Table 3.1 lists the approximate density classifications commonly associated with ranges of relative density values.

TABLE 3.1 Density Classifications Associated With Relative Density

D_r (%)	Density Classification
<15	Very loose
15–35	Loose
35–65	Medium loose
65–85	Dense
>85	Very dense

Compaction data are typically presented using a so-called *compaction curve*. In such a curve, the moisture content is plotted as the abscissa, and the dry unit weight or dry density is plotted as the ordinate. Fig. 3.1 shows a hypothetical compaction curve. The maximum dry unit weight and the optimum moisture content are identified in Fig. 3.1.

The peak point of the compaction curve is idntified. The dry density or dry unit weight at this point is a *maximum*. The moisture content corresponding to the peak point is referred to as the *optimum moisture content* or *optimum water content*.

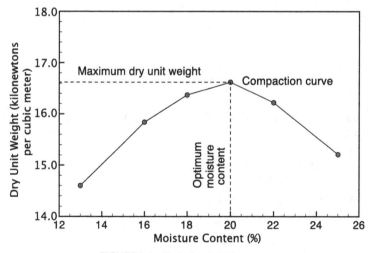

FIGURE 3.1 Typical compaction curve.

A very important compaction *end product* or *performance specification* is the relative compaction (*RC*), which is defined as follows:

$$RC = \left(\frac{\rho_{d_{field}}}{\rho_{d_{max}}}\right) * 100\% = \left(\frac{\gamma_{d_{field}}}{\gamma_{d_{max}}}\right) * 100\% \qquad (3.19)$$

where $\rho_{d_{field}}$ and $\rho_{d_{max}}$ are the field dry density and the laboratory maximum dry density, respectively. In a similar manner, $\gamma_{d_{field}}$ and $\gamma_{d_{max}}$ are the field dry unit weight and the laboratory maximum dry unit weight, respectively. Typical values for *RC* are 90% or 95% of the laboratory maximum.

EXAMPLE PROBLEM 3.1

General Remarks

This example problem illustrates the manner in which the relative density is computed.

Problem Statement

For a given sandy soil, the maximum and minimum void ratios were determined to be 0.80 and 0.25, respectively. The bulk (moist) density (ρ) of a field specimen of the soil at a moisture content (w) of 12% was 1.85 Mg/m³. Assuming $G_s = 2.68$, determine the relative density (D_r) and degree of saturation (S) of this specimen.

Solution

From the definition of the relative density, with e_{min} and e_{max} given, it remains to determine the actual void ratio (e) of the soil. In light of the values given in the problem, recall that, from Case 1.4 of Chapter 1

$$\rho = \frac{G_s \rho_w (1 + w)}{1 + e} \qquad (3.1.1)$$

Solving Eq. (3.1.1) for the void ratio and substituting the given values into the resulting expression gives

$$e = \frac{G_s \rho_w (1 + w)}{\rho} - 1 = \frac{(2.68)(1.0 \text{ Mg/cm}^3)(1 + 0.12)}{1.85 \text{ Mg/cm}^3} - 1 = 0.622$$

$$(3.1.2)$$

The relative density is thus

$$D_r = \left(\frac{e_{max} - e}{e_{max} - e_{min}}\right) * 100\% = \left(\frac{0.80 - 0.622}{0.80 - 0.25}\right) * 100\% = \mathbf{32.3\%} \quad (3.1.3)$$

In light of this relative density value, the soil is considered to be "loose" (Table 3.1).

Finally, the degree of saturation is computed using Case 1.3 of Chapter 1, i.e.,

$$S = \left(\frac{G_s w}{e}\right) * 100\% = \left[\frac{(2.68)(0.12)}{0.622}\right] * 100\% = \mathbf{51.7\%} \qquad (3.1.4)$$

EXAMPLE PROBLEM 3.2

General Remarks

This example problem illustrates the manner in which the maximum and minimum void ratios, as well as the relative density are computed.

Problem Statement

A certain cohesionless soil has a specific gravity of solids (G_s) of 2.67. A 1000 cm^3 container is just filled with a dry sample of this soil in its loosest possible state. Later, the container is filled at the densest state obtainable. The total weights for the loosest and densest samples are 1550 and 1720 g, respectively.

a) Determine the maximum and minimum void ratios.

b) If the dry unit weight of the soil in situ is equal to 103.5 lb/ft^3, compute the relative density (D_r).

Solution

a) In general, the volume of the solid phase is computed from the definition of G_s, i.e.,

$$G_s = \frac{W_s}{V_s \gamma_w} \quad \Rightarrow \quad V_s = \frac{W_s}{G_s \gamma_w} \qquad (3.2.1)$$

Since the soil is dry, the voids are filled only with air (consequently $W_w = 0$). The void ratio is then computed as follows:

$$e = \frac{V_v}{V_s} = \frac{V - V_s}{V_s} \qquad (3.2.2)$$

In the loosest possible state, $W_s = 1550$ g. Using Eq. (3.2.1), the associated volume of the solid phase is thus

$$V_{s_{loose}} = \frac{W_s}{G_s \gamma_w} = \frac{1550 \text{ g}}{(2.67)(1.0 \text{ g/cm}^3)} = 580.5 \text{ cm}^3 \quad (3.2.3)$$

Using Eq. (3.2.2), the maximum void ratio is thus

$$e_{max} = \frac{V - V_s}{V_s} = \frac{(1000 - 580.5) \text{ cm}^3}{580.5 \text{ cm}^3} = 0.723 \quad (3.2.4)$$

In the densest possible state, $W_s = 1720$ g. Using Eq. (3.2.1), the associated volume of the solid phase is thus

$$V_{s_{dense}} = \frac{W_s}{G_s \gamma_w} = \frac{1720 \text{ g}}{(2.67)(1.0 \text{ g/cm}^3)} = 644.2 \text{ cm}^3 \quad (3.2.5)$$

Using Eq. (3.2.2), the minimum void ratio is thus

$$e_{min} = \frac{V - V_s}{V_s} = \frac{(1000 - 644.4) \text{ cm}^3}{644.2 \text{ cm}^3} = 0.552 \quad (3.2.6)$$

b) The void ratio in situ is computed from the dry unit weight using the expression derived in Case 1.4 of Chapter 1, i.e.,

$$\gamma_d = \frac{G_s \gamma_w}{1 + e} \quad \Rightarrow \quad e = \frac{G_s \gamma_w}{\gamma_d} - 1 = \frac{(2.67)(62.4 \text{ lb/ft}^3)}{103.5 \text{ lb/ft}^3} - 1 = 0.610 \quad (3.2.7)$$

The relative density is thus

$$D_r = \left(\frac{e_{max} - e}{e_{max} - e_{min}}\right) * 100\% = \left(\frac{0.723 - 0.610}{0.723 - 0.552}\right) * 100\% = 66.1\% \quad (3.2.8)$$

The soil in situ would thus be considered to be "dense" (Table 3.1).

EXAMPLE PROBLEM 3.3

General Remarks

This example problem illustrates the manner in which ranges in unit weights are computed from known values of the minimum and maximum void ratio.

Problem Statement

The values of the minimum (e_{min}) and maximum void ratio (e_{max}) for a sample of pure silica sand were found to be 0.42 and 0.67, respectively. The specific gravity of solids (G_s) for the sand is 2.69.

a) What is the corresponding *range* in dry unit weights? b) What is the corresponding range in the *saturated* unit weights? c) If a relative density (D_r) of 78% is desired, what will be the moisture content associated with *full* saturation? d) Finally, compute the density index (I_d).

Solution

a) The dry unit weight can be written in terms of G_s and void ratio as follows (recall Case 1.4 of Chapter 1):

$$\gamma_d = \frac{G_s \gamma_w}{1 + e} \tag{3.3.1}$$

It follows that

$$\gamma_{d_{min}} = \frac{G_s \gamma_w}{1 + e_{max}} \tag{3.3.2}$$

and

$$\gamma_{d_{max}} = \frac{G_s \gamma_w}{1 + e_{min}} \tag{3.3.3}$$

Substituting all given values into Eqs. (3.3.2) and (3.3.3) leads to the following range in dry unit weights:

$$\gamma_{d_{min}} = \frac{G_s \gamma_w}{1 + e_{max}} = \frac{(2.69)(9.81 \text{ kN/m}^3)}{1 + 0.67} = \textbf{15.80 kN/m}^3 \tag{3.3.4}$$

and

$$\gamma_{d_{max}} = \frac{G_s \gamma_w}{1 + e_{min}} = \frac{(2.69)(9.81 \text{ kN/m}^3)}{1 + 0.42} = \textbf{18.58 kN/m}^3 \tag{3.3.5}$$

b) The saturated unit weight is next written as derived in Case 1.8 of Chapter 1, i.e.,

$$\gamma_{sat} = \frac{\gamma_w (G_s + e)}{1 + e} \tag{3.3.6}$$

Since $G_s > 1$, it follows that $(G_s + e) > (1 + e)$. Thus,

$$\gamma_{sat_{min}} = \frac{\gamma_w (G_s + e_{max})}{1 + e_{max}} \tag{3.3.7}$$

and

$$\gamma_{sat_{max}} = \frac{\gamma_w (G_s + e_{min})}{1 + e_{min}} \tag{3.3.8}$$

Substituting all given values into Eqs. (3.3.7) and (3.3.8) leads to the following range in saturated unit weights:

$$\gamma_{sat_{min}} = \frac{\gamma_w(G_s + e_{max})}{1 + e_{max}} = \frac{(9.81 \text{ kN/m}^3)(2.69 + 0.67)}{1 + 0.67} = \textbf{19.74 kN/m}^3$$

$$(3.3.9)$$

and

$$\gamma_{sat_{max}} = \frac{\gamma_w(G_s + e_{min})}{1 + e_{min}} = \frac{(9.81 \text{ kN/m}^3)(2.69 + 0.42)}{1 + 0.42} = \textbf{21.49 kN/m}^3$$

$$(3.3.10)$$

c) The void ratio of the soil is written in terms of the relative density as follows:

$$D_r = \left(\frac{e_{max} - e}{e_{max} - e_{min}}\right) * 100\% \quad \Rightarrow \quad e = e_{max} - \frac{D_r}{(100\%)}(e_{max} - e_{min})$$

$$(3.3.11)$$

Since for full saturation $e = G_s w \Rightarrow w = e/G_s$, the desired moisture content is thus

$$w = \left[\frac{e_{max} - \dfrac{D_r}{(100\%)}(e_{max} - e_{min})}{G_s}\right] * 100\%$$

$$(3.3.12)$$

$$= \left[\frac{0.67 - (0.78)(0.67 - 0.42)}{2.69}\right] * 100\% = \textbf{17.7\%}$$

d) The first step in computing the *density index* is the determination of the void ratio. This can be done in one of two ways. For example,

$$e = G_s w = (2.69)(0.177) = 0.475 \tag{3.3.13}$$

or

$$e = e_{max} - \frac{D_r}{(100\%)}(e_{max} - e_{min}) = 0.67 - (0.78)(0.67 - 0.42) = 0.475$$

$$(3.3.14)$$

In either case, the dry unit weight of the soil is thus

$$\gamma_d = \frac{G_s \gamma_w}{1 + e} = \frac{(2.69)(9.81 \text{ kN/m}^3)}{1 + 0.475} = 17.89 \text{ kN/m}^3 \tag{3.3.15}$$

Finally, the *density index* is computed as follows:

$$I_d = \left(\frac{\gamma_d - \gamma_{d_{min}}}{\gamma_{d_{max}} - \gamma_{d_{min}}}\right) * 100\% = \left(\frac{17.89 - 15.80}{18.58 - 15.80}\right) * 100\% = \textbf{75.2\%}$$

$$(3.3.16)$$

As a check, compute the following product:

$$I_d \left(\frac{\gamma_{d_{\max}}}{\gamma_d} \right) = (75.2\%) \left(\frac{18.58 \text{ kN/m}^3}{17.89 \text{ kN/m}^3} \right) = 78.0\% \qquad (3.3.17)$$

which is equal to the relative density (D_r) and thus confirms the result.

EXAMPLE PROBLEM 3.4

General Remarks

This example problem applies several of the expressions developed in Chapter 1 to the topic of compaction.

Problem Statement

In a standard Proctor test a sample of silty sand was compacted in a mold whose volume is $1/30$ ft^3. The moist weight of the sample was 4.01 lb. When dried, the soil weighted 3.52 lb. If the specific gravity of solids (G_s) for the soil is 2.69, compute the following:

a) The void ratio (e), b) the moisture content (w), c) the dry unit weight (γ_d), d) the degree of saturation (S), e) the moist unit weight (γ), and f) the saturated unit weight (γ_{sat}).

Solution

From the given information,

$$W_s = 3.52 \text{ lb} \qquad (3.4.1)$$

$$W_w = 4.01 - 3.52 = 0.49 \text{ lb} \qquad (3.4.2)$$

The volume of solids is computed from the definition of the specific gravity of solids, i.e.,

$$G_s = \frac{W_s}{V_s \gamma_w} \quad \Rightarrow \quad V_s = \frac{W_s}{G_s \gamma_w} = \frac{3.52 \text{ lb}}{(2.69)(62.4 \text{ lb/ft}^3)} = 2.097 \times 10^{-2} \text{ ft}^3$$

$$(3.4.3)$$

The volume of the pore fluid is computed from the definition of the unit weight of water, i.e.,

$$\gamma_w = \frac{W_w}{V_w} \quad \Rightarrow \quad V_w = \frac{W_w}{\gamma_w} = \frac{0.49 \text{ lb}}{62.4 \text{ lb/ft}^3} = 7.853 \times 10^{-3} \text{ ft}^3 \qquad (3.4.4)$$

The volume of air contained in the pores (voids) is thus

$$V_a = V - V_w - V_s = \frac{1}{30} - (7.853 \times 10^{-3}) - (2.097 \times 10^{-2})$$
$$= 4.510 \times 10^{-3} \text{ ft}^3 \qquad (3.4.5)$$

a) The void ratio is computed as follows:

$$e = \frac{V_v}{V_s} = \frac{V - V_s}{V_s} = \frac{\frac{1}{30} - (2.097 \times 10^{-2})}{2.097 \times 10^{-2}} = \mathbf{0.590} \qquad (3.4.6)$$

or

$$e = \frac{V_v}{V_s} = \frac{V_w + V_a}{V_s} = \frac{(7.853 \times 10^{-3}) + (4.510 \times 10^{-3})}{2.097 \times 10^{-2}} = \mathbf{0.590} \qquad (3.4.7)$$

b) The moisture content is next computed:

$$w = \left(\frac{W_w}{W_s}\right) * 100\% = \left(\frac{0.49 \text{ lb}}{3.52 \text{ lb}}\right) * 100\% = \mathbf{13.9\%} \qquad (3.4.8)$$

c) The dry unit weight is next computed, i.e.,

$$\gamma_d = \frac{W_s}{V} = \frac{(3.52 \text{ lb})}{\left(\frac{1}{30} \text{ ft}^3\right)} = \mathbf{105.6 \text{ lb/ft}^3} \qquad (3.4.9)$$

d) The degree of saturation is thus

$$S = \left(\frac{V_w}{V_v}\right) * 100\% = \left(\frac{V_w}{V_a + V_w}\right) * 100\%$$

$$= \left(\frac{7.853 \times 10^{-3}}{4.510 \times 10^{-3} + 7.853 \times 10^{-3}}\right) * 100\% = \mathbf{63.4\%} \qquad (3.4.10)$$

or, using the expression developed in Case 1.3 in Chapter 1,

$$S = \left(\frac{G_s w}{e}\right) * 100\% = \frac{(2.69)(0.139)}{0.590} * 100\% = \mathbf{63.4\%} \qquad (3.4.11)$$

e) The moist unit weight is next computed as

$$\gamma = \frac{W_s + W_w}{V} = \frac{4.01 \text{ lb}}{\left(\frac{1}{30} \text{ ft}^3\right)} = \mathbf{120.3 \text{ lb/ft}^3} \qquad (3.4.12)$$

or, using the expression developed in Case 1.7 of Chapter 1,

$$\gamma = \frac{\gamma_w (G_s + Se)}{1 + e} = \frac{(62.4 \text{ lb/ft}^3)[2.69 + (0.634)(0.590)]}{1 + 0.590} = \mathbf{120.3 \text{ lb/ft}^3}$$

$$(3.4.13)$$

f) Finally, the saturated unit weight is computed

$$\gamma = \frac{\gamma_w(G_s + e)}{1 + e} = \frac{(62.4 \text{ lb/ft}^3)(2.69 + 0.590)}{1 + 0.590} = \textbf{128.7 lb/ft}^3 \quad (3.4.14)$$

Figure Ex. 3.4 shows the phase diagram associated with this soil.

EXAMPLE PROBLEM 3.5

General Remarks

This example problem investigates the calculation of certain quantities associated with the process of compaction.

Problem Statement

Laboratory tests on a cohesionless soil show that the minimum and maximum dry unit weights that can be obtained for this material are 100.5 and 115.2 lb/ft^3, respectively. The specific gravity of solids (G_s) for the soil is 2.69. Field tests indicate that, in situ, the moist unit weight of the soil is 121.7 lb/ft^3 at a moisture content of 10.7%.

a) Determine the relative density (D_r) in situ. b) The soil is to be compacted to a dry unit weight that is 97% of the maximum dry density without adding or removing water. Determine the volume of compacted soil that will be obtained for each cubic foot of soil in situ. c) What is the degree of saturation in the compacted soil?

Solution

a) The requisite void ratio values are computed from the general expression for the dry unit weight derived in Case 1.4 of Chapter 1, i.e.,

$$\gamma_d = \frac{G_s \gamma_w}{1 + e} \quad \Rightarrow \quad e = \frac{G_s \gamma_w}{\gamma_d} - 1 \quad (3.5.1)$$

Thus, applying Eq. (3.5.1) for $\gamma_d = \gamma_{d_{min}}$ gives

$$e_{max} = \frac{G_s \gamma_w}{\gamma_{d_{min}}} - 1 = \frac{(2.69)(62.4 \text{ lb/ft}^3)}{100.5 \text{ lb/ft}^3} - 1 = 0.670 \quad (3.5.2)$$

Similarly, for $\gamma_d = \gamma_{d_{max}}$,

$$e_{min} = \frac{G_s \gamma_w}{\gamma_{d_{max}}} - 1 = \frac{(2.69)(62.4 \text{ lb/ft}^3)}{115.2 \text{ lb/ft}^3} - 1 = 0.457 \quad (3.5.3)$$

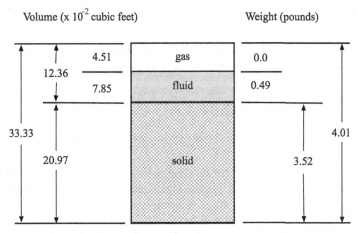

FIGURE EX. 3.4 Phase diagram showing relationship between volume and weight for the silty sand.

Knowing the moist unit weight of the field soil, the void ratio is computed from the expression derived in Case 1.5 of Chapter 1, i.e.,

$$\gamma = \frac{G_s \gamma_w (1 + w)}{1 + e} \quad \Rightarrow \quad e = \frac{G_s \gamma_w (1 + w)}{\gamma} - 1$$

$$= \frac{(2.69)(62.4 \text{ lb/ft}^3)(1 + 0.107)}{121.7 \text{ lb/ft}^3} - 1 = 0.527$$

$$(3.5.4)$$

Using the values obtained in Eqs. (3.5.2)–(3.5.4), the relative density of the field soil is thus

$$D_r = \left(\frac{e_{max} - e}{e_{max} - e_{min}} \right) * 100\% = \left(\frac{0.670 - 0.527}{0.670 - 0.457} \right) * 100\% = \mathbf{67.3\%}$$

$$(3.5.5)$$

The field soil is thus considered to be "dense" (Table 3.1).

b) During the compaction process, the weight (W_s) and volume (V_s) of the solid phase remain unchanged. The former quantity is computed from the definition of the dry unit weight, i.e.,

$$\gamma_d = \frac{W_s}{V} = \frac{\gamma}{1 + w} \quad \Rightarrow \quad W_s = \frac{\gamma V}{1 + w} = \frac{(62.4 \text{ lb/ft}^3)V}{1 + 0.107} = 109.9V \text{ lb} \quad (3.5.6)$$

Since no volume has explicitly been specified in the problem, assume that $V = 1 \text{ ft}^3$. Thus, $W_s = 109.9(1) = 109.9$ lb. The volume of the solid phase is then computed from the definition of the specific gravity of solids as follows:

$$G_s = \frac{W_s}{V_s \gamma_w} \implies V_s = \frac{W_s}{G_s \gamma_w} = \frac{(109.9 \text{ lb})}{(2.69)(62.4 \text{ lb/ft}^3)} = 0.655 \text{ ft}^3 \quad (3.5.7)$$

With the volume of the solid phase known, the volume of the voids (V_v) is computed from the definition of the void ratio as follows:

$$e = \frac{V_v}{V_s} \implies V_v = eV_s = (0.527)(0.655 \text{ ft}^3) = 0.345 \text{ ft}^3 \quad (3.5.8)$$

The final quantity that needs to be computed is the volume of the pore water. The first step in this process is the computation of the weight of the water from the definition of the moisture content, i.e.,

$$w = \frac{W_w}{W_s} \implies W_w = wW_s = (0.107)(109.9 \text{ lb}) = 11.76 \text{ lb} \quad (3.5.9)$$

The volume of the pore water is next computed from the definition of the unit weight of water, i.e.,

$$\gamma_w = \frac{W_w}{V_w} \implies V_w = \frac{W_w}{\gamma_w} = \frac{11.76 \text{ lb}}{62.4 \text{ lb/ft}^3} = 0.188 \text{ ft}^3 \quad (3.5.10)$$

For completeness, the volume of air in the pore space is

$$V_a = V - V_w = 0.345 - 0.188 = 0.157 \text{ ft}^3 \quad (3.5.11)$$

Following compaction,

$$\gamma_{d_{new}} = 0.97\gamma_{d_{max}} = 0.97(115.2 \text{ lb/ft}^3) = 111.7 \text{ lb/ft}^3 = \frac{W_s}{V_{new}} \quad (3.5.12)$$

Since the weight of the solid phase is unchanged during compaction, it follows that

$$V_{new} = \frac{W_s}{\gamma_{d_{new}}} = \frac{109.9 \text{ lb}}{111.7 \text{ lb/ft}^3} = \mathbf{0.984 \text{ ft}^3} \quad (3.5.13)$$

Since the volume of the solid phase is also unchanged during compaction, it follows that

$$V_{v_{new}} = V_{new} - V_s = 0.984 - 0.655 = 0.329 \text{ ft}^3 \quad (3.5.14)$$

c) Since the volume of water is also unchanged during compaction, it follows that the degree of saturation after compaction will be

$$S = \left(\frac{V_w}{V_{v_{new}}}\right) * 100\% = \left(\frac{0.188}{0.329}\right) * 100\% = \mathbf{57.1\%} \quad (3.5.15)$$

EXAMPLE PROBLEM 3.6

General Remarks

This example problem investigates the calculation of certain quantities associated with the process of compaction.

Problem Statement

The results of a set of standard Proctor compaction tests on a particular soil give a maximum dry density of 19.6 kN/m^3 and an optimum moisture content of 12.5%. The specific gravity of solids for the soil is 2.68. Determine a) the degree of saturation and the void ratio at optimum moisture content and b) what percentage of the voids is occupied by air at this condition.

Solution

a) Recalling Eq. (3.6), i.e.,

$$\gamma_d = \frac{G_s \gamma_w}{1 + \left(\dfrac{G_s w}{S}\right)} \tag{3.6.1}$$

and solving for the degree of saturation gives

$$S = \frac{w \, G_s}{\dfrac{G_s \gamma_w}{\gamma_d} - 1} = \frac{(0.125)(2.68)}{\left[\dfrac{(2.68)(9.91 \text{ kN/m}^3)}{(19.6 \text{ kN/m}^3)} - 1\right]} = 0.981 = \textbf{98.1\%} \tag{3.6.2}$$

The void ratio is then computed using Eq. (3.5), i.e.,

$$e = \frac{G_s w}{S} = \frac{(2.68)(0.125)}{0.981} = \textbf{0.341} \tag{3.6.3}$$

b) The ratio of the volume of air to the volume of voids can be computed in one of two ways. The first approach uses the degree of saturation (written as a decimal rather than a percentage) as follows:

$$S = \frac{V_w}{V_v} = \frac{V_v - V_a}{V_v} = 1 - \frac{V_a}{V_v} \quad \Rightarrow \quad \frac{V_a}{V_v} = (1 - S) \tag{3.6.4}$$

In the second approach, $V_s = 1 \text{ m}^3$ is assumed, implying that $V_v = e$. From the definition of the specific gravity of solids,

$$W_s = G_s \gamma_w V_s = G_s \gamma_w (1) \tag{3.6.5}$$

From the definition of the moisture content,

$$w = \frac{W_w}{W_s} \quad \Rightarrow \quad W_w = wW_s = wG_s\gamma_w \tag{3.6.6}$$

where w is represented as a decimal number. From the definition of the unit weight of water,

$$\gamma_w = \frac{W_w}{V_w} \quad \Rightarrow \quad V_w = \frac{W_w}{\gamma_w} = wG_s \tag{3.6.7}$$

The volume of air contained in the voids is thus

$$V_a = V_v - V_w = e - wG_s \tag{3.6.8}$$

From Eq. (3.5), $wG_s = Se$. Substituting this result into Eq. (3.6.8) gives

$$V_a = e - Se = (1 - S)e = (1 - S)V_v \quad \Rightarrow \quad \frac{V_a}{V_v} = (1 - S) \tag{3.6.9}$$

Thus,

$$\frac{V_a}{V_v} = (1 - S) = 1 - 0.981 = 0.0190 = \mathbf{1.90\%} \tag{3.6.10}$$

EXAMPLE PROBLEM 3.7

General Remarks

This example problem illustrates the manner in which a compaction curve is created.

Problem Statement

The first two columns in Table Ex. 3.7a list values of moisture content and moist unit weight that were determined in the laboratory. The specific gravity of solids (G_s) for the soil is equal to 2.71. a) Compute the associated dry unit weight, b) draw a compaction curve for the data, and c) draw the associated ZAV curve. d) If the contractor is required to obtain 90% RC, what is the range of admissible moisture content values?

Solution

a) The dry unit weight, which is listed in the third column of Table Ex. 3.7a, is computed as follows (recall Case 1.6 in Chapter 1):

$$\gamma_d = \frac{\gamma}{1 + w} \tag{3.7.1}$$

TABLE EX. 3.7a Compaction Data

Moisture Content (%)	Moist Unit Weight (kN/m³)	Dry Unit Weight (kN/m³)
10	15.24	13.85
13	16.49	14.60
16	18.38	15.84
18	19.32	16.37
20	19.95	16.62
22	19.79	16.22
25	19.01	15.21

b) Plotting the values in the first and third columns of Table Ex. 3.7a gives the compaction curve shown in Figure Ex. 3.7a.

From this figure the *maximum dry unit weight* is approximately **16.6 kN/m³**. The corresponding *optimum moisture content* is **20%**.

c) In light of the given data, the unit weight associated with the ZAV curve is computed from Eq. (3.7), i.e.,

$$\gamma_{d_{ZAV}} = \frac{\gamma_w}{\dfrac{1}{G_s} + w} \qquad (3.7.2)$$

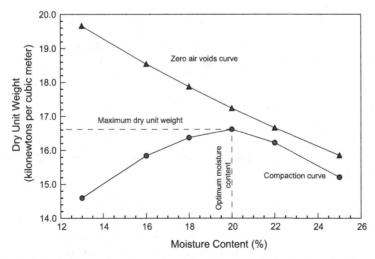

FIGURE EX. 3.7A Compaction curve and zero air voids curve for laboratory data of Table Ex. 3.7.

TABLE EX. 3.7b Data Used to Draw the Zero Air Voids Curve

Moisture Content (%)	Unit Weight for Zero Air Voids Curve (kN/m³)
10	20.92
13	19.66
16	18.54
18	17.87
20	17.24
22	16.66
25	15.85

For example, when $w = 10\%$,

$$\gamma_{d_{ZAV}} = \frac{9.81 \text{ kN/m}^3}{\left(\dfrac{1}{2.71}\right) + 0.10} = 20.92 \text{ kN/m}^3 \tag{3.7.3}$$

The unit weight for the other moisture contents is computed in a similar manner. Table Ex. 3.7b lists the resulting values.

Figure Ex. 3.7a shows the ZAV curve and its proximity to the compaction curve.

d) Since the contractor is required to obtain 90% RC, the minimal allowable dry unit weight is thus

$$\gamma_{d_{min}} = 0.90\gamma_{d_{max}} = 0.90\left(16.6 \text{ kN/m}^3\right) = 14.9 \text{ kN/m}^3 \tag{3.7.4}$$

From the compaction curve, the range of acceptable moisture contents is thus approximately **13.8–26.0%** (Figure Ex. 3.7b).

EXAMPLE PROBLEM 3.8

General Remarks

This example problem illustrates how information is obtained from a compaction curve.

Problem Statement

Figure Ex. 3.8a shows the laboratory field compaction curve for a given soil. Specifications call for the compacted unit weight to be at least 95% of the

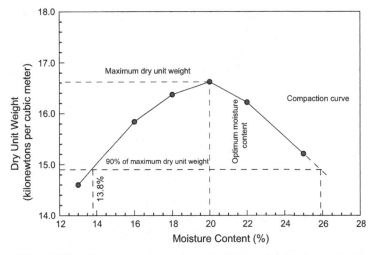

FIGURE EX. 3.7B   Compaction curve and acceptable range of moisture content values.

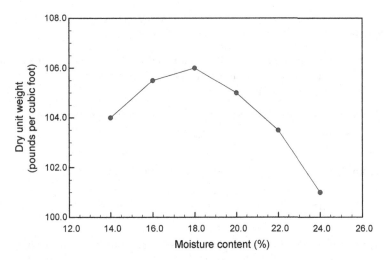

FIGURE EX. 3.8A   Results of laboratory standard proctor compaction test.

standard Proctor maximum and within ±2% of the optimum moisture content. When a sample of the soil was excavated from the field, it had a volume of 57.6 in³. The sample weighed 4.00 lbs wet and 3.36 lbs dry.

Determine a) the field moisture content, b) the compacted dry unit weight, and c) the *RC*. d) Does the soil meet specifications? e) If $G_s = 2.70$, what is the degree of saturation of the field sample? f) If the field sample was saturated at constant total volume (i.e., the air in the voids was replaced by water), what would be the resulting moisture content? g) It is desired to plot the *ZAV* curve. For this purpose, compute the *ZAV* unit weight associated with moisture contents of 20%, 22%, and 24%.

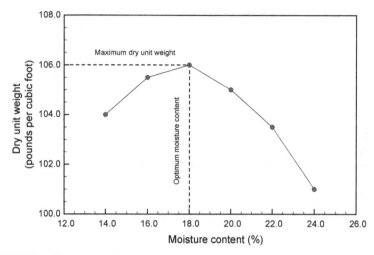

FIGURE EX. 3.8B Results of laboratory standard proctor compaction test with maximum values identified.

Solution

The compaction curve of Figure Ex. 3.8a is redrawn only with the maximum dry unit weight and optimum moisture content identified.

From Figure Ex. 3.8b it is evident that

- The maximum dry unit weight ($\gamma_{d_{max}}$) is approximately 106.0 lb/ft^3.
- The optimum moisture content (w_{opt}) is approximately 18%.

a) From the given information, field the moisture content is

$$w = \left(\frac{W_w}{W_s}\right) * 100\% = \left(\frac{4.00 - 3.36}{3.36}\right) * 100\% = \mathbf{19.05\%} \qquad (3.8.1)$$

b) The compacted dry unit weight is likewise computed from the given information, i.e.,

$$\gamma_d = \frac{W_s}{V} = \left(\frac{3.36 \text{ lb}}{57.6 \text{ in}^3}\right)\left(\frac{12 \text{ in}}{\text{ft}}\right)^3 = \mathbf{100.8 \text{ lb/ft}^3} \qquad (3.8.2)$$

This can also be computed from the moist unit weight (γ) and the field moisture content, i.e.,

$$\gamma = \frac{W}{V} = \left(\frac{4.00 \text{ lb}}{57.6 \text{ in}^3}\right)\left(\frac{12 \text{ in}}{\text{ft}}\right)^3 = 120.0 \text{ lb/ft}^3 \qquad (3.8.3)$$

and then

$$\gamma_d = \frac{\gamma}{1+w} = \frac{(120.0 \text{ lb/ft}^3)}{1+0.1905} = \textbf{100.8 lb/ft}^3 \qquad (3.8.4)$$

c) Since the maximum dry unit weight is approximately 106.0 lb/ft^3, the relative compaction (RC) is

$$RC = \left(\frac{\gamma_d}{\gamma_{d_{max}}}\right) * 100\% = \left(\frac{100.0 \text{ lb/ft}^3}{106.0 \text{ lb/ft}^3}\right) * 100\% = \textbf{95.1\%} \qquad (3.8.5)$$

d) Since the optimum moisture content (w_{opt}) is approximately 18%, the acceptable range in moisture content is thus $16.0\% \le w \le 20.0\%$. The moisture content of 19.05% falls in this range. In addition, since $RC > 95\%$, it follows that *the soil meets specifications*.

e) The determination of the degree of saturation of the field sample requires knowledge of the void ratio. This is computed as follows:

$$\gamma_d = \frac{G_s\gamma_w}{1+e} \quad\Rightarrow\quad e = \frac{G_s\gamma_w}{\gamma_d} - 1 = \frac{(2.70)(62.4 \text{ lb/ft}^3)}{100.8 \text{ lb/ft}^3} - 1 = 0.671$$

$$(3.8.6)$$

The desired degree of saturation is thus

$$S = \left(\frac{G_s w}{e}\right) * 100\% = \left[\frac{(2.70)(0.1905)}{0.671}\right] * 100\% = \textbf{76.7\%} \qquad (3.8.7)$$

This can also be computed from the moist unit weight (γ) and the field moisture content, i.e.,

$$\gamma = \frac{G_s\gamma_w(1+w)}{1+e} = \frac{G_s\gamma_w(1+w)}{1+\dfrac{G_s w}{S}} \quad\Rightarrow\quad S = \frac{G_s w}{\dfrac{(1+w)G_s\gamma_w}{\gamma} - 1} * 100\%$$

$$(3.8.8)$$

Substituting the given quantities gives

$$S = \frac{(2.70)(0.1905)}{\dfrac{(1+0.1905)(2.70)(62.4 \text{ lb/ft}^3)}{120.0 \text{ lb/ft}^3} - 1} * 100\% = \textbf{76.6\%} \qquad (3.8.9)$$

f) To determine the moisture content if the field sample was saturated (i.e., $S = 1.0$) at constant total volume, note that the void ratio remains unchanged at $e = 0.671$. Thus, using Eq. (3.5) gives

$$w = \left(\frac{Se}{G_s}\right) * 100\% = \frac{(1.0)(0.671)}{2.70} * 100\% = \textbf{24.9\%} \qquad (3.8.10)$$

TABLE EX. 3.8 Summary of Zero Air Voids Calculations

w (%)	$\gamma_{d_{ZAV}}$ (lb/ft³)
20.0	109.4
22.0	105.7
24.0	102.2

g) To generate data for the ZAV curve, recall that for this special case $S = 100\%$, giving $e = wG_s$. The first part of Eq. (3.8.2) thus becomes

$$\gamma_{d_{ZAV}} = \frac{G_s \gamma_w}{1 + G_s w} = \frac{(2.70)(62.4 \text{ lb/ft}^3)}{1 + (2.70)w} \qquad (3.8.11)$$

where w is understood to be a decimal number. Table Ex. 3.8 summarizes the values of $\gamma_{d_{ZAV}}$ for the specified values of w.

EXAMPLE PROBLEM 3.9

General Remarks

This example problem illustrates the manner in which a compaction curve is created from actual laboratory data.

Problem Statement

Table Ex. 3.9a lists the results of standard Proctor compaction tests that were performed on a brown inorganic silty clay of medium plasticity. In all five tests, the volume of the compaction mold was equal to 1/30 ft³. Table Ex. 3.9b lists the results of moisture content tests that were performed on the same soil. In addition, the specific gravity of solids (G_s) for the silty clay was found to be 2.71.

a) Compute the dry densities and moisture content for each of the samples tested. b) Draw the compaction curve and determine the maximum dry unit weight and the optimum moisture content. c) Finally, draw the ZAV curve.

Solution

a) Consider the data for sample 1 given in the first two rows in Table Ex. 3.9a. The weight of the moist soil is thus

$$W = 3688 - 1978 = \mathbf{1710 \text{ g}} \qquad (3.9.1)$$

TABLE EX. 3.9a Standard Proctor Compaction Data for Determining Unit Weight

Sample	1	2	3	4	5
Weight of moist sample + mold (g)	3688	3801	3913	3897	3865
Weight of mold (g)	1978	1978	2030	2030	2030
Weight of moist sample (g)	1710	1823	1883	1867	1835
Moist unit weight (lb/ft³)	113.1	120.6	124.5	123.5	121.4
Dry unit weight (lb/ft³)	101.2	104.5	105.4	101.9	99.1

TABLE EX. 3.9b Data for Determining Moisture Content

Sample	1	2	3	4	5
Weight of moist sample + container (g)	83.13	89.20	96.13	132.01	122.90
Weight of dry sample + container (g)	78.05	81.95	86.72	115.12	106.77
Weight of container (g)	34.80	34.90	34.80	35.40	35.10
Weight of pore fluid (g)	5.08	7.25	9.41	16.89	16.13
Weight of dry soil (g)	43.25	47.05	51.92	79.72	71.67
Moisture content (%)	11.75	15.41	18.12	21.19	22.51

Choosing units of *pounds* and *feet*, the moist unit weight is thus

$$\gamma = \frac{(1710 \text{ g})(2.205 \times 10^{-3} \text{ lb/g})}{\left(\dfrac{1}{30} \text{ ft}^3\right)} = \textbf{113.1 lb/ft}^3 \qquad (3.9.2)$$

Performing similar calculations gives the values listed in the last two rows of Table Ex. 3.9a.

Next consider the data for sample 1 given in the first three rows in Table Ex. 3.9b. The weight of the pore fluid is computed as follows:

$$W_w = 83.13 - 78.05 = \textbf{5.08 g} \qquad (3.9.3)$$

The weight of the dry soil is next computed

$$W_s = 78.05 - 34.80 = \textbf{43.25 g} \qquad (3.9.4)$$

The moisture content is thus

$$w = \left(\frac{W_w}{W_s}\right) * 100\% = \left(\frac{5.08 \text{ g}}{43.25 \text{ g}}\right) * 100\% = \mathbf{11.75\%} \tag{3.9.5}$$

Performing similar calculations gives the values listed in the last three rows of Table Ex. 3.9b.

The dry unit weight can now be computed from the moist unit weight and the moisture content, i.e.,

$$\gamma_d = \frac{\gamma}{1+w} = \frac{113.1 \text{ lb/ft}^3}{1+0.1175} = \mathbf{101.2 \text{ lb/ft}^3} \tag{3.9.6}$$

Performing similar calculations gives the values listed in the final row of Table Ex. 3.9a.

Remark: Additional information regarding relative amounts of the three phases can easily be quantified using the equations presented in Chapter 1.

For example, consider sample 1 in Table Ex. 3.9b. From Case 1.4 of Chapter 1,

$$\gamma_d = \frac{G_s \gamma_w}{1+e} \quad \Rightarrow \quad e = \frac{G_s \gamma_w}{\gamma_d} - 1 = \frac{(2.71)(62.4 \text{ lb/ft}^3)}{101.2 \text{ lb/ft}^3} = \mathbf{0.671} \tag{3.9.7}$$

From Case 1.3 of Chapter 1,

$$w = \left(\frac{Se}{G_s}\right) * 100\% \quad \Rightarrow \quad S = \left(\frac{wG_s}{e}\right) * 100\%$$
$$= \left[\frac{(0.1175)(2.71)}{0.671}\right] * 100\% = 47.5\% \tag{3.9.8}$$

The volume of the solid phase is computed from the definition of the specific gravity of solids, i.e.,

$$G_s = \frac{W_s}{V_s \gamma_w} \quad \Rightarrow \quad V_s = \frac{W_s}{G_s \gamma_w} = \frac{(43.25 \text{ g})}{(2.71)(1.0 \text{ g/cm}^3)} = \mathbf{15.96 \text{ cm}^3} \tag{3.9.9}$$

The volume of the voids is thus

$$e = \frac{V_v}{V_s} \quad \Rightarrow \quad V_v = eV_s = (0.671)(15.96 \text{ cm}^3) = \mathbf{10.71 \text{ cm}^3} \tag{3.9.10}$$

The volume of the pore fluid is computed from the definition of the unit weight of water, i.e.,

$$\gamma_w = \frac{W_w}{V_w} \quad \Rightarrow \quad V_w = \frac{W_w}{\gamma_w} = \frac{5.08 \text{ g}}{1.0 \text{ g/cm}^3} = \mathbf{5.08 \text{ cm}^3} \tag{3.9.11}$$

The volume of air occupying the voids is thus

$$V_a = V_v - V_w = (10.71 - 5.08) \text{ cm}^3 = \textbf{5.63 cm}^3 \tag{3.9.12}$$

As a check on the aforesaid results, compute the degree of saturation (S) in an alternate way than was used earlier. In particular,

$$S = \left(\frac{V_w}{V_v}\right) * 100\% = \left(\frac{5.08 \text{ cm}^3}{10.71 \text{ cm}^3}\right) * 100\% = \textbf{47.5\%} \tag{3.9.13}$$

which is identical to the value computed earlier using Case 1.3 of Chapter 1. Figure Ex. 3.9a shows the phase diagram associated with sample 1.

b) Plotting the dry unit weight from Table Ex. 3.9a versus the moisture content from Table Ex. 3.9b gives the compaction curve shown in Figure Ex. 3.9b.

The maximum dry unit weight is approximately 105.7 lb/ft^3 and the optimum moisture content is about 17.5%.

c) In light of the given data, the unit weight associated with the ZAV curve is again computed from Eq. (3.7), i.e.,

$$\gamma_{d_{ZAV}} = \frac{\gamma_w}{\dfrac{1}{G_s} + w} \tag{3.9.14}$$

For example, when $w = 12\%$,

$$\gamma_{d_{ZAV}} = \frac{62.4 \text{ lb/ft}^3}{\left(\dfrac{1}{2.71}\right) + 0.12} = 127.6 \text{ kN/m}^3 \tag{3.9.15}$$

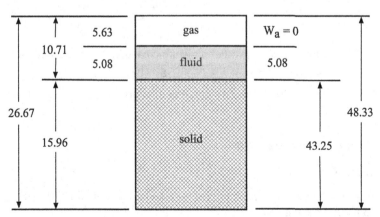

FIGURE EX. 3.9A Phase diagram showing relationship between volume and mass for soil sample 1.

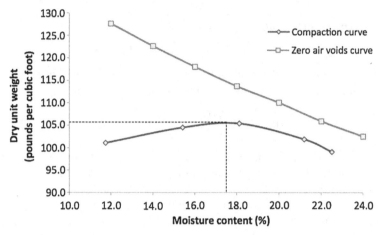

FIGURE EX. 3.9B Compaction curve and zero air voids curve for laboratory data of Tables Ex. 3.9a and Ex. 3.9b.

TABLE EX. 3.9c Data Used to Draw the Zero Air Voids Curve

Moisture Content (%)	Unit Weight for Zero Air Voids Curve (lb/ft³)
12	127.6
14	122.6
16	118.0
18	113.7
20	110.0
22	105.9
24	102.5

The unit weight for the other moisture contents is computed in a similar manner. Table Ex. 3.9c lists the resulting values.

Figure Ex. 3.9b shows the *ZAV* curve and its proximity to the compaction curve.

EXAMPLE PROBLEM 3.10

General Remarks

In this problem the compaction curves for standard and modified Proctor tests are compared for a given soil.

Problem Statement

Both standard and modified Proctor compaction tests were performed on a soil. Table Ex. 3.10a summarizes the moisture contents and dry densities. The former were determined in the laboratory; the latter were computed from moist unit weights in the manner shown in Example Problem 3.8. The specific gravity of solids (G_s) for the soil is equal to 2.68.

Given the data in Table Ex. 3.10a, draw the compaction curves for the two sets of data and the ZAV curve.

Solution

Plotting the values in the first and third columns of Table Ex. 3.10a gives the compaction curve shown in Figure Ex. 3.10.

In light of the given data, the unit weight associated with the ZAV curve is again computed from Eq. (3.7)

$$\gamma_{d_{ZAV}} = \frac{\gamma_w}{\dfrac{1}{G_s} + w} \qquad (3.10.1)$$

where $G_s = 2.68$. Table Ex. 3.10b lists the resulting values.

Figure Ex. 3.10 shows the ZAV curve and its proximity to the compaction curves.

Remark: The maximum dry unit weight associated with the modified Proctor method *increases* by 1.4 kN/m³. The moisture content *decreases* by 1.5%.

TABLE EX. 3.10a Data for Standard and Modified Proctor Compaction Tests

Standard Proctor Method		Modified Proctor Method	
Moisture Content (%)	Dry Unit Weight (kN/m³)	Moisture Content (%)	Dry Unit Weight (kN/m³)
6	16.02	6	16.81
9	16.65	9	17.74
12	16.97	12	18.54
14	17.12	13	18.54
16	16.97	14	18.38
19	16.49	16	17.59
22	15.71	18	16.97

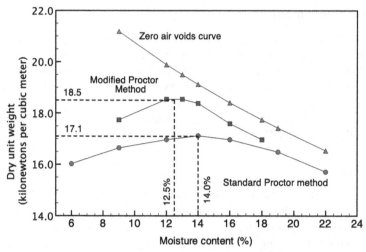

FIGURE EX. 3.10 Compaction curves and zero air voids curve for laboratory data of Table Ex. 3.10.

TABLE EX. 3.10b Data Used to Draw the Zero Air Voids Curve

Moisture Content (%)	Unit Weight for Zero Air Voids Curve (kN/m^3)
6	22.65
9	21.18
12	19.89
13	19.50
14	19.12
16	18.40
18	17.74
19	17.72
22	16.54

EXAMPLE PROBLEM 3.11

General Remarks

This example problem illustrates the manner in which the volume of soil required from a borrow area is computed.

Problem Statement

Specifications for a road embankment require the soil to be compacted to a moist unit weight of 19.2 kN/m^3 at a moisture content of 17.5%. The soil is to be obtained from a borrow area where the in situ dry unit weight is 15.9 kN/m^3. The volume of the finished embankment is to be 1520 m^3. What volume of borrow material is required for the road embankment? It is assumed that the volume of the fill be essentially *unchanged* when it is transported from the borrow area to the site of the embankment.

Solution

The specified dry density is first computed from the given information as

$$\gamma_d = \frac{\gamma}{1+w} = \frac{19.2 \text{ kN/m}^3}{1+0.175} = 16.34 \text{ kN/m}^3 \tag{3.11.1}$$

Denoting the volume required for the embankment by V_{emb}, the associated weight of solids is thus

$$\gamma_d = \frac{W_s}{V_{emb}} \quad \Rightarrow \quad W_s = \gamma_d \, V_{emb} = \left(16.34 \text{ kN/m}^3\right)\left(1520 \text{ m}^3\right) = 24,837 \text{ kN} \tag{3.11.2}$$

Since the weight of the solid phase remains unchanged, it follows that at the borrow area,

$$\gamma_d = \frac{W_s}{V_{borrow}} \quad \Rightarrow \quad V_{borrow} = \frac{W_s}{\gamma_d} = \frac{24,837 \text{ kN}}{15.9 \text{ kN/m}^3} = \mathbf{1562 \text{ m}^3} \tag{3.11.3}$$

Thus, 1562 m^3 of soil must be transported to the embankment site. Once compacted at the site, this amount of soil will give the required volume for the finished embankment.

EXAMPLE PROBLEM 3.12

General Remarks

This example problem illustrates the manner in which certain quantities for soil taken from a borrow area are computed.

Problem Statement

The in situ moisture content of a soil at a borrow area is 16% and its moist unit weight is 17.3 kN/m^3. The specific gravity of solids for the soil is 2.72.

The soil is to be excavated and transported to a construction site for use in a compacted fill. a) If the specifications call for the soil to be compacted to a dry unit weight of 18.1 kN/m^3 at the *same* moisture content of 16%, how many cubic meters of soil from the excavation are needed to produce 2000 m^3 of compacted fill? b) What is the void ratio and degree of saturation of the compacted soil? c) If a dump truck can carry 20 tons of soil in a given trip, how many trips are required to transport the soil to the construction site?

Solution

a) The required weight of the solid phase is computed from the definition of the dry unit weight. Denoting the volume of the compacted fill by V_{CF}, it follows that

$$\gamma_d = \frac{W_s}{V_{CF}} \quad \Rightarrow \quad W_s = \gamma_d\, V_{CF} = \left(18.1 \text{ kN/m}^3\right)\left(2000 \text{ m}^3\right) = 36,200 \text{ kN}$$

$$(3.12.1)$$

For the soil at the borrow area,

$$\gamma_d = \frac{\gamma}{1+w} = \frac{17.3 \text{ kN/m}^3}{1+0.16} = 14.91 \text{ kN/m}^3 \qquad (3.12.2)$$

The required volume of soil from the borrow area is thus

$$V_{borrow} = \frac{W_s}{\gamma_d} = \frac{36,200 \text{ kN}}{14.91 \text{ kN/m}^3} = \textbf{2427.3 m}^3 \qquad (3.12.3)$$

b) The void ratio of the compacted soil is computed from the dry unit weight of this soil as follows:

$$\gamma_d = \frac{G_s \gamma_w}{1+e} \quad \Rightarrow \quad e = \frac{G_s \gamma_w}{\gamma_d} - 1 = \frac{(2.72)(9.81 \text{ kN/m}^3)}{18.1 \text{ kN/m}^3} - 1 = \textbf{0.474}$$

$$(3.12.4)$$

With the void ratio known, the degree of saturation is then computed using Eq. (3.5), i.e.,

$$S = \left(\frac{G_s w}{e}\right) * 100\% = \left[\frac{(2.72)(0.16)}{0.474}\right] * 100\% = \textbf{91.8\%} \qquad (3.12.5)$$

c) The number of trips required to transport the soil to the construction site is computed by noting that the required moist weight of the solid phase is

$$\gamma = \frac{W}{V} \quad \Rightarrow \quad W = \gamma\, V = \left(17.3 \text{ kN/m}^3\right)\left(2427.3 \text{ m}^3\right) = 41,992 \text{ kN}$$

$$(3.12.6)$$

Since a dump truck can carry 20 tons of soil per trip, it follows that

$$(41,992 \text{ kN})\left(\frac{1000 \text{ N}}{\text{kN}}\right)\left(\frac{\text{lb}}{4.448 \text{ N}}\right)\left(\frac{\text{ton}}{2000 \text{ lb}}\right)\left(\frac{\text{trip}}{20 \text{ ton}}\right) = \textbf{236 trips}$$

$$(3.12.7)$$

EXAMPLE PROBLEM 3.13

General Remarks

This example presents a somewhat more complicated compaction problem.

Problem Statement

An embankment is to be constructed as part of a roadway improvement project. The dimensions of the embankment are 8 m wide by 0.70 m compacted thickness by 0.75 mile long.

To create the embankment, soil will be excavated and transported from a borrow area to the construction site. Construction specifications require the soil to be compacted to a moisture content (w) of 19.6%, achieving a dry unit weight (γ_d) of 17.0 kN/m^3. At the borrow area the soil has a specific gravity of solids (G_s) equal to 2.70, a moisture content of $w = 15\%$ and a dry unit weight of 15 kN/m^3. When loaded on dump trucks for transport, the soil *loosens* and its dry unit weight drops to 14 kN/m^3.

Determine a) the volume of soil to be excavated at the borrow area, b) the number of trips of trucks between the borrow area and the construction site assuming that each truck can carry 10 yd^3 of loose soil, c) The volume of water, in cubic meters, to be added at the construction site to achieve the desired moisture content before compaction, d) The degree of saturation of soil at the construction site after compaction, and e) the moisture content of the compacted soil if it is saturated after construction due to rainfall.

Solution

The following notation is used in the solution: *stage A* refers to the excavation at the borrow area, *stage B* refers to transport on trucks, and *stage C* refers to the construction phase. The following quantities are thus known for the respective stages:

> *For stage A*: $G_s = 2.70$, $w = 15\%$, $\gamma_d = 15$ kN/m^3.
> *For stage B*: $\gamma_d = 14$ kN/m^3, volume per dump truck $= 10$ m^3.
> *For stage C*: $\gamma_d = 17$ kN/m^3, $w = 19.6\%$.

Since it is 8.0 m wide, 0.70 m thick, and 0.75 miles long, the required volume of the embankment is

$$V = (8.0 \text{ m})(0.70 \text{ m})(0.75 \text{ mile})\left(\frac{5280 \text{ ft}}{\text{mile}}\right)\left(\frac{\text{m}}{3.281 \text{ ft}}\right) = 6759 \text{ m}^3 \quad (3.13.1)$$

Remark: Before continuing with the solution, it is important to note that the weight of solids (W_s) and specific gravity of solids (G_s) remain *unchanged* throughout the aforesaid three stages.

---

a) For *stage C* the dry unit weight is 17.0 kN/m^3. Thus,

$$\gamma_{dc} = \frac{W_s}{V_C} \quad \Rightarrow \quad W_s = \gamma_{dc} V_C \tag{3.13.2}$$

Since the weight of solids remains unchanged throughout the construction of the embankment, it follows that the volume of soil to be excavated at the borrow area (V_A) is thus

$$\gamma_{dc} V_C = \gamma_{d_A} V_A \quad \Rightarrow \quad V_A = \frac{\gamma_{dc} V_C}{\gamma_{d_A}} = \frac{(17.0 \text{ kN/m}^3)(6759 \text{ m}^3)}{15.0 \text{ kN/m}^3} = \mathbf{7660 \text{ m}^3} \tag{3.13.3}$$

b) To determine the number of trips of trucks between the borrow area and the construction site requires knowledge of the volume of material (V_B) to be transported. This is computed as follows:

$$\gamma_{dc} V_C = \gamma_{d_B} V_B \quad \Rightarrow \quad V_B = \frac{\gamma_{dc} V_C}{\gamma_{d_B}} = \frac{(17.0 \text{ kN/m}^3)(6759 \text{ m}^3)}{14.0 \text{ kN/m}^3} = 8207 \text{ m}^3 \tag{3.13.4}$$

The volume of soil carried by a dump truck in a single trip is

$$V_{truck} = (10 \text{ yd}^3)\left(\frac{3 \text{ ft}}{\text{yd}}\right)^3 \left(\frac{\text{m}}{3.281 \text{ ft}}\right)^3 = 7.644 \text{ m}^3 \tag{3.13.5}$$

The total number of dump truck trips required is thus

$$(8207 \text{ m}^3)\left(\frac{\text{truck trip}}{7.644 \text{ m}^3}\right) = \mathbf{1074} \text{ truck trips} \tag{3.13.6}$$

c) To compute the volume of water to be added at the construction site to achieve the desired moisture content before compaction, note that, from the definition of the unit weight of water and the moisture content, $\gamma_w = W_w/V_w$ and $w = W_w/W_s$. Thus,

$$V_w = \frac{W_w}{\gamma_w} = \frac{w W_s}{\gamma_w} \tag{3.13.7}$$

Thus, for stages A and C:

$$V_{W_A} = \frac{w_A W_s}{\gamma_w}, \quad V_{w_C} = \frac{w_C W_s}{\gamma_w} \tag{3.13.8}$$

The volume of water to be added at the construction site to achieve the desired moisture content before compaction is thus

$$\Delta V_w = V_{w_C} - V_{W_A} = \frac{W_s}{\gamma_w}(w_C - w_A) = \frac{\gamma_{dc} V_C}{\gamma_w}(w_C - w_A) \tag{3.13.9}$$

Substituting for the known quantities gives

$$\Delta V_w = \frac{(17.0 \text{ kN/m}^3)(6759 \text{ m}^3)}{(9.81 \text{ kN/m}^3)}(0.196 - 0.150) = \mathbf{538.8 \text{ m}^3} \quad (3.13.10)$$

d) The determination of the degree of saturation of soil at the construction site after compaction requires knowledge of the void ratio of the compacted soil at the site. This is computed by recalling the expression for the dry unit weight given in Case 1.4 of Chapter 1, i.e.,

$$\gamma_d = \frac{G_s\gamma_w}{1+e} \quad \Rightarrow \quad e_C = \frac{G_s\gamma_w}{\gamma_{dc}} - 1 = \frac{(2.70)(9.81 \text{ kN/m}^3)}{17.0 \text{ kN/m}^3} - 1 = \mathbf{0.558}$$

$$(3.13.11)$$

Using Eq. (3.5), the desired degree of saturation is thus

$$S_C = \left(\frac{G_s w}{e_C}\right) * 100\% = \left[\frac{(2.70)(0.196)}{0.558}\right] * 100\% = \mathbf{94.8\%} \quad (3.13.12)$$

e) If the compacted soil is saturated after construction due to rainfall, it is assumed that the void ratio remains constant during this process. The moisture content of the compacted soil is thus,

$$w = \left(\frac{e}{G_s}\right) * 100\% = \left(\frac{0.558}{2.70}\right) * 100\% = \mathbf{20.7\%} \quad (3.13.13)$$

which is quite close to the specified moisture content of 19.6%.

EXAMPLE PROBLEM 3.14

General Remarks

This example illustrates the inclusion of financial considerations into a compaction problem.

Problem Statement

A proposed embankment fill requires 9000 m³ of compacted soil. The void ratio of the compacted fill is specified to be 0.72. Five potential borrow pits are under consideration for use in this project. Table Ex. 3.14a lists the respective void ratios of the soils and the cost per cubic meter for moving the soil to the embankment construction site. The specific gravity of solids is assumed to be the *same* for the soils found at all of the borrow pits. Make the necessary calculations so as to select the pit from which the soil should be brought to *minimize* costs.

TABLE EX. 3.14a Data for Potential Borrow Pits

Borrow Pit	Void Ratio	Cost for Transport (per m³)
A	1.15	$8.00
B	0.85	$7.50
C	0.76	$12.50
D	0.92	$9.00
E	1.20	$9.50

Solution

The required weight of solids is obtained from the dry density according to

$$\gamma_d = \frac{G_s\gamma_w}{1+e} = \frac{W_s}{V} \quad \Rightarrow \quad W_s = \frac{G_s\gamma_w V}{1+e} = \frac{9000\,G_s\gamma_w}{1+0.720} \tag{3.14.1}$$

Since the weight of solids must remain unchanged, it follows that for each of the borrow pits,

$$\frac{G_s\gamma_w}{1+e} = \frac{W_s}{V} \quad \Rightarrow \quad V = \frac{W_s(1+e)}{G_s\gamma_w} = \left(\frac{9000\,G_s\gamma_w}{1.720}\right)\frac{(1+e)}{G_s\gamma_w} = \frac{9000(1+e)}{1.720} \tag{3.14.2}$$

where V has units of cubic meters. For example, for borrow pit A,

$$V = \frac{9000(1+1.15)}{1.720} = \mathbf{11,250\ m^3} \tag{3.14.3}$$

Once the total volume of soil from a given borrow pit is known, it is multiplied by the cost of transport to get the total cost. The cost of transporting soil from borrow pit A to the embankment site is thus

$$\text{cost} = \left(11,250\ m^3\right)\left(\$8.00/m^3\right) = \mathbf{\$90,000} \tag{3.14.4}$$

Similar calculations are performed for the other four borrow pits. Table Ex. 3.14b summarizes the results obtained. Using **borrow pit B** thus constitutes the most economical solution.

EXAMPLE PROBLEM 3.15

General Remarks

This example illustrates the effect that uncertainty has on compaction calculations.

TABLE EX. 3.14b Summary of Results Obtained for Potential Borrow Pits

Borrow Pit	Volume Required (m³)	Total Cost
A	11,250	$90,000
B	9,680	$72,602
C	9,209	$115,116
D	10,047	$90,419
E	11,512	$109,360

Problem Statement

Specifications require that 750 yd³ of a particular well-graded granular fill be placed at a relative density (D_r) of 90%. The fill material is to be obtained from a borrow pit. In situ, the material at the borrow pit has a moisture content of 13.7% and a moist unit weight of 115.7 lb/ft³. Estimate the volume of soil (in cubic yards) that must be removed from the borrow pit for the fill.

Solution

The dry unit weight of the fill at the job site is

$$\gamma_{d_{site}} = \frac{G_s \gamma_w}{1 + e_{site}} = \frac{W_s}{V_{site}} \tag{3.15.1}$$

where $V_{site} = 750$ yd³. The required weight of solids is thus

$$W_s = \left(\frac{G_s \gamma_w}{1 + e_{site}}\right) V_{site} \tag{3.15.2}$$

The dry unit weight of the material at the borrow pit is

$$\gamma_{d_{borrow}} = \frac{W_s}{V_{borrow}} \quad \Rightarrow \quad V_{borrow} = \frac{W_s}{\gamma_{d_{borrow}}} \tag{3.15.3}$$

Since the weight of the solid phase remains unchanged from the borrow pit to the site, Eq. (3.15.2) is substituted into Eq. (3.15.3) to give

$$V_{borrow} = \frac{1}{\gamma_{d_{borrow}}} \left(\frac{G_s \gamma_w}{1 + e_{site}}\right) V_{site} \tag{3.15.4}$$

The dry unit weight of the borrow material is computed from the given moist unit weight and moisture content

$$\gamma_{d_{borrow}} = \frac{\gamma_{borrow}}{1+w_{borrow}} = \frac{(115.7 \text{ lb/ft}^3)}{1+0.137} = 101.8 \text{ lb/ft}^3 \qquad (3.15.5)$$

The void ratio at the job site is computed from the definition of the relative density, i.e.,

$$D_r = \frac{e_{max}-e_{site}}{e_{max}-e_{min}} \quad \Rightarrow \quad e_{site} = e_{max} - (D_r)(e_{max}-e_{min}) \qquad (3.15.6)$$

where D_r is specified to be 0.90.

Some uncertainty exists regarding the values of minimum and maximum void ratio for the borrow pit material. In general, for cohesionless soils, these extreme values of void ratio depend on (1) grain size, (2) grain shape, (3) nature of the grain size distribution curve, and (4) the fines content[1].

One grain size distribution curve indicates that the well-graded granular fill is essentially free of gravel size particles and has relatively little fines. As such, e_{min} and e_{max} are estimated to be 0.38 and 0.80, respectively. Substituting these values into Eq. (3.15.6) gives

$$e_{site} = 0.80 - (0.90)(0.80 - 0.38) = 0.422 \qquad (3.15.7)$$

A second grain size distribution curve, however, indicates that the gravel content is approximately 20%. In this case, e_{min} and e_{max} are estimated to be 0.30 and 0.58, respectively. Substituting these values into Eq. (3.15.6) gives

$$e_{site} = 0.58 - (0.90)(0.58 - 0.30) = 0.328 \qquad (3.15.8)$$

Another uncertainty associated with quantities appearing in Eq. (3.15.4) is the value of G_s. Tests performed on the borrow pit material indicate that G_s varies between 2.67 and 2.71.

Eq. (3.15.4) is first evaluated assuming $G_s = 2.67$. For $e_{site} = 0.422$,

$$V_{borrow} = \frac{1}{(101.8 \text{ lb/ft}^3)}\left[\frac{(2.67)(62.4 \text{ lb/ft}^3)}{1+0.422}\right](750 \text{ yd}^3) = \textbf{863 yd}^3 \quad (3.15.9)$$

For $e_{site} = 0.328$,

$$V_{borrow} = \frac{1}{(101.8 \text{ lb/ft}^3)}\left[\frac{(2.67)(62.4 \text{ lb/ft}^3)}{1+0.328}\right](750 \text{ yd}^3) = \textbf{924 yd}^3$$

$$(3.15.10)$$

1. Perloff, W.H., Baron, W., 1976. Soil Mechanics, Principles and Applications. The Ronald Press Company, New York, NY.

Eq. (3.15.4) is first evaluated assuming $G_s = 2.71$. For $e_{site} = 0.422$,

$$V_{borrow} = \frac{1}{(101.8 \text{ lb/ft}^3)} \left[\frac{(2.71)(62.4 \text{ lb/ft}^3)}{1 + 0.422} \right] (750 \text{ yd}^3) = \textbf{876 yd}^3$$

$$(3.15.11)$$

For $e_{site} = 0.328$,

$$V_{borrow} = \frac{1}{(101.8 \text{ lb/ft}^3)} \left[\frac{(2.71)(62.4 \text{ lb/ft}^3)}{1 + 0.328} \right] (750 \text{ yd}^3) = \textbf{938 yd}^3$$

$$(3.15.12)$$

From the aforesaid results, it is evident that changes in the specific gravity of solids have a relatively minor effect on the computed volumes of borrow pit material. Changes in the extreme void ratio values, however, have a much more pronounced affect on the results. As such, if additional laboratory tests on the borrow pit material are to be performed, they should focus on determining additional values of e_{min} and e_{max}.

Chapter 4

Stresses, Strains, and Elastic Response of Soils

4.0 INTRODUCTORY COMMENTS

Mechanics is defined as the study of the motion of matter and the forces that cause such motion. Mechanics is based on the concepts of time, space, force, energy, and matter.

One of the most important facts learned in an undergraduate mechanics education is that a valid solution to any problem in solid mechanics must satisfy three sets of equations, namely (1) the equations of equilibrium of body, surface, and inertia forces and stresses; (2) the equations of compatibility of strains and displacements; and (3) the constitutive equations for the material.

This chapter reviews key aspects associated with the aforesaid three sets of equations. The presentation is not meant to be exhaustive.

4.1 GENERAL DEFINITIONS

The following general definitions, included here for completeness, facilitate understanding of this chapter.

4.1.1 The Continuum Concept

In formulating the aforementioned sets of equations, the material is assumed to be a *continuum*. In general, a continuum is defined as an entity that has continuity; i.e., an unbroken connection or sequence.

The adoption of a continuum concept for geomaterials (i.e., soils and rock) is complicated by the particulate nature of this class of materials. As noted in Chapter 1, a porous material is treated as the superimposition of two continua: the solid matrix continuum (i.e., the solid skeleton) and the pore continuum (i.e., the voids).[1]

As noted in Section 1.1, if a two-phase continuum consisting of a coherent solid matrix (skeleton) with fluid-filled pore space is assumed, the geomaterial

---

1. Coussy, O., 2004. Poromechanics. John Wiley & Sons, Ltd., Chichester, UK.

Soil Mechanics. http://dx.doi.org/10.1016/B978-0-12-804491-9.00004-5
131

is *saturated*. If, on the other hand, the pore space is not fully occupied by fluid, the geomaterial is *unsaturated*.

4.1.2 Homogeneity

A material is said to be *homogeneous* if the matter of the body is the same throughout the body and is continuously distributed over the volume occupied by the body. This may not be the case with materials exhibiting localized behavior (e.g., cracks, shear banding, etc.). Although natural geomaterials are generally heterogeneous, some assumptions concerning homogeneity are typically required. For example, a soil profile will typically be assumed to consist of one or more discrete layers; each such layer will then be assumed to be homogeneous.

4.1.3 Isotropy

An *isotropic* material has identical mechanical properties in *all* directions. If due to some technological process (such as the rolling of metals), natural growth (such as trees), or natural deposition (such as geomaterials) the properties of the material are *different* in different directions, the material is said to be *anisotropic*.

4.2 CONCEPT OF STRESS

The subject of internal forces and moments acting in a body is commonly introduced in a statics course. Obtaining the *distribution* of internal forces and their *intensity* is of primary importance in mechanics of solids. To properly study this issue, it is necessary to establish the concept of *state stress at a point*. In discussing the concept of stress at a point in a body, a material *continuum* is assumed.

4.2.1 Definition of Stress at a Point

Consider a general body subjected to several applied (external) forces. Figure 4.1 shows an *internal* force increment ΔF acting on the area $\Delta A = (\Delta y)(\Delta z)$ at a typical interior point O in the body. The normal to this plane is assumed to point in the positive x-axis direction.

Resolve ΔF into *components* along the coordinate axes in the manner shown in Figure 4.2. According to the approach proposed by Cauchy[2], define the *normal* stress in the x-direction in the following manner:

$$\sigma_x = \lim_{\Delta A \to 0} \frac{\Delta F_x}{\Delta A} \tag{4.1}$$

---

2. Named in honor of Augustin Cauchy (1789–1857).

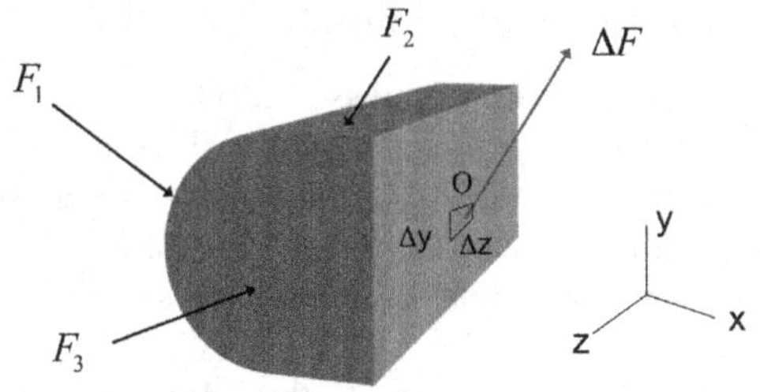

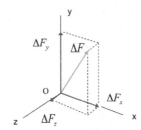

FIGURE 4.1 Internal force increment acting in a body subjected to applied forces.

FIGURE 4.2 Internal force increment resolved into components along coordinate axes.

The designation "normal stress" is used since it denotes the intensity of the force acting *normal* (i.e., at right angles) to ΔA. As shown, σ_x is a tensile normal stress. In soil mechanics, *compressive* normal stresses are typically taken as being *positive*.

Two additional stresses are defined on ΔA; i.e.,

$$\tau_{xy} = \lim_{\Delta A \to 0} \frac{\Delta F_y}{\Delta A}; \quad \tau_{xz} = \lim_{\Delta A \to 0} \frac{\Delta F_z}{\Delta A} \qquad (4.2)$$

Since these quantities involve forces in the plane being considered, they are called *shear stresses.*

Two subscripts must be used to describe shear stresses. The first subscript refers to the direction of the outward normal vector (in this case x) associated with ΔA. The second subscript refers to the direction (in this case y or z) in which the force component is acting.

Before defining positive and negative stresses in a general manner, it is necessary to define the *positive* and *negative* faces of a infinitesimal cube of material at a given point (e.g., point O) in a body. A face will be defined as *positive* when its outwardly directed normal vector points in the direction of the *positive* coordinate axis. A face will be defined as *negative* when its outwardly directed normal vector points in the direction of the *negative* coordinate axis.

By definition, positive shear stresses act in the *positive* coordinate direction on a *positive* face or in the *negative* coordinate direction on a *negative* face.

4.2.2 Definition of the State of Stress at a Point

To completely describe the state of stress at a point requires the force components on *three mutually perpendicular planes*. Figure 4.3 shows the stress components associated with a three-dimensional state of stress at a point (the components acting on the x-y plane are omitted for clarity). In accordance with the aforementioned sign convention for normal and shear stresses, all of these components are positive.

The following matrix of components thus defines the general state of Cauchy stress at a point in a body:

$$\boldsymbol{\sigma} = \begin{bmatrix} \sigma_x & \tau_{xy} & \tau_{xz} \\ \tau_{yx} & \sigma_y & \tau_{yz} \\ \tau_{zx} & \tau_{zy} & \sigma_z \end{bmatrix} \tag{4.3}$$

From moment equilibrium, it is found that

$$\tau_{xy} = \tau_{yx}; \quad \tau_{xz} = \tau_{zx}; \quad \tau_{yz} = \tau_{zy} \tag{4.4}$$

Thus, only *six* stress components are independent, rendering the stress matrix given in Eq. (4.3) symmetric.

4.2.3 Mean Stress

For a three-dimensional state of stress, the mean stress is defined as follows:

$$\sigma_m = \frac{1}{3}(\sigma_x + \sigma_y + \sigma_z) \tag{4.5}$$

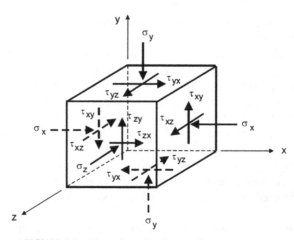

FIGURE 4.3 Three-dimensional state of stress at a point.

4.2.4 State of Plane Stress

If a relatively thin body is loaded by forces applied at the boundary, parallel to the plane of the plate and distributed uniformly over the thickness (Figure 4.4), the stress component σ_z will be negligible (often zero) through the thickness of the body. The shear stresses $\tau_{xz} = \tau_{zx}$ and $\tau_{yz} = \tau_{zy}$ will likewise be negligible. Under these assumptions, it follows from the balance of linear momentum that under static conditions the body force in the z-direction must be zero.

The state of stress is thus completely determined by the values of σ_x, σ_y, and $\tau_{xy} = \tau_{yx}$ and is referred to as *plane stress*. Eq. (4.6) gives the associated stress matrix.

$$\boldsymbol{\sigma} = \begin{bmatrix} \sigma_x & \tau_{xy} \\ \tau_{xy} & \sigma_y \end{bmatrix} \tag{4.6}$$

Figure 4.5 shows the stress components associated with a state of plane stress. In accordance with the aforementioned sign convention for normal and shear stresses, all of these components are positive.

Remark: Although $\sigma_z = 0$ under conditions of plane stress, ε_z will not be zero. Section 4.4 presents additional details regarding this issue.

4.2.5 Stress Transformations

In general, both normal and shear stresses simultaneously act on a small element of a body (Figure 4.6A). In many instances, it is necessary to determine the state of stress acting on an inclined plane through the element (Figure 4.6B). An infinite number of such planes can be chosen. In all cases the stresses acting on the inclined plane are *equivalent* to the original stress

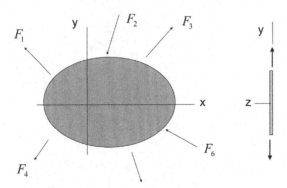

FIGURE 4.4 Schematic illustration of plane stress.

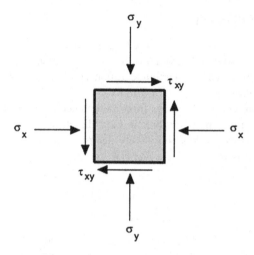

FIGURE 4.5 Stress components associated with a state of plane stress.

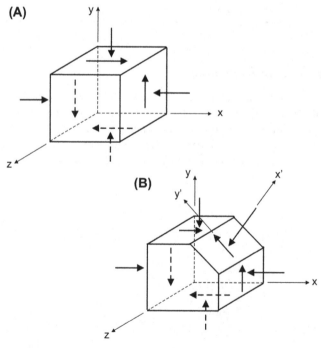

FIGURE 4.6 Schematic illustration of the (A) original stress state, (B) equivalent transformed stress state.

state. This is due to the fact that both the original and transformed stress states must maintain the equilibrium of the element.

The equations for transforming a given stress state into an equivalent one acting on any plane through an element in a body are now derived. Rather than developing such equations for a general three-dimensional stress state, the simpler case of plane stress will be considered. Figure 4.5 shows the three nonzero stress components associated with a state of plane stress.

4.2.5.1 Equations for Plane Stress Transformation

Figure 4.7 shows a state of general plane stress and the original $(x - y)$ and rotated $(x' - y')$ axes. The angle θ quantifies the rotation with counterclockwise angles taken as being positive.

Compressive normal stresses are assumed to be positive; tensile normal stresses are thus negative. A positive shear stress is defined as acting in a positive coordinate direction on a face whose outward normal is directed in a positive coordinate direction, or in a negative coordinate direction on a face whose outward normal is directed in a negative coordinate direction. For example, the shear stress on the face PS acts in the positive y-direction; the outward normal to the face acts parallel to the positive x-direction. The shear stress acting on face PS is thus positive. Following the same logic, all of the shear stresses shown in Figures 4.7 and 4.8 are seen to be positive.

The equations for plane stress transformation are developed by analyzing a small element of a body. Figure 4.8 shows such an element.

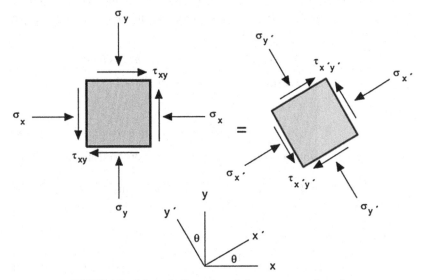

FIGURE 4.7 Schematic illustration of plane stress transformation.

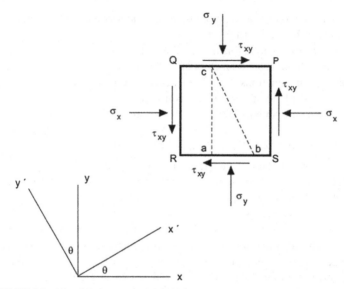

FIGURE 4.8 Material element used in deriving plane stress transformation equations.

Figure 4.9A shows the wedge that is created by passing a plane bc normal to the x'-axis through the element shown in Figure 4.8. The plane bc makes an angle θ with the vertical (y) axis. If the area of this inclined plane is dA, then the areas of the faces ac and ab will be $dA\cos\theta$ and $dA\sin\theta$, respectively.

Figure 4.9B shows the forces that are obtained by multiplying the stresses from Figure 4.9A by their corresponding areas. Writing the equations of force equilibrium in the x' direction gives

$$\sum F_{x'} = -\sigma_{x'}dA + (\sigma_x \, dA \cos\theta)\cos\theta - (\tau_{xy} \, dA \cos\theta)\sin\theta$$
$$- (\tau_{xy} \, dA \sin\theta)\cos\theta + (\sigma_y \, dA \sin\theta)\sin\theta = 0 \tag{4.7}$$

or

$$\sigma_{x'} = \sigma_x \cos^2\theta + \sigma_y \sin^2\theta - 2\tau_{xy}\sin\theta\cos\theta \tag{4.8}$$

Substituting the following double angle trigonometric identities

$$\cos^2\theta = \frac{1}{2}(1 + \cos 2\theta); \quad \sin^2\theta = \frac{1}{2}(1 - \cos 2\theta); \quad \sin\theta\cos\theta = \frac{1}{2}\sin 2\theta \tag{4.9}$$

into Eq. (4.8) gives

$$\sigma_{x'} = \frac{1}{2}(\sigma_x + \sigma_y) + \frac{1}{2}(\sigma_x - \sigma_y)\cos 2\theta - \tau_{xy}\sin 2\theta \tag{4.10}$$

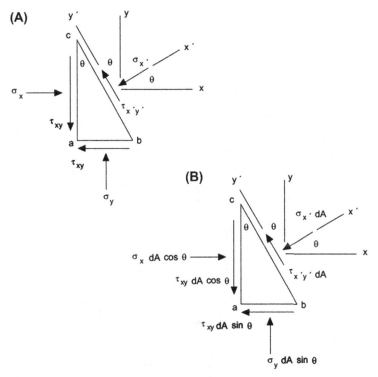

FIGURE 4.9   Portion of a material element used in deriving plane stress transformation: (A) stresses and (B) forces acting on inclined face.

Next, writing the equations of force equilibrium in the y' direction gives

$$\sum F_{y'} = \tau_{x'y'}\, dA - (\sigma_x\, dA \cos \theta)\sin \theta - (\tau_{xy}\, dA \cos \theta)\cos \theta$$
$$+ (\tau_{xy}\, dA \sin \theta)\sin \theta + (\sigma_y\, dA \sin \theta)\cos \theta = 0$$
(4.11)

or

$$\tau_{x'y'} = (\sigma_x - \sigma_y)\sin \theta \cos \theta - \tau_{xy}\left(\sin^2 \theta - \cos^2 \theta\right) \qquad (4.12)$$

Substituting the trigonometric identities given in Eq. (4.9) into Eq. (4.12) gives

$$\tau_{x'y'} = \frac{1}{2}(\sigma_x - \sigma_y)\sin 2\theta + \tau_{xy}\cos 2\theta \qquad (4.13)$$

To obtain an equation for the transformed normal stress $\sigma_{y'}$, consider the material wedge shown in Figure 4.10.

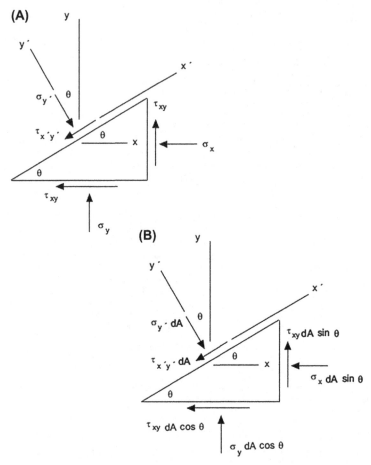

FIGURE 4.10 Portion of a material element used in deriving plane stress transformation: (A) stresses and (B) forces acting on inclined face.

Writing the equations of force equilibrium in the y' direction gives

$$\sum F_{y'} = -\sigma_{y'} \, dA + (\sigma_x \, dA \sin \theta)\sin \theta + (\tau_{xy} \, dA \sin \theta)\cos \theta$$
$$+ (\tau_{xy} \, dA \cos \theta)\sin \theta + (\sigma_y \, dA \cos \theta)\cos \theta = 0 \tag{4.14}$$

or

$$\sigma_{y'} = \sigma_x \sin^2 \theta + \sigma_y \cos^2 \theta + 2\tau_{xy} \sin \theta \cos \theta \tag{4.15}$$

Substituting the trigonometric identities given in Eq. (4.9) into Eq. (4.15) gives

$$\sigma_{y'} = \frac{1}{2}(\sigma_x + \sigma_y) - \frac{1}{2}(\sigma_x - \sigma_y)\cos 2\theta + \tau_{xy} \sin 2\theta \tag{4.16}$$

The angle θ, appearing in Eqs. (4.8)–(4.16), is positive when measured *counterclockwise*.

Remark: Adding Eqs. (4.10) and (4.16) gives the relation $\sigma_{x'} + \sigma_{y'} = \sigma_x \sigma_y$, which holds for all transformation angles.

4.2.5.2 Determination of Principal Stress

In many cases, the maximum and minimum normal stresses associated with a given stress state, and the planes on which they act, are of interest. To determine the orientation of the plane associated with a minimum or a maximum normal stress, Eq. (4.10) is differentiated with respect to θ and the resulting expression is set equal to zero, giving

$$\frac{d\sigma_{x'}}{d\theta} = -\frac{1}{2}(\sigma_x - \sigma_y)(2 \sin 2\theta) - \tau_{xy}(2 \cos 2\theta) = 0$$

Thus,

$$\tan 2\theta_p = \frac{-\tau_{xy}}{\frac{1}{2}(\sigma_x - \sigma_y)} \tag{4.17}$$

where the subscript on θ indicates that this angle defines the maximum or minimum normal stress. Eq. (4.17) has two roots, since the value of the tangent of an angle in diametrically opposite quadrants is the same. These roots are thus 180 degrees apart.

The angles θ_{p1} and θ_{p2} locate the planes on which the maximum and minimum normal stresses act, respectively. Since $2\theta_{p1}$ and $2\theta_{p2}$ are 180 degrees apart, θ_{p1} and θ_{p2} will be 90 degrees apart.

These planes are called the *principal planes* of stress. The stresses acting on these planes—the maximum and minimum normal stresses—are called the *principal stresses*.

Remark: For three-dimensional stress states, the three principal stresses (σ_1, σ_2, and σ_3) are commonly ordered such that $\sigma_1 \geq \sigma_2 \geq \sigma_3$, where σ_1 is the *major* principal stress, σ_2 is the *intermediate* principal stress, and σ_3 is the *minor* principal stress. There are three principal planes that are perpendicular to each other.

Remark: The sum of the normal stresses is invariant, i.e., it is independent of the coordinate system. Consequently, the mean stress defined in Eq. (4.5) is expanded to

$$\sigma_m = \frac{1}{3}(\sigma_x + \sigma_y + \sigma_z) = \frac{1}{3}(\sigma_1 + \sigma_2 + \sigma_3)$$

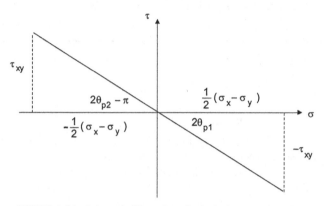

FIGURE 4.11 Schematic illustration of principal stress orientation.

Figure 4.11 schematically illustrates the orientation of the principal stress directions in σ-τ space. From the triangles shown in this figure,

$$\sin 2\theta_{p_1} = \frac{-\tau_{xy}}{\sqrt{\left(\dfrac{\sigma_x - \sigma_y}{2}\right)^2 + \left(\tau_{xy}\right)^2}} \tag{4.18}$$

$$\cos 2\theta_{p_1} = \frac{\dfrac{1}{2}(\sigma_x - \sigma_y)}{\sqrt{\left(\dfrac{\sigma_x - \sigma_y}{2}\right)^2 + \left(\tau_{xy}\right)^2}} \tag{4.19}$$

and

$$\sin 2\theta_{p_2} = \frac{\tau_{xy}}{\sqrt{\left(\dfrac{\sigma_x - \sigma_y}{2}\right)^2 + \left(\tau_{xy}\right)^2}} = -\sin 2\theta_{p_1} \tag{4.20}$$

$$\cos 2\theta_{p_2} = \frac{-\dfrac{1}{2}(\sigma_x - \sigma_y)}{\sqrt{\left(\dfrac{\sigma_x - \sigma_y}{2}\right)^2 + \left(\tau_{xy}\right)^2}} = -\cos 2\theta_{p_1} \tag{4.21}$$

Substituting Eqs. (4.18) and (4.19) into Eq. (4.10) gives the magnitude of the maximum normal stress, which corresponds to the *major* principal stress σ_1, i.e.,

$$\sigma_{x'} = \frac{1}{2}(\sigma_x + \sigma_y) + \frac{1}{2}(\sigma_x - \sigma_y)\left[\frac{\frac{1}{2}(\sigma_x - \sigma_y)}{\sqrt{\left(\frac{\sigma_x - \sigma_y}{2}\right)^2 + (\tau_{xy})^2}}\right] - \tau_{xy}\left[\frac{-\tau_{xy}}{\sqrt{\left(\frac{\sigma_x - \sigma_y}{2}\right)^2 + (\tau_{xy})^2}}\right]$$

$$= \frac{1}{2}(\sigma_x + \sigma_y) + \sqrt{\left(\frac{\sigma_x - \sigma_y}{2}\right)^2 + (\tau_{xy})^2} \equiv \sigma_1$$

(4.22)

Next, substituting Eqs. (4.18) and (4.19) into Eq. (4.16) gives the magnitude of the minimum normal stress, which corresponds to the *minor* principal stress σ_2, i.e.,

$$\sigma_{y'} = \frac{1}{2}(\sigma_x + \sigma_y) - \frac{1}{2}(\sigma_x - \sigma_y)\left[\frac{\frac{1}{2}(\sigma_x - \sigma_y)}{\sqrt{\left(\frac{\sigma_x - \sigma_y}{2}\right)^2 + (\tau_{xy})^2}}\right] + \tau_{xy}\left[\frac{-\tau_{xy}}{\sqrt{\left(\frac{\sigma_x - \sigma_y}{2}\right)^2 + (\tau_{xy})^2}}\right]$$

$$= \frac{1}{2}(\sigma_x + \sigma_y) - \sqrt{\left(\frac{\sigma_x - \sigma_y}{2}\right)^2 + (\tau_{xy})^2} \equiv \sigma_2$$

(4.23)

Finally, substituting Eqs. (4.18) and (4.19) into Eq. (4.13) gives

$$\tau_{x'y'} = \frac{1}{2}(\sigma_x - \sigma_y)\left[\frac{-\tau_{xy}}{\sqrt{\left(\frac{\sigma_x - \sigma_y}{2}\right)^2 + (\tau_{xy})^2}}\right] + \tau_{xy}\left[\frac{\frac{1}{2}(\sigma_x - \sigma_y)}{\sqrt{\left(\frac{\sigma_x - \sigma_y}{2}\right)^2 + (\tau_{xy})^2}}\right] = 0$$

(4.24)

The shear stress is thus *zero* on the principal planes of stress.[3] Figure 4.12 shows the principal stress state for conditions of plane stress.

4.2.5.3 Determination of Maximum In-Plane Shear Stress

The planes on which the shear stress is maximum are next sought. Differentiating Eq. (4.13) with respect to θ and setting the resulting expression equal to zero gives

$$\frac{d\tau_{x'y'}}{d\theta} = \frac{1}{2}(\sigma_x - \sigma_y)(2\cos 2\theta) + \tau_{xy}(-2\sin 2\theta) = 0$$

---

3. This result can also be realized by noting that Eq. (4.17) is obtained by setting Eq. (4.13) equal to zero.

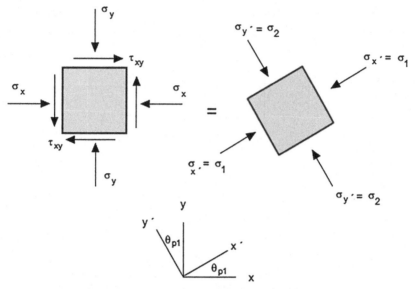

FIGURE 4.12 Schematic illustration of principal stress state.

Thus,

$$\tan 2\theta_s = \frac{\frac{1}{2}(\sigma_x - \sigma_y)}{\tau_{xy}} \qquad (4.25)$$

Eq. (4.25) has two roots, θ_{s_1} and θ_{s_2}. Similar to the roots of Eq. (4.17), $2\theta_{s_1}$ and $2\theta_{s_2}$ are 180 degrees apart (Figure 4.13). The angles θ_{s_1} and θ_{s_2} will thus be 90 degrees apart.

From the triangles shown in Figure 4.13,

$$\sin 2\theta_{s_1} = \frac{\frac{1}{2}(\sigma_x - \sigma_y)}{\sqrt{\left(\frac{\sigma_x - \sigma_y}{2}\right)^2 + (\tau_{xy})^2}} \qquad (4.26)$$

$$\cos 2\theta_{s_1} = \frac{\tau_{xy}}{\sqrt{\left(\frac{\sigma_x - \sigma_y}{2}\right)^2 + (\tau_{xy})^2}} \qquad (4.27)$$

and

$$\sin 2\theta_{s_2} = \frac{-\frac{1}{2}(\sigma_x - \sigma_y)}{\sqrt{\left(\frac{\sigma_x - \sigma_y}{2}\right)^2 + (\tau_{xy})^2}} = -\sin 2\theta_{s_1} \qquad (4.28)$$

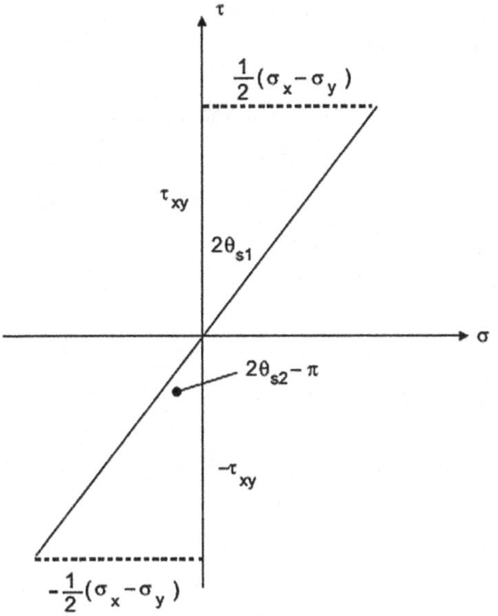

FIGURE 4.13 Schematic illustration of maximum shear stress orientation.

$$\cos 2\theta_{s_2} = \frac{-\tau_{xy}}{\sqrt{\left(\dfrac{\sigma_x - \sigma_y}{2}\right)^2 + (\tau_{xy})^2}} = -\cos 2\theta_{s_1} \qquad (4.29)$$

Substituting Eqs. (4.26) and (4.27) into Eq. (4.13) gives the magnitude of the maximum shear stress, i.e.,

$$\tau_{x'y'} \equiv \tau_{max} = \frac{1}{2}(\sigma_x - \sigma_y) \left[\frac{\frac{1}{2}(\sigma_x - \sigma_y)}{\sqrt{\left(\dfrac{\sigma_x - \sigma_y}{2}\right)^2 + (\tau_{xy})^2}} \right]$$

$$+ \tau_{xy} \left[\frac{\tau_{xy}}{\sqrt{\left(\dfrac{\sigma_x - \sigma_y}{2}\right)^2 + (\tau_{xy})^2}} \right] = \sqrt{\left(\dfrac{\sigma_x - \sigma_y}{2}\right)^2 + (\tau_{xy})^2}$$

$$(4.30)$$

The quantity τ_{max} is the *maximum in-plane shear stress* at a point. The associated normal stresses are determined by substituting Eqs. (4.26) and (4.27) into Eqs. (4.10) and (4.16), giving

$$\sigma_{x'} = \sigma_{y'} = \frac{1}{2}(\sigma_x + \sigma_y) \qquad (4.31)$$

The normal stresses associated with the maximum in-plane shear stress are thus equal to the *average* of σ_x and σ_y.

4.2.5.4 Mohr's Circle of Stress for Plane Stress

Let C denote the aforementioned average stress, i.e.,

$$C = \frac{1}{2}(\sigma_x + \sigma_y) \tag{4.32}$$

The plane stress transformation Eq. (4.10) is then rewritten as

$$\sigma_{x'} - C = \frac{1}{2}(\sigma_x - \sigma_y)\cos 2\theta - \tau_{xy}\sin 2\theta \tag{4.33}$$

Squaring Eq. (4.33) as well as the plane stress transformation Eq. (4.13) gives

$$(\sigma_{x'} - C)^2 = \frac{1}{4}(\sigma_x - \sigma_y)^2 \cos^2 2\theta - (\sigma_x - \sigma_y)\tau_{xy}\sin 2\theta \cos 2\theta + (\tau_{xy})^2 \sin^2 2\theta \tag{4.34}$$

$$(\tau_{x'y'})^2 = \frac{1}{4}(\sigma_x - \sigma_y)^2 \sin^2 2\theta + (\sigma_x - \sigma_y)\tau_{xy}\sin 2\theta \cos 2\theta + (\tau_{xy})^2 \cos^2 2\theta \tag{4.35}$$

Adding Eqs. (4.34) and (4.35) gives

$$(\sigma_{x'} - C)^2 + (\tau_{x'y'})^2 = \left(\frac{\sigma_x - \sigma_y}{2}\right)^2 + (\tau_{xy})^2 \tag{4.36}$$

The equation of a circle centered at $x = a$ and $y = b$ and having a radius equal to R is

$$(x - a)^2 + (y - b)^2 = R^2 \tag{4.37}$$

If $\sigma_{x'}$ is taken as the abscissa and $\tau_{x'y'}$ as the ordinate, then Eq. (4.36) is seen to be the equation of a circle centered at $(C,0)$ and having a radius of

$$R = \sqrt{\left[\frac{1}{2}(\sigma_x - \sigma_y)\right]^2 + (\tau_{xy})^2} \tag{4.38}$$

Such a circle is called Mohr's circle of stress, in honor of the German engineer[4] who first proposed its use to graphically represent the state of stress at a point. Every point on the circumference of such a circle corresponds to a specific direction in the plane of a stress element. The circle is constructed in accordance with the following sign convention:

1. Compressive normal stresses (Figure 4.14) are considered positive and are plotted to the right of the τ-axis.
2. Tensile normal stresses are *negative* and are plotted to the *left* of the τ-axis.

---

4. Otto Mohr (1835–1918).

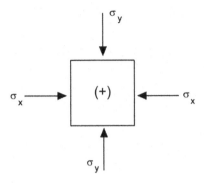

FIGURE 4.14 Positive normal stress components associated with two-dimensional stress states.

3. Shear stresses that produce a *positive* (counterclockwise) couple (Figure 4.15) are plotted *above* the σ-axis.
4. Shear stresses that produce a *negative* (clockwise) couple (Figure 9.3) are plotted *below* the σ-axis.
5. Counterclockwise rotations θ (from one direction in the plane of the stress element to another or from one point on the circumference of Mohr's circle to another) are considered positive.

Given σ_x, σ_y, and τ_{xy}, Figure 4.16 shows a typical Mohr's circle.

The center of the circle always lies along the σ-axis; it is located at the point $(C,0)$, where C is given by Eq. (4.32).

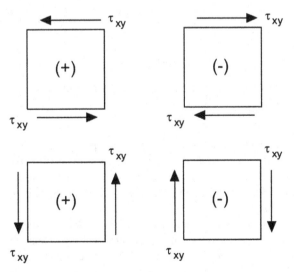

FIGURE 4.15 Sign convention for shear stress components associated with two-dimensional stress states.

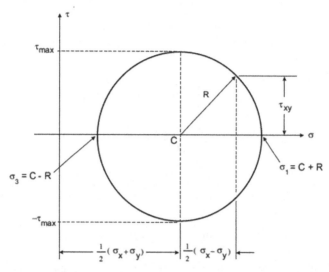

FIGURE 4.16 Typical Mohr's circle for plane stress.

The radius of the circle (R) is given by Eq. (4.38). When measured along the σ-axis, $\tau_{xy} = 0$ and $R = (\sigma_x - \sigma_y)/2$.

In light of the definition of the center (Eq. 4.32) and radius (Eq. 4.38) of a Mohr's circle, the major and minor principal stresses given by Eqs. (4.22) and (4.23), respectively, can be computed as follows (Figure 4.16):

$$\sigma_1 = C + R; \quad \sigma_2 = C - R \tag{4.39}$$

Finally, the magnitude of the maximum in-plane shear stress is equal to the radius of the Mohr's circle, i.e.,

$$|\tau_{max}| = R \tag{4.40}$$

The associated normal stresses are equal to C (Figure 4.16).

4.2.5.5 The Pole Method

The equivalent plane stress state acting on any plane can also be determined using the so-called *pole method*. The *pole* is an especially useful point on the Mohr's circle of stress. Any straight line drawn through this point will intersect the circle at a point representing the stress on a plane inclined at the *same* orientation as the line.

When the stresses acting on any one plane are known, the location of the pole is determined by drawing a line parallel to the plane and passing through the point on the circle corresponding to the stresses on the plane. The intersection of this line with the circle is the pole.

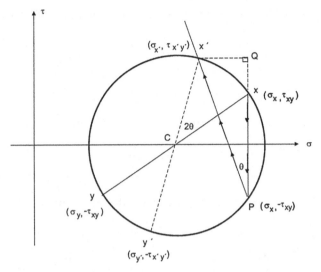

FIGURE 4.17 Determination of pole location using a vertical plane.

Figure 4.17 shows a typical Mohr's circle associated with a state of plane stress. The points on the circle labeled x and y correspond to the (σ, τ) stress states associated with the x and y faces of the material element, respectively. The points on the circle labeled x' and y' correspond to the transformed (rotated) stress state characterized by the counterclockwise angle θ, as determined using Eqs. (4.10), (4.13), and (4.16).

The locations of the points x' and y' can also be determined using the pole method. Since the plane associated with the face having an outward normal parallel to the x-axis is vertical, the pole is located by passing a vertical line through point x as shown in Figure 4.17. The intersection of this line with the Mohr's circle is the pole P. A line oriented θ degrees in a counterclockwise direction is next drawn through the pole. The intersection of this line with the circle is the point x'.

To verify that the location of point x' obtained using Eqs. (4.10) and (4.13) will be identical to that obtained using the pole method, extend the line passing through points P and x until it intersects the horizontal line passing through points x' and Q. Denote the angle $x'PQ$ by γ; it remains to show that $\gamma = \theta$. From the right triangle PQx',

$$\tan \gamma = \frac{\sigma_x - \sigma'_x}{\tau_{x'y'} - \tau_{xy}} \tag{4.41}$$

Substituting Eqs. (4.10) and (4.13) for $\sigma_{x'}$ and $\tau_{x'y'}$, respectively, into Eq. (4.41) gives

$$\tan \gamma = \frac{\sigma_x - \sigma'_x}{\tau_{x'y'} - \tau_{xy}} = \frac{\sigma_x - \left[\frac{1}{2}(\sigma_x - \sigma_y) + \frac{1}{2}(\sigma_x - \sigma_y)\cos 2\theta - \tau_{xy} \sin 2\theta\right]}{\left[\frac{1}{2}(\sigma_x - \sigma_y)\sin 2\theta + \tau_{xy} \cos 2\theta\right] - \tau_{xy}}$$

$$= \frac{\frac{1}{2}(\sigma_x + \sigma_y)(1 - \cos 2\theta) + \tau_{xy} \sin 2\theta}{\frac{1}{2}(\sigma_x - \sigma_y)\sin 2\theta + \tau_{xy}(\cos 2\theta + 1)}$$

(4.42)

Using the double-angle trigonometric identities given by Eq. (4.9), Eq. (4.42) becomes

$$\tan \gamma = \frac{(\sigma_x - \sigma_y)\sin^2 \theta + 2\tau_{xy} \sin \theta \cos \theta}{(\sigma_x - \sigma_y)\sin \theta \cos \theta + 2\tau_{xy} \cos^2 \theta} = \frac{\left[(\sigma_x - \sigma_y)\sin \theta + 2\tau_{xy} \cos \theta\right]\sin \theta}{\left[(\sigma_x - \sigma_y)\sin \theta + 2\tau_{xy} \cos \theta\right]\cos \theta}$$

$$= \tan \theta \Rightarrow \gamma = \theta$$

(4.43)

which thus verifies the validity of using the pole method to determine the location of the point x'.

The location of the pole can likewise be determined from the point labeled y on the Mohr's circle. Since the plane associated with the face having an outward normal parallel to the y-axis is horizontal, the pole is located by passing a horizontal line through point y as shown in Figure 4.18. The intersection of this line with the Mohr's circle is the pole P. A line oriented θ degrees in a counterclockwise direction is next drawn through the pole. The intersection of this line with the circle is the point y'.

4.3 DEFORMATION AND STRAIN

Deformations and strains are kinematic quantities. The subject of kinematics deals with purely geometrical quantities.

Perhaps the most fundamental kinematic quantity is *displacement*, which is defined as the movement of a point from its original or previous location to the current one.

By contrast, deformation refers to a change in shape, size, or both of the continuum between some initial (typically undeformed) configuration and a subsequent (deformed) configuration. More precisely, deformation is characterized by the change in the relative position of pairs of points in the body. When materials are loaded (stressed), they deform. Two types of deformation are possible, namely (1) dilatational or contractual changes in the geometry that effect the *size* (e.g., area or volume) of the body and (2) distortional changes in the geometry associated with changes in the *shape* of the body. Figure 4.19 shows both types of deformation.

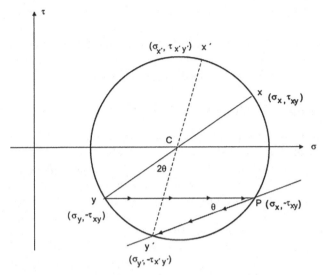

FIGURE 4.18 Determination of pole location using a vertical plane.

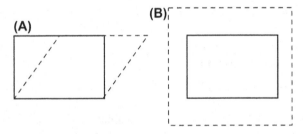

FIGURE 4.19 Deformation in a body: (A) change in shape, (B) change in size.

A continuous body is said to be *strained* when the relative position of points in the body is altered. Consequently, the study of deformations is directly related to the analysis of *strain* in the body. Strain, a normalized deformation, is the quantity used to measure the intensity of deformation.

Strains may be the result of mechanical loading (e.g., an applied force causing stress), of a change in temperature, or other physical phenomena (e.g., shrinkage, settlement). Strains are dimensionless (e.g., in/in, mm/mm, etc.).

Prior to deriving actual strain measures, it is timely to recall the definition of a *rigid* or *nondeformable* body.[5] A rigid body is an ideal body in which the distance between every pair of its points remains *unchanged* throughout the loading history applied to the body. The possible motion in a rigid body consists of translations and rotations.

---

5. The equilibrium of rigid bodies is studied in Statics, which is typically the first course in mechanics taken by undergraduates.

A fundamental requirement of proper strain measures is that they vanish for any rigid body motion, that is, for a rigid body translation or rotation. If a strain measure fails to satisfy this requirement, it will predict nonzero strains, and thus nonzero stresses, for a rigid body motion. This will, however, give results that violate the physics of deformable bodies.

4.3.1 Normal and Shear Strains

As noted in Section 4.3, two types of deformation are possible; i.e., dilatational or contractual changes in the geometry that effect the *size* of a body and distortional changes in the geometry that change a body's *shape*.

Two types of strains are thus defined. *Normal strains* quantify the change in size (elongation or contraction) of an arbitrary line segment during deformation. Figure 4.20 shows a horizontal bar of length L fixed at its left end and subjected to a uniaxial stress state. The relative elongation of the bar is denoted by δ. Axial strain will thus be $\varepsilon = \delta/L$.

Shear strains quantify the change in shape (i.e., change in angle that occurs between two lines that were initially perpendicular). Figure 4.21 shows a rectangular body subjected to a state of pure shear. The associated shear strain is equal to the change in initial right angle (in radians). Section 4.3.2 presents more general definitions of normal and shear strains.

4.3.2 Infinitesimal Strains

If both the displacements and the displacement gradients (rotations) are small compared to unity, there is very little difference between the initial and deformed coordinates of a particle in the continuum. The strains are then said to be *infinitesimal*. Denoting by u, v, and w the displacement components in the x, y, and z directions of the coordinate axes, the infinitesimal normal strains are defined as follows:

$$\varepsilon_x = \frac{\partial u}{\partial x}, \quad \varepsilon_y = \frac{\partial v}{\partial y}, \quad \varepsilon_z = \frac{\partial w}{\partial z} \tag{4.44}$$

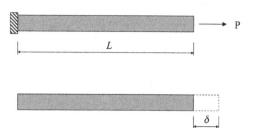

FIGURE 4.20   Bar subjected to uniaxial stress state and resulting relative elongation.

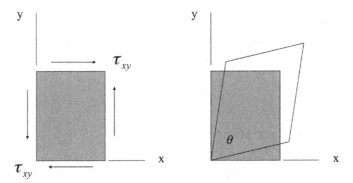

FIGURE 4.21   Rectangular body subjected to state of pure shear.

The infinitesimal *engineering* shear strains associated with the x-y, x-z, and y-z coordinates are

$$\gamma_{xy} = \gamma_{yx} = \frac{\partial v}{\partial x} + \frac{\partial u}{\partial y}, \quad \gamma_{xz} = \gamma_{zx} = \frac{\partial w}{\partial x} + \frac{\partial u}{\partial z}, \quad \gamma_{yz} = \gamma_{zy} = \frac{\partial w}{\partial y} + \frac{\partial v}{\partial z}$$

(4.45)

In examining Eqs. (4.44) and (4.45), it is evident that the six strain–displacement equations depend on only three displacements (i.e., u, v, and w). It follows that the equations cannot be independent. Additional independent equations, which interrelate the strain components, are thus required. The derivation and application of such equations, known as the *equations of compatibility*, are given in standard textbooks on the theory of elasticity.[6]

Normal strains are positive when a line segment *contracts*. They are negative when a line segment *elongates*. Shear strains are positive if the angle between two perpendicular reference lines *decreases*.

4.3.3 Definition of State of Strain at a Point

The state of infinitesimal strain at a point is completely defined by the values of the normal and shear strains. Written in matrix form, the state of strain at a point is thus

$$\varepsilon = \begin{bmatrix} \varepsilon_x & \gamma_{xy} & \gamma_{xz} \\ \gamma_{yx} & \varepsilon_y & \gamma_{yz} \\ \gamma_{zx} & \gamma_{zy} & \varepsilon_z \end{bmatrix}$$

(4.46)

Since $\gamma_{xy} = \gamma_{yx}$, $\gamma_{xz} = \gamma_{zx}$, and $\gamma_{yz} = \gamma_{zy}$, only six of the nine strain components are independent.

---

6. See, for example, Timoshenko, S.P., Goodier, J.N., 1970. Theory of Elasticity, third ed. McGraw-Hill, New York, NY.

4.3.4 Volumetric Strain

For infinitesimal strains, the volumetric strain associated with a three-dimensional state of strain is defined as follows:

$$\varepsilon_{vol} = \varepsilon_x + \varepsilon_y + \varepsilon_z \tag{4.47}$$

4.3.5 State of Plane Strain

Consider a body that is substantially longer in one coordinate direction as compared to the other two directions. Take the z-axis parallel to the "long" direction. The body is assumed to be loaded by forces acting in the planes of the cross section, that is, normal to the z-axis, and not varying along its length. The stresses and deformations of all cross sections in the body are only functions of the x and y coordinates. The deformation is then said to be *plane strain*. For such a condition, $\varepsilon_z = \gamma_{xz} = \gamma_{yz} \approx 0$. The following strain matrix thus represents the state of plane strain:

$$\varepsilon = \begin{bmatrix} \varepsilon_x & \gamma_{xy} \\ \gamma_{xy} & \varepsilon_y \end{bmatrix} \tag{4.48}$$

4.3.6 Strain Transformations

Since the state of stress (Eq. 4.3) and strain (Eq. 4.46) at a point is represented by a square matrix (or more generally, by a rank-two tensor), the equations for transforming a given strain state into an equivalent one acting on any plane through an element in a body are obtained by analogy to the stress transformation equations.

4.3.6.1 Equations for Plane Strain Transformation

The transformation equations for a state of plane strain are obtained by analogy to the transformation equations for a state of plane stress given by Eqs. (4.10), (4.16), and (4.13). As such,

$$\varepsilon_{x'} = \frac{1}{2}(\varepsilon_x + \varepsilon_y) + \frac{1}{2}(\varepsilon_x - \varepsilon_y)\cos 2\theta - \frac{\gamma_{xy}}{2}\sin 2\theta \tag{4.49}$$

$$\varepsilon_{y'} = \frac{1}{2}(\varepsilon_x + \varepsilon_y) - \frac{1}{2}(\varepsilon_x - \varepsilon_y)\cos 2\theta + \frac{\gamma_{xy}}{2}\sin 2\theta \tag{4.50}$$

$$\frac{1}{2}\gamma_{x'y'} = \frac{1}{2}(\varepsilon_x - \varepsilon_y)\sin 2\theta + \frac{\gamma_{xy}}{2}\cos 2\theta \tag{4.51}$$

Similar to the case of plane stress transformation, the angle θ, appearing in Eqs. (4.49)–(4.51), is positive when measured *counterclockwise*.

4.3.6.2 Determination of Principal Strains

By analogy to Eq. (4.39), the principal strains are given by

$$\varepsilon_1 = C + R; \quad \varepsilon_2 = C - R \tag{4.52}$$

where, by analogy to Eqs. (4.32) and (4.38),

$$C = \frac{1}{2}(\varepsilon_x + \varepsilon_y) \tag{4.53}$$

$$R = \sqrt{\left[\frac{1}{2}(\varepsilon_x - \varepsilon_y)\right]^2 + \left(\frac{\gamma_{xy}}{2}\right)^2} \tag{4.54}$$

and $\gamma_{xy} = 0$. The principal strain direction is computed from

$$\tan 2\theta_p = \frac{-\gamma_{xy}}{\varepsilon_x - \varepsilon_y} \tag{4.55}$$

4.3.6.3 Determination of Maximum In-Plane Shear Strain

The magnitude of the maximum in-plane shearing strain (γ_{max}) is

$$\gamma_{max} = 2R = 2\sqrt{\left[\frac{1}{2}(\varepsilon_x - \varepsilon_y)\right]^2 + \left(\frac{\gamma_{xy}}{2}\right)^2} \tag{4.56}$$

The associated normal strains are the *average* of ε_x and ε_y, i.e.,

$$\varepsilon_x = \varepsilon_y = \varepsilon_{avg} = \frac{1}{2}(\varepsilon_x + \varepsilon_y) \tag{4.57}$$

The orientation of the maximum in-plane shear strain is computed from

$$\tan 2\theta_s = \frac{\varepsilon_x - \varepsilon_y}{\gamma_{xy}} \tag{4.58}$$

4.3.6.4 Mohr's Circle of Stress for Plane Strain

A Mohr's circle of strain is constructed in the space of ε versus $\gamma/2$. The steps to constructing the circle are identical to those listed in Section 4.2.5.4 for the case of a Mohr's circle of stress. Eqs. (4.53) and (4.54) give the ε-coordinate of the center and the radius of the circle, respectively.

4.4 CONSTITUTIVE RELATIONS

As noted in Section 4, a valid solution to any problem in solid mechanics must satisfy (1) the equations of equilibrium of body, surface, and inertia forces and stresses; (2) the equations of compatibility of strains and displacements; and (3) the constitutive equations for the material.

The equilibrium equations and the equations of compatibility are valid for *all* materials, irrespective of their internal constitution. Thus, a unique solution to a boundary value problem cannot be obtained only with these two sets of equations. Instead, a unique solution requires additional considerations that account for the nature of the material. The quantification of material response is realized through *constitutive relations.*[7]

4.4.1 General Form of Constitutive Relations

The general form of a constitutive relation for solids relates stresses to strains. This section presents the vector–matrix form of the constitutive relations in both direct and inverse form.

In "direct" form, the constitutive relations are written as follows:

$$\boldsymbol{\varepsilon} = \mathbf{A}\boldsymbol{\sigma} \tag{4.59}$$

where $\boldsymbol{\varepsilon}$ is the vector of infinitesimal strains, $\boldsymbol{\sigma}$ is the Cauchy stress vector, and $\mathbf{A}$ is a square symmetric matrix of compliance coefficients characterizing the material.

Written in "inverse" form, the constitutive relations become

$$\boldsymbol{\sigma} = \mathbf{C}\boldsymbol{\varepsilon} \tag{4.60}$$

where $\boldsymbol{\sigma}$ and $\boldsymbol{\varepsilon}$ are as previously defined. The square symmetric matrix of material coefficients $\mathbf{C}$ is the inverse of $\mathbf{A}$.

In the most general case of a three-dimensional analysis, $\mathbf{A}$ and $\mathbf{C}$ are (6*6) in size. The strain and stress vectors are (6*1) in size and contain the following entries:

$$\boldsymbol{\varepsilon} = \left\{ \varepsilon_x \quad \varepsilon_y \quad \varepsilon_z \quad \gamma_{xy} \quad \gamma_{xz} \quad \gamma_{yz} \right\}^T \tag{4.61}$$

$$\boldsymbol{\sigma} = \left\{ \sigma_x \quad \sigma_y \quad \sigma_z \quad \tau_{xy} \quad \tau_{xz} \quad \tau_{yz} \right\}^T \tag{4.62}$$

In the given expressions the superscript T denotes the operation of matrix transposition.

4.4.2 Insight Into the Constitutive Matrices

To make better sense of material idealizations based on entries in $\mathbf{C}$, partition this matrix into the following (3*3) submatrices:

$$\mathbf{C} = \begin{bmatrix} \mathbf{C}_{11} & \mathbf{C}_{12} \\ \mathbf{C}_{21} & \mathbf{C}_{22} \end{bmatrix} \tag{4.63}$$

where, due to symmetry, $\mathbf{C}_{12} = \mathbf{C}_{21}^T$. The following observations are then pertinent:

- Nonzero entries in $\mathbf{C}_{11}$ mean that the normal stresses σ_x, σ_y, and σ_z are functions of the strains ε_x, ε_y, and ε_z. It follows that $\mathbf{C}_{11}$ will *never* be the zero matrix.

---

7. Constitutive relations are also referred to as "constitutive laws" and "constitutive models."

- If $\mathbf{C_{12}} \neq \mathbf{0}$, the normal stresses are also functions of the shear strains γ_{xy}, γ_{xz}, and γ_{yz}. If $\mathbf{C_{12}} \neq \mathbf{0}$, then $\mathbf{C_{21}} \neq \mathbf{0}$, implying that the shear stresses τ_{xy}, τ_{xz}, and τ_{yz} are functions of ε_x, ε_y, and ε_z.
- Finally, if $\mathbf{C_{22}} \neq \mathbf{0}$, the shear stresses are also functions of the shear strains. It follows that $\mathbf{C_{22}}$ will *never* be the zero matrix. Analyzing $\mathbf{C_{22}}$ more closely, it is evident that a diagonal form of this submatrix means that shear stresses are only functions of the corresponding shear strains. Once nonzero diagonal entries are present in $\mathbf{C_{22}}$, the shear stresses become functions of two or more shear strains.

Although the aforesaid discussion focused on $\mathbf{C}$, it could have likewise been presented in terms of $\mathbf{A}$.

4.4.3 General Classes of Material Idealizations

The discussion of constitutive relations is typically organized by the type of material idealization. The following general classes of such idealizations are commonly used:

- Linear elasticity
- Nonlinear elasticity
- Linear viscoelasticity
- Nonlinear viscoelasticity
- Elastoplasticity (i.e., time- and rate-independent inelastic response)
- Viscoplasticity (i.e., time- and rate-dependent inelastic response).

As noted in Chapter 1, soils consist of a porous skeleton whose voids are filled with fluid (typically air and water). Macroscopically, soils exhibit an anisotropic, inelastic, strain hardening (and softening), and time- and temperature-dependent behavior. The proper mathematical characterization of soils thus requires rather sophisticated constitutive relations that account for some or all of the aforementioned behavioral characteristics.

However, since introductory courses in soil mechanics largely consider elastic material response, this focus in the remainder of this section is on elastic constitutive relations.

4.4.4 Elastic Material Idealizations

Elastic materials exhibit the following characteristics:

- The state of stress is a function only of the current state of deformation; it does *not* depend on the history of straining or loading.
- Upon removal of the applied loads, the material completely recovers to the undeformed configuration.
- When loaded, an elastic material stores 100% of the energy due to deformation (i.e., the strain energy).

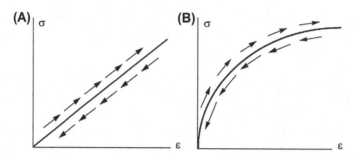

FIGURE 4.22 Uniaxial loading and unloading of an elastic material: (A) linear, (B) nonlinear.

- When unloaded, such a material releases 100% of the stored energy and returns to its initial state. No permanent deformation is realized.
- The response is *rate-independent*, i.e., the rate at which the loading is applied has no effect upon the material response.

In a *linear* elastic material (Figure 4.22A), stress is proportional to strain. In general, however, elastic materials are *nonlinear*. This is particularly true in the case of geomaterials (soil and rock). The loading and unloading paths for such materials essentially coincide (Figure 4.22B).

4.4.4.1 Linear Elastic Material Idealizations

The most general linear elastic constitutive relation, which pertains to *anisotropic* linear elastic materials, is generalized Hooke's law.[8] The general form of this relation, in direct and inverse form, is given by Eqs. (4.59) and (4.60), respectively. Since the strain and stress vectors in Eqs. (4.61) and (4.62), respectively, are (6*1) in size, it follows that the square matrices **A** and **C** are (6*6) in size. Due to the symmetry of the strain and stress matrices (recall Eqs. 4.46 and 4.3), as well as **A** and **C** themselves, only 21 of the 36 coefficients of **A** and **C** are independent.

From the point of view of material characterization, the analysis of bodies made of anisotropic materials is quite complicated, especially in three dimensions. This is because of the need to experimentally determine the aforementioned 21 entries in **A** or **C**. Fortunately, many of the important engineering materials possess some internal structure that exhibits certain symmetries and thus simplifies the composition of **A** and **C**.

For example, in the case of materials possessing three planes of elastic symmetry (i.e., orthotropic materials), the number of independent entries reduces to 9. For materials possessing a plane of isotropy (i.e., transversely isotropic materials), the number of independent entries further reduces to 5. Although transversely isotropic material idealizations are particularly well

---

8. Named in honor of Robert Hooke (1635–1703).

suited to soils because of their method of deposition, the prospect of determining the values of five material parameters often precludes their use.

Instead, the elastic response of soils is typically assumed to be *isotropic* (recall the discussion in Section 4.1.3). This represents the simplest elastic material for which the elastic behavior is *independent* of the orientation of the coordinate axes. For an isotropic elastic material the constitutive relations simplify in that only *two* independent constants are required to completely describe the material behavior, for example, the elastic modulus E and Poisson's ratio[9] v. The compliance matrix $\mathbf{A}$ then has the following entries:

$$\mathbf{A} = \frac{1}{E} \begin{bmatrix} 1 & -v & -v & 0 & 0 & 0 \\ -v & 1 & -v & 0 & 0 & 0 \\ -v & -v & 1 & 0 & 0 & 0 \\ 0 & 0 & 0 & 2(1+v) & 0 & 0 \\ 0 & 0 & 0 & 0 & 2(1+v) & 0 \\ 0 & 0 & 0 & 0 & 0 & 2(1+v) \end{bmatrix} \tag{4.64}$$

Recalling the direct form of the constitutive relations given by Eq. (4.59) and the strain and stress vectors given by Eqs. (4.61) and (4.62), respectively, the constitutive relations for an isotropic linear elastic material are

$$\varepsilon_x = \frac{1}{E}\left[\sigma_x - v(\sigma_y + \sigma_z)\right] \tag{4.65}$$

$$\varepsilon_y = \frac{1}{E}\left[\sigma_y - v(\sigma_x + \sigma_z)\right] \tag{4.66}$$

$$\varepsilon_z = \frac{1}{E}\left[\sigma_z - v(\sigma_x + \sigma_y)\right] \tag{4.67}$$

$$\gamma_{xy} = \frac{2(1+v)}{E}\tau_{xy} = \frac{1}{G}\tau_{xy} \tag{4.68}$$

$$\gamma_{xz} = \frac{1}{G}\tau_{xz} \tag{4.69}$$

$$\gamma_{yz} = \frac{1}{G}\tau_{yz} \tag{4.70}$$

where G is the elastic shear modulus.

In inverse form, the nonzero entries in $\mathbf{C}$ are

$$C_{11} = C_{22} = C_{33} = \frac{E(1-v)}{(1+v)(1-2v)} \tag{4.71}$$

---

9. Named in honor of S.D. Poisson (1781–1840).

$$C_{12} = C_{13} = C_{23} = \frac{E\nu}{(1+\nu)(1-2\nu)} \qquad (4.72)$$

$$C_{44} = C_{55} = C_{66} = \frac{E}{2(1+\nu)} = G \qquad (4.73)$$

where G is as defined previously.

4.5 STRESSES IN SOIL DUE TO SURFACE LOADS

When a load, such as the weight of a structure, is applied to the surface of a soil mass, the vertical total, and possibly effective, stress within the mass will increase. Load applied at one point will be transferred vertically and laterally throughout the mass.

If the load is applied over a large areal extent, the vertical total stress on a horizontal section at a given depth will be uniformly distributed. Such will not be the case if the load is applied over a smaller area such as a footing. In these cases the distribution of vertical total stress is determined using various formulas based on the theory of elasticity. Such formulas are similar in that they assume the soil to be homogeneous (at least in a given layer), elastic and isotropic with constant elastic modulus and Poisson's ratio. In addition, the load is typically assumed to be applied at the ground surface over a flexible loading area. The aforementioned formulas differ only in the specific assumptions made to represent the elastic soil mass and the extent of the solution domain (e.g., infinite versus semiinfinite).

Once the elastic stresses in a soil mass are known, Eq. (4.59) is used to compute the associated strains. If the strain distribution is integrated over the depth of the soil mass, it yields the displacement field in the soil. Among the three displacement components, it is the vertical one that is typically of most interest in that it is used to quantify settlement of structures. Chapter 8 gives additional details pertaining to the subject of settlement.

4.6 SUPERPOSITION PRINCIPLE

One of the benefits of using a linear elastic material idealization is that the superposition principle can be used. This principle states that for all linear systems, the net response at a given location and time caused by two or more excitations (e.g., applied loads, applied displacements, etc.) is the sum of the responses that would have been caused by each excitation individually.

EXAMPLE PROBLEM 4.1

General Remarks

This example problem illustrates the manner in which average normal and shear strains are computed.

Problem Statement

The rectangular plate *PQRS* is deformed into the shape shown by dashed lines in Figure Ex. 4.1A. Compute (a) the average normal strain along diagonal *PR*, (b) the average normal strain along diagonal *QS*, (c) the average engineering shear strain (γ_{xy}) at corner *P*, and (d) the average engineering shear strain (γ_{xy}) at corner *Q* and the average engineering shear strain (γ_{xy}) at corner *R*.

Solution

a) The initial length of diagonal *PR* is

$$PR = \sqrt{(300)^2 + (250)^2} = 390.512 \text{ mm} \qquad (4.1.1)$$

The deformed length of diagonal *PR'* is

$$PR' = \sqrt{(300 + 6.5)^2 + (250 + 6.0)^2} = 399.347 \text{ mm} \qquad (4.1.2)$$

Using Eqs. (4.1.1) and (4.1.2), the average normal strain along diagonal *PR* is thus

$$\varepsilon_{PR} = \frac{PR' - PR}{PR} = -\left(\frac{399.347 - 390.512}{390.512}\right) = -2.262 \times 10^{-2} \text{ mm/mm}$$

$$(4.1.3)$$

The negative sign indicates that the strain is extensional.

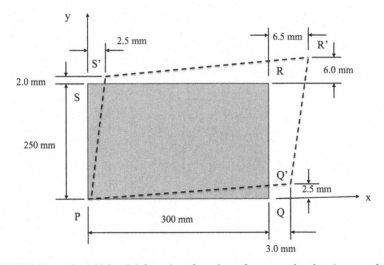

FIGURE EX. 4.1A Initial and deformed configurations of a rectangular plate (not to scale).

b) Proceeding in a similar manner, the initial length of diagonal QS is equal to the length of PR. The deformed length of diagonal $Q'S'$ is

$$Q'S' = \sqrt{(303 - 2.5)^2 + (2.5 - 252.0)^2} = 390.577 \text{ mm} \qquad (4.1.4)$$

Using Eqs. (4.1.1) and (4.1.4), the average normal strain along diagonal QS is thus

$$\varepsilon_{QS} = \frac{Q'S' - QS}{QS} = -\left(\frac{390.577 - 390.512}{390.512}\right) = \mathbf{-1.656 \times 10^{-4} \text{ mm/mm}} \qquad (4.1.5)$$

The negative sign indicates that the strain is also extensional.

c) To facilitate the computation of the average engineering shear strain at corner P, refer to Figure Ex. 4.1B. Recalling that small angle geometry is assumed, the values of the angles α and β are computed as follows:

$$\tan \alpha = \frac{2.5 \text{ mm}}{(250 + 2.0) \text{ mm}} \Rightarrow \alpha = \tan^{-1}\left(\frac{2.5}{252.0}\right) = 0.010 \text{ rad} \qquad (4.1.6)$$

$$\tan \beta = \frac{2.5 \text{ mm}}{(300 + 3.0) \text{ mm}} \Rightarrow \beta = \tan^{-1}\left(\frac{2.5}{303.0}\right) = 0.008 \text{ rad} \qquad (4.1.7)$$

The average engineering shear strain at corner P is thus

$$\gamma_{xy_P} = \alpha + \beta = 0.010 + 0.008 = \mathbf{0.018 \text{ rad}} \qquad (4.1.8)$$

The positive shear strain indicates that the initial right angle at point P has decreased in magnitude.

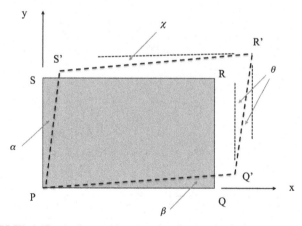

FIGURE EX. 4.1B   Angles used in computing shear strains in the rectangular plate.

d) The average engineering shear strain at corner Q requires the value of the angle θ (Figure Ex. 4.1B), which is computed as follows:

$$\tan \theta = \frac{(6.5 - 3.0)\text{mm}}{(250 + 6.0 - 2.5)\text{mm}} \Rightarrow \theta = \tan^{-1}\left(\frac{3.5}{253.5}\right) = 0.014 \text{ rad} \quad (4.1.9)$$

Thus,

$$\gamma_{xy_Q} = \theta + \beta = -(0.014 + 0.008) = \mathbf{-0.022 \text{ rad}} \quad (4.1.10)$$

The negative shear strain indicates that the initial right angle at point Q has increased in magnitude due to shearing.

e) In addition to the value of θ, the average engineering shear strain at corner R also requires the value of the angle χ (Figure Ex. 4.1B), which is computed as follows:

$$\tan \chi = \frac{(6.0 - 2.0)\text{mm}}{(300 + 6.5 - 2.5)\text{mm}} \Rightarrow \chi = \tan^{-1}\left(\frac{4.0}{304.0}\right) = 0.013 \text{ rad} \quad (4.1.11)$$

Thus,

$$\gamma_{xy_R} = \theta + \chi = 0.014 + 0.013 = \mathbf{0.027 \text{ rad}} \quad (4.1.12)$$

EXAMPLE PROBLEM 4.2

General Remarks

This example problem illustrates the manner in which normal strains are computed and then used to compute other kinematic quantities.

Problem Statement

During testing of a sand sample in axisymmetric triaxial compression, the original diameter of 36 mm was increased by 2.403×10^{-3} mm and the original length of 76 mm was decreased by 1.810×10^{-2} mm. Compute (a) the axial normal strain ε_z, (b) the lateral normal strain $(\varepsilon_x = \varepsilon_y)$, (c) the volumetric strain ε_{vol}, and (d) the Poisson's ratio ν.

Solution

a) Assuming the direction of load application to be parallel to the z-axis, the axial normal strain is computed as follows:

$$\varepsilon_z = \frac{1.810 \times 10^{-2} \text{ mm}}{76 \text{ mm}} = \mathbf{2.382 \times 10^{-4} \text{ mm/mm}} \quad (4.2.1)$$

b) Similarly, the lateral normal strain is

$$\varepsilon_x = \varepsilon_y = \frac{-2.403 \times 10^{-3} \text{ mm}}{36 \text{ mm}} = \mathbf{-6.675 \times 10^{-5} \text{ mm/mm}} \qquad (4.2.2)$$

c) Using the previous results, the infinitesimal volumetric strain is thus

$$\varepsilon_{vol} = \varepsilon_x + \varepsilon_y + \varepsilon_z = 2\left(-6.675 \times 10^{-5}\right) + 2.382 \times 10^{-4} = \mathbf{1.047 \times 10^{-4} \text{ mm/mm}}$$
$$(4.2.3)$$

d) Finally, assuming a linear elastic material idealization for the sand, the Poisson's ratio is

$$\nu = -\frac{\varepsilon_x}{\varepsilon_z} = -\frac{\left(-6.675 \times 10^{-5} \text{ mm/mm}\right)}{2.382 \times 10^{-4} \text{ mm/mm}} = \mathbf{0.28} \qquad (4.2.4)$$

EXAMPLE PROBLEM 4.3

General Remarks

This example problem illustrates the manner in which uniaxial stress and axial and lateral strains are related through a linear elastic constitutive relation.

Problem Statement

When testing a concrete cylinder in compression, the original diameter of 150 mm was increased by 0.01016 mm and the original length of 300 mm was decreased by 0.16510 mm under the action of a compressive axial force of 224 kN. Assuming a linear elastic material idealization for the concrete, determine (a) the volumetric strain, (b) the Poisson's ratio (ν), and (c) the modulus of elasticity (Young's modulus) E.

Solution

a) The axial strain developed in the concrete cylinder as a result of the applied load is

$$\varepsilon_{axial} = \frac{0.16510 \text{ mm}}{300 \text{ mm}} = 5.503 \times 10^{-4} \text{ mm/mm} \qquad (4.3.1)$$

The lateral strain resulting from the applied load is

$$\varepsilon_{lateral} = -\frac{0.01016 \text{ mm}}{150 \text{ mm}} = -6.777 \times 10^{-5} \text{ mm/mm} \qquad (4.3.2)$$

The volumetric strain resulting from the applied load is thus

$$\varepsilon_{vol} = \varepsilon_{axial} + 2\varepsilon_{lateral} = \left(5.503 \times 10^{-4}\right) + 2\left(-6.777 \times 10^{-5}\right)$$
$$= \mathbf{4.148 \times 10^{-4} \ mm/mm} \tag{4.3.3}$$

b) The Poisson's ratio for the concrete is computed from Eqs. (4.3.1) and (4.3.2) as follows:

$$\nu = -\frac{\varepsilon_{lateral}}{\varepsilon_{axial}} = -\frac{\left(-6.777 \times 10^{-5} \ mm/mm\right)}{5.503 \times 10^{-4} \ mm/mm} = \mathbf{0.12} \tag{4.3.4}$$

c) The modulus of elasticity is the slope of the initial (linear) portion of the stress–strain curve. As such,

$$E = \frac{\sigma_{axial}}{\varepsilon_{axial}} \tag{4.3.5}$$

The axial stress is computed as follows:

$$\sigma_{axial} = \frac{P_{axial}}{A} = \frac{224 \ kN}{\frac{\pi}{4}(150 \ mm)^2} * \left(\frac{1000 \ mm}{m}\right)^2 = 1.268 \times 10^4 \ kPa \tag{4.3.6}$$

Substituting Eqs. (4.3.6) and (4.3.1) into Eq. (4.3.5) gives the desired value of the modulus of elasticity

$$E = \frac{\sigma_{axial}}{\varepsilon_{axial}} = \frac{1.268 \times 10^4 \ kPa}{5.503 \times 10^{-4} \ mm/mm} = \mathbf{2.303 \times 10^7 \ kPa = 23.0 \ GPa} \tag{4.3.7}$$

EXAMPLE PROBLEM 4.4

General Remarks

This example problem illustrates the manner in which uniaxial stress and axial and lateral strains are related through a linear elastic constitutive relation.

Problem Statement

The rectangular specimen shown in Figure Ex. 4.4 is subjected to an axial tensile force of $P = 28.5$ kN. As a result of this force, the specimen elongates 0.0562 mm over its length.

The specimen's response is known to be in the linear elastic range, and the Poisson's ratio (ν) is equal to 0.34. Compute (a) the elastic modulus (E) for the material from which the specimen is made, (b) the axial strain (ε_x), (c) the transverse normal strains ε_y and ε_z, (d) the changes in length in the y and z directions, and (e) the volumetric strain (ε_{vol}).

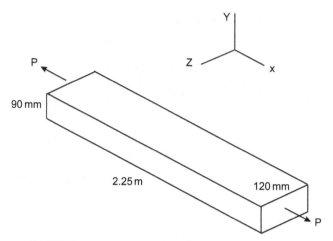

FIGURE EX. 4.4 Rectangular specimen subjected to axial load.

Solution

a) Since the response is known to be in the linear elastic range, the uniaxial stress and strain are related through Hooke's law, i.e.,

$$\sigma_x = E\varepsilon_x \Rightarrow \frac{P}{A} = E\frac{\delta_x}{L} \Rightarrow E = \frac{PL}{A\delta_x} \tag{4.4.1}$$

where δ_x is the deformation in the x-direction and A is the cross-sectional area. Substituting all known quantities into Eq. (4.4.1) and recalling that extensional strains and tensile stresses are negative gives

$$E = \frac{PL}{A\delta_x} = \frac{(-28.5 \text{ kN})(2.25 \text{ m})}{(0.120 \text{ m})(0.090 \text{ m})(-0.0562 \text{ mm})(\text{m}/1000 \text{ mm})} \tag{4.4.2}$$
$$= 1.056 \times 10^8 \text{ kPa} = 105.6 \text{ GPa}$$

b) The axial strain in the specimen is

$$\varepsilon_x = \frac{\delta_x}{L} = \frac{-0.0562 \text{ mm}}{2.25 \text{ m}} \left(\frac{1 \text{ m}}{1000 \text{ mm}}\right) = -2.498 \times 10^{-5} \text{ mm/mm} \tag{4.4.3}$$

c) The transverse normal strains are computed from the definition of Poisson's ratio, i.e.,

$$\nu = -\frac{\varepsilon_y}{\varepsilon_x} = -\frac{\varepsilon_z}{\varepsilon_x} \Rightarrow \varepsilon_y = \varepsilon_z = -\nu\varepsilon_x \tag{4.4.4}$$

Substituting all known quantities into Eq. (4.4.4) and recalling that compressive strains are positive gives

$$\varepsilon_y = \varepsilon_z = -(0.34)(-2.498 \times 10^{-5} \text{ mm/mm}) = 8.492 \times 10^{-6} \text{ mm/mm}$$

(4.4.5)

d) The changes in length in the y and z directions are computed from the lateral strains as follows:

$$\varepsilon_y = \frac{\delta_y}{L_y} \Rightarrow \delta_y = \varepsilon_y L_y = (8.492 \times 10^{-6} \text{ mm/mm})(90 \text{ mm})$$
$$= 7.643 \times 10^{-4} \text{ mm}$$

(4.4.6)

$$\varepsilon_z = \frac{\delta_z}{L_z} \Rightarrow \delta_z = \varepsilon_z L_z = (8.492 \times 10^{-6} \text{ mm/mm})(120 \text{ mm})$$
$$= 1.019 \times 10^{-3} \text{ mm}$$

(4.4.7)

e) Finally, the volumetric strain is

$$\varepsilon_{vol} = \varepsilon_x + \varepsilon_y + \varepsilon_z = (-2.498 \times 10^{-5} \text{ mm/mm}) + 2(8.492 \times 10^{-6} \text{ mm/mm})$$
$$= -8.000 \times 10^{-6} \text{ mm/mm}$$

(4.4.8)

EXAMPLE PROBLEM 4.5

General Remarks

This example problem illustrates the manner in which elastic constants and strain components are computed for a soil specimen subjected to plane strain conditions.

Problem Statement

A 200 mm cube of sand is subjected to conditions of plane strain with the strain increments $\Delta\varepsilon_z = \Delta\gamma_{xz} = \Delta\gamma_{yz} = 0$. The loading induces the following stress state: $\sigma_x = 150$ kPa (compressive), $\sigma_y = 800$ kPa (compressive), and $\tau_{xy} = 0$.

The applied stresses result in a 0.70 mm expansion in the x-direction and a 6.5 mm compression in the y-direction. Assuming that the sand exhibits isotropic, linear elastic response, determine the following:

a) The magnitude of the elastic modulus (E) and Poisson's ratio (ν) for the sand.

b) The magnitude of the normal stress σ_z.

Solution

a) Begin the solution with the linear elastic constitutive relations in direct form as given by Eqs. (4.65)–(4.7), i.e.,

$$\varepsilon_x = \frac{1}{E}\left[\sigma_x - \nu(\sigma_y + \sigma_z)\right] \tag{4.5.1}$$

$$\varepsilon_y = \frac{1}{E}\left[\sigma_y - \nu(\sigma_x + \sigma_z)\right] \tag{4.5.2}$$

$$\varepsilon_z = \frac{1}{E}\left[\sigma_z - \nu(\sigma_x + \sigma_y)\right] \tag{4.5.3}$$

Since ε_x, ε_y, σ_x, and σ_y are either given or easily computed from the information provided, and since $\varepsilon_z = 0$ because of plane strain conditions, the three unknowns associated with the problem are E, ν, and σ_z.

Begin by solving Eqs. (4.5.1) and (4.5.2) for E and then equating the resulting expressions. This leads to the following result:

$$\frac{1}{\varepsilon_x}\left[\sigma_x - \nu(\sigma_y + \sigma_z)\right] = \frac{1}{\varepsilon_y}\left[\sigma_y - \nu(\sigma_x + \sigma_z)\right] \tag{4.5.4}$$

To eliminate σ_z, note that from Eq. (4.5.3),

$$\varepsilon_z = \frac{1}{E}\left[\sigma_z - \nu(\sigma_x + \sigma_y)\right] = 0 \Rightarrow \sigma_z = \nu(\sigma_x + \sigma_y) \tag{4.5.5}$$

Substituting this into Eq. (4.5.4) gives a single equation in ν, i.e.,

$$(\sigma_x + \sigma_y)\left(\frac{1}{\varepsilon_y} - \frac{1}{\varepsilon_x}\right)\nu^2 + \left(\frac{\sigma_x}{\varepsilon_y} - \frac{\sigma_y}{\varepsilon_x}\right)\nu + \left(\frac{\sigma_x}{\varepsilon_x} - \frac{\sigma_y}{\varepsilon_y}\right) = 0 \tag{4.5.6}$$

From the information given in the problem, $\sigma_x = 150$ kPa, $\sigma_y = 800$ kPa, and

$$\varepsilon_x = \frac{-0.70 \text{ mm}}{200 \text{ mm}} = -3.500 \times 10^{-3} \text{ mm/mm} \tag{4.5.7}$$

$$\varepsilon_y = \frac{6.50 \text{ mm}}{200 \text{ mm}} = 3.250 \times 10^{-2} \text{ mm/mm} \tag{4.5.8}$$

Substituting these values into Eq. (4.5.6) gives

$$(3.007 \times 10^5)\nu^2 + (2.332 \times 10^5)\nu + (-6.747 \times 10^4) = 0 \tag{4.5.9}$$

The solution to this quadratic equation is $\nu = -1.000, 0.224$. Although the first root is theoretically possible, the second one is the desired quantity. Thus,

$$\nu = \mathbf{0.224} \tag{4.5.10}$$

To determine the elastic modulus E, subtract Eq. (4.5.2) from Eq. (4.5.1), giving

$$\varepsilon_x - \varepsilon_y = \left(\frac{1+\nu}{E}\right)(\sigma_x - \sigma_y) \Rightarrow E = (1+\nu)\frac{(\sigma_x - \sigma_y)}{\varepsilon_x - \varepsilon_y} \qquad (4.5.11)$$

Substituting the known values into Eq. (4.5.11) gives

$$E = (1+0.224)\left[\frac{150 - 800}{(-3.500 \times 10^{-3}) - (3.250 \times 10^{-2})}\right] = \mathbf{2.210 \times 10^4 \ kPa}$$
$$(4.5.12)$$

b) The normal stress in the z-direction is computed from Eq. (4.5.5), i.e.,

$$\sigma_z = \nu(\sigma_x + \sigma_y) = (0.224)(150 + 800) = \mathbf{2.218 \times 10^2 \ kPa} \qquad (4.5.13)$$

Thus, similar to σ_x and σ_y, the normal stress acting in the thickness direction is compressive.

EXAMPLE PROBLEM 4.6

General Remarks

This example problem investigates quantities associated with confined (one-dimensional) compression.

Problem Statement

In an isotropic linear elastic material idealization, the constrained modulus (E_{1d}) is defined as the ratio of the axial stress to the axial strain for confined (uniaxial) compression such as that realized in an oedometer. Taking the z-direction as the direction of loading and deformation, it follows that $\varepsilon_z \neq 0$ and $\varepsilon_x = \varepsilon_y = 0$.

a) Express the lateral stresses σ_x and σ_y in terms of σ_z and Poisson's ratio (ν).

b) Derive an expression for the coefficient of earth pressure at rest ($K_0 = \sigma_x/\sigma_z$) in terms of ν.

c) Derive an expression for E_{1d} in terms of the elastic modulus E and ν.

d) Besides changes in volume, uniaxial loading and confined compression also involve shear strains. As such, determine the volumetric strain ($\varepsilon_{vol} = \varepsilon_x + \varepsilon_y + \varepsilon_z$) and maximum shear strain (γ_{max}) developed during confined compression. Express the results in terms of E, ν, and σ_z.

Solution

a) Specializing the constitutive relations (recall Eqs. 4.59 and 4.64), written in direct form, for the case of one-dimensional compression gives

$$\varepsilon_x = \frac{1}{E}\left[\sigma_x - \nu(\sigma_y + \sigma_z)\right] = 0 \tag{4.6.1}$$

$$\varepsilon_y = \frac{1}{E}\left[\sigma_y - \nu(\sigma_x + \sigma_z)\right] = 0 \tag{4.6.2}$$

$$\varepsilon_z = \frac{1}{E}\left[\sigma_z - \nu(\sigma_x + \sigma_y)\right] \tag{4.6.3}$$

From Eq. (4.6.1),

$$\sigma_x = \nu(\sigma_y + \sigma_z) \tag{4.6.4}$$

Substituting this result into Eq. (4.6.2) gives

$$\sigma_y = \nu(\sigma_x + \sigma_z) = \nu\left[\nu\sigma_y + (1 + \nu)\sigma_z\right] \tag{4.6.5}$$

Solving Eq. (4.6.5) for σ_y gives

$$\sigma_y = \left(\frac{\nu}{1 - \nu}\right)\sigma_z \tag{4.6.6}$$

Substituting this result into Eq. (4.6.1) gives

$$\sigma_x = \nu(\sigma_y + \sigma_z) = \left(\frac{\nu^2}{1 - \nu}\right)\sigma_z + \nu\sigma_z = \left(\frac{\nu}{1 - \nu}\right)\sigma_z = \sigma_y \tag{4.6.7}$$

Thus,

$$\sigma_x = \sigma_y = \left(\frac{\nu}{1 - \nu}\right)\sigma_z \tag{4.6.8}$$

The result makes sense, as the lateral stresses would be expected to be equal under the uniaxial (one-dimensional) conditions imposed in an oedometer.

b) From Eq. (4.6.7) it follows that

$$K_0 = \frac{\sigma_x}{\sigma_z} = \left(\frac{\nu}{1 - \nu}\right) \tag{4.6.9}$$

Before leaving this part of the problem, it is timely to note that if K_0 is known in a test, then the aforesaid expression can be inverted and solved for Poisson's ratio, giving

$$\nu = \frac{K_0}{1 + K_0} \tag{4.6.10}$$

where it is understood that the response is assumed to be elastic.

c) Noting that $\sigma_x = \sigma_y = \nu\sigma_z/(1-\nu)$, Eq. (4.6.3) is rewritten as

$$
\varepsilon_z = \frac{1}{E}[\sigma_z - \nu(\sigma_x + \sigma_y)] = \frac{1}{E}\left[\sigma_z - 2\nu\left(\frac{\nu}{1-\nu}\right)\sigma_z\right]
$$
$$
= \frac{1}{E}\left[\frac{1-\nu-2\nu^2}{1-\nu}\right]\sigma_z = \frac{1}{E}\left[\frac{(1+\nu)(1-2\nu)}{1-\nu}\right]\sigma_z
$$

(4.6.11)

Thus,

$$
E_{1d} = \frac{\sigma_z}{\varepsilon_z} = \frac{(1-\nu)}{(1+\nu)(1-2\nu)}E
$$

(4.6.12)

For example, if $\nu = \frac{1}{3}$, then $E_{1d} = \frac{3}{2}E$.

d) For confined (uniaxial) compression, $\varepsilon_x = \varepsilon_y = 0$. The volumetric strain is thus given by Eq. (4.6.11), i.e.,

$$
\varepsilon_{vol} = \varepsilon_z = \frac{\sigma_z}{E}\left[\frac{(1+\nu)(1-2\nu)}{1-\nu}\right] = \frac{\sigma_z}{E_{1d}}
$$

(4.6.13)

The maximum shear strain in the x-z plane is given by

$$
\tau_{max} = \frac{1}{2}(\sigma_z - \sigma_x) = \frac{1}{2}\left(\sigma_z - \frac{\nu}{1-\nu}\sigma_z\right) = \frac{1}{2}\left(\frac{1-2\nu}{1-\nu}\right)\sigma_z
$$

(4.6.14)

where Eq. (4.6.7) has been used. The maximum engineering shear strain is thus

$$
\gamma_{max} = \frac{1}{G}\tau_{max} = \left[\frac{2(1+\nu)}{E}\right]\frac{1}{2}\left(\frac{1-2\nu}{1-\nu}\right)\sigma_z
$$

(4.6.15)

or

$$
\gamma_{max} = \frac{\sigma_z}{E}\left[\frac{(1+\nu)(1-2\nu)}{1-\nu}\right] = \frac{\sigma_z}{E_{1d}}
$$

(4.6.16)

Remark: For elastic response during confined (uniaxial) compression, the maximum engineering shear strain is thus equal to the volumetric strain (which itself is equal to the axial strain).

EXAMPLE PROBLEM 4.7

General Remarks

This example problem reinforces that an isotropic linear elastic material idealization is defined by the values of *two* independent material parameters.

The possible parameters include the elastic modulus E, Poisson's ratio ν, the bulk modulus K, the shear modulus G, and the Lame' parameters[10] λ and μ. Any one of these parameters can be expressed in terms of two of other parameters.

Problem Statement

For an isotropic linear elastic material the relation between the Cauchy stress vector and the infinitesimal strain vector can be written in terms of the elastic modulus E and Poisson's ratio ν as

$$\sigma_x = \frac{E}{1+\nu}\left[\varepsilon_x + \left(\frac{\nu}{1-2\nu}\right)\varepsilon_{vol}\right]; \quad \sigma_y = \frac{E}{1+\nu}\left[\varepsilon_y + \left(\frac{\nu}{1-2\nu}\right)\varepsilon_{vol}\right];$$

$$\sigma_z = \frac{E}{1+\nu}\left[\varepsilon_z + \left(\frac{\nu}{1-2\nu}\right)\varepsilon_{vol}\right] \tau_{xy} = \frac{E}{2(1+\nu)}\gamma_{xy};$$

$$\tau_{xz} = \frac{E}{2(1+\nu)}\gamma_{xz}; \quad \tau_{yz} = \frac{E}{2(1+\nu)}\gamma_{yz}$$

where $\varepsilon_{vol} = \varepsilon_x + \varepsilon_y + \varepsilon_z$. The aforementioned relation between stress and strain can likewise be written in terms of the Lame' parameters λ and μ as

$$\sigma_x = \lambda\varepsilon_{vol} + 2\mu\varepsilon_x; \quad \sigma_y = \lambda\varepsilon_{vol} + 2\mu\varepsilon_y; \quad \sigma_z = \lambda\varepsilon_{vol} + 2\mu\varepsilon_z$$

$$\tau_{xy} = \mu\gamma_{xy}; \quad \tau_{xz} = \mu\gamma_{xz}; \quad \tau_{yz} = \mu\gamma_{yz}$$

a) Using the aforementioned equations, verify that

$$\lambda = \frac{\nu E}{(1+\nu)(1-2\nu)}; \quad \mu = G = \frac{E}{2(1+\nu)}; \quad K = \frac{E}{3(1-2\nu)} = \frac{3\lambda + 2\mu}{3}$$

Then, using the aforementioned relations between material parameters, express the following:

b) λ in terms of K and G.

c) ν in terms of λ and μ, and

d) E in terms of K and G.

---

10. Named in honor of Gabriel Lame' (1795–1870).

Solution

a) Comparing like terms in the following equations:

$$\sigma_x = \frac{E}{1+v}\left[\varepsilon_x + \left(\frac{v}{1-2v}\right)\varepsilon_{vol}\right]; \quad \sigma_y = \frac{E}{1+v}\left[\varepsilon_y + \left(\frac{v}{1-2v}\right)\varepsilon_{vol}\right];$$

$$\sigma_z = \frac{E}{1+v}\left[\varepsilon_z + \left(\frac{v}{1-2v}\right)\varepsilon_{vol}\right]$$

(4.7.1)

$$\sigma_x = \lambda\varepsilon_{vol} + 2\mu\varepsilon_x; \quad \sigma_y = \lambda\varepsilon_{vol} + 2\mu\varepsilon_y; \quad \sigma_z = \lambda\varepsilon_{vol} + 2\mu\varepsilon_z \qquad (4.7.2)$$

it follows that

$$\lambda = \frac{E}{1+v}\left(\frac{v}{1-2v}\right) = \frac{Ev}{(1+v)(1-2v)} \qquad (4.7.3)$$

Similarly,

$$\frac{E}{1+v} = 2\mu \Rightarrow \mu = \frac{E}{2(1+v)} \qquad (4.7.4)$$

Comparing the equations for shear stresses, i.e.,

$$\tau_{xy} = \frac{E}{2(1+v)}\gamma_{xy}; \quad \tau_{xz} = \frac{E}{2(1+v)}\gamma_{xz}; \quad \tau_{yz} = \frac{E}{2(1+v)}\gamma_{yz}$$

$$\tau_{xy} = \mu\gamma_{xy}; \quad \tau_{xz} = \mu\gamma_{xz}; \quad \tau_{yz} = \mu\gamma_{yz}$$

and recalling that

$$\tau_{xy} = G\gamma_{xy}; \quad \tau_{xz} = G\gamma_{xz}; \quad \tau_{yz} = G\gamma_{yz}$$

it follows that

$$\mu = \frac{E}{2(1+v)} = G \qquad (4.7.5)$$

For an isotropic linear elastic material, the mean stress is related to the volumetric strain as follows:

$$\frac{1}{3}(\sigma_x + \sigma_y + \sigma_z) = K\varepsilon_{vol} \qquad (4.7.6)$$

Substituting Eq. (4.7.1) for the normal stresses gives

$$\frac{1}{3}(\sigma_x + \sigma_y + \sigma_z) = \frac{1}{3}\left(\frac{E}{1+v}\right)\left[(\varepsilon_x + \varepsilon_y + \varepsilon_z) + 3\left(\frac{v}{1-2v}\right)\varepsilon_{vol}\right]$$

$$= \frac{E}{3(1+v)}\left[1 + \frac{3v}{1-2v}\right]\varepsilon_{vol} = \frac{E}{3(1+v)}\frac{(1+v)}{(1-2v)}\varepsilon_{vol}$$

$$= \frac{E}{3(1-2v)}\varepsilon_{vol}$$

(4.7.7)

Comparing Eqs. (4.7.6) and (4.7.7), it is evident that

$$K = \frac{E}{3(1 - 2\nu)} \qquad (4.7.8)$$

Returning to Eq. (4.7.6) and substituting Eq. (4.7.2) for the normal stresses gives

$$\frac{1}{3}(\sigma_x + \sigma_y + \sigma_z) = \frac{1}{3}\left[(\lambda\varepsilon_{vol} + 2\mu\varepsilon_x) + (\lambda\varepsilon_{vol} + 2\mu\varepsilon_y) + (\lambda\varepsilon_{vol} + 2\mu\varepsilon_z)\right]$$

$$(4.7.9)$$

Expanding the equation and collecting like terms gives

$$\frac{1}{3}(\sigma_x + \sigma_y + \sigma_z) = \frac{1}{3}\left[3\lambda\varepsilon_{vol} + 2\mu(\varepsilon_x + \varepsilon_y + \varepsilon_z)\right] = \frac{1}{3}(3\lambda + 2\mu)\varepsilon_{vol} \quad (4.7.10)$$

Comparing Eqs. (4.7.6) and (4.7.10), it is evident that

$$K = \frac{3\lambda + 2\mu}{3} \qquad (4.7.11)$$

b) Solving Eq. (4.7.11) for λ gives the desired result, i.e.,

$$\lambda = \frac{3K - 2\mu}{3} = K - \frac{2}{3}\mu \qquad (4.7.12)$$

c) Rewriting Eq. (4.7.3) gives

$$\lambda = \frac{E\nu}{(1 + \nu)(1 - 2\nu)} = \frac{E}{2(1 + \nu)}\frac{2\nu}{(1 - 2\nu)} = \mu\frac{2\nu}{(1 - 2\nu)} \qquad (4.7.13)$$

Solving for ν gives the desired expression, i.e.,

$$\nu = \frac{\lambda}{2(\lambda + \mu)} \qquad (4.7.14)$$

d) From Eq. (4.7.8), it follows that

$$E = 3K(1 - 2\nu) \qquad (4.7.15)$$

Solving Eq. (4.7.5) for Poisson's ratio gives

$$G = \frac{E}{2(1 + \nu)} \Rightarrow \nu = \frac{E}{2G} - 1 \qquad (4.7.16)$$

Substituting Eq. (4.7.16) into Eq. (4.7.15) gives

$$E = 3K\left[1 - 2\left(\frac{E}{2G} - 1\right)\right] = 3K - 6K\left(\frac{E}{2G} - 1\right) \qquad (4.7.17)$$

Solving for E gives the desired expression, i.e.,

$$E = \frac{9KG}{3K + G}$$

(4.7.18)

EXAMPLE PROBLEM 4.8

General Remarks

This example problem investigates a state of uniaxial compression.

Problem Statement

Draw the Mohr's circle associated with a state of uniaxial compression with the normal stress applied (a) in the x-direction and (b) in the y-direction.

Solution

a) Figure Ex. 4.8A shows an example of uniaxial compression that is imposed by normal stresses applied in the x-direction.

The state of plane stress is thus

$$\sigma_x = \sigma^*; \quad \sigma_y = 0; \quad \tau_{xy} = 0$$

(4.8.1)

The points on the circle labeled x and y correspond to the (σ, τ) stress states associated with the x and y faces of the material element, respectively. Using Eq. (4.32), the coordinates of the center of the circle are

$$C = \left(\frac{\sigma^*}{2}, 0\right)$$

(4.8.2)

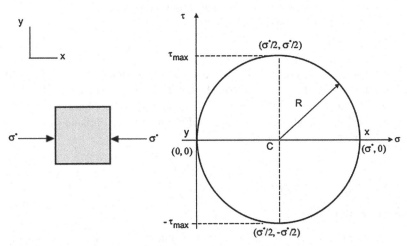

FIGURE EX. 4.8A   State of uniaxial compression applied in the x-direction and the associated Mohr's circle.

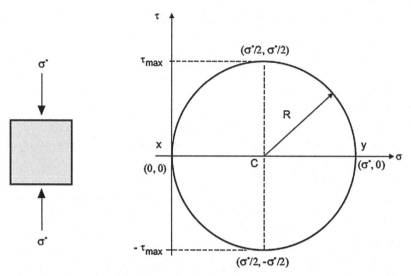

FIGURE EX. 4.8B State of uniaxial compression applied in the y-direction and the associated Mohr's circle.

The radius of the circle is computed from Eq. (4.38), giving

$$R = \frac{\sigma^*}{2} \qquad (4.8.3)$$

The magnitude of the maximum in-plane shear stress associated with this state of uniaxial stress is thus $|\tau_{max}| = R = \sigma^*/2$ (recall Eq. 4.40).

b) Figure Ex. 4.8B shows an example of uniaxial compression that is now imposed by normal stresses applied in the y-direction.

The state of plane stress is now

$$\sigma_x = 0; \quad \sigma_y = \sigma^*; \quad \tau_{xy} = 0 \qquad (4.8.4)$$

The points on the circle labeled x and y are simply interchanged with those shown in Figure Ex. 4.8A. The coordinates of the center of the circle and the magnitude of the radius are again given by Eqs. (4.8.2) and (4.8.3), respectively. Finally, the magnitude of the maximum in-plane shear stress is again $|\tau_{max}| = R = \sigma^*/2$.

EXAMPLE PROBLEM 4.9

General Remarks

This example problem investigates state of uniaxial extension.

Problem Statement

Draw the Mohr's circle associated with a state of uniaxial extension with the normal stress applied a) in the x-direction and b) in the y-direction.

Solution

a) Figure Ex. 4.9A shows an example of uniaxial extension that is imposed by normal stresses applied in the x-direction.

The state of plane stress is thus

$$\sigma_x = -\sigma^*; \quad \sigma_y = 0; \quad \tau_{xy} = 0 \tag{4.9.1}$$

The points on the circle labeled x and y correspond to the (σ, τ) stress states associated with the x and y faces of the material element, respectively. Using Eq. (4.32), the coordinates of the center of the circle are

$$C = \left(-\frac{\sigma^*}{2}, 0\right) \tag{4.9.2}$$

The radius of the circle is computed from Eq. (4.38), giving

$$R = \frac{\sigma^*}{2} \tag{4.9.3}$$

The magnitude of the maximum in-plane shear stress associated with this state of uniaxial stress is thus $|\tau_{max}| = R = \sigma^*/2$ [recall Eq. 4.40].

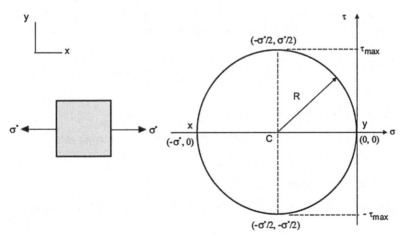

FIGURE EX. 4.9A State of uniaxial extension applied in the x-direction and the associated Mohr's circle.

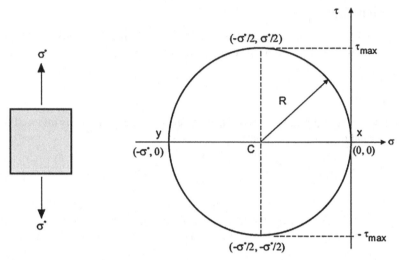

FIGURE EX. 4.9B State of uniaxial extension applied in the y-direction and the associated Mohr's circle.

b) Figure Ex. 4.9B shows an example of uniaxial extension that is now imposed by normal stresses applied in the y-direction.

The state of plane stress is now

$$\sigma_x = 0; \quad \sigma_y = -\sigma^*; \quad \tau_{xy} = 0 \tag{4.9.4}$$

The points on the circle labeled x and y are simply interchanged as compared to those shown in Figure Ex. 4.9A. The coordinates of the center of the circle and the magnitude of the radius are again given by Eqs. (4.9.2) and (4.9.3), respectively. Finally, the maximum in-plane shear stress is again $|\tau_{max}| = R = |\sigma^*|/2$.

EXAMPLE PROBLEM 4.10

General Remarks

This example problem investigates a state of hydrostatic compression.

Problem Statement

Draw the Mohr's circle associated with a state of hydrostatic compression.

FIGURE EX. 4.10   State of hydrostatic compression and the associated Mohr's circle.

Solution

Figure Ex. 4.10 shows an example of hydrostatic compression applied to a material element.

The state of plane stress is thus

$$\sigma_x = \sigma_y = \sigma^*; \quad \tau_{xy} = 0 \tag{4.10.1}$$

Using Eq. (4.32), the coordinates of the center of the circle are

$$C = (\sigma^*, 0) \tag{4.10.2}$$

The radius of the circle, computed using Eq. (4.38), is $R = 0$. The Mohr's circle thus *reduces to a point* (Figure Ex. 4.10).

EXAMPLE PROBLEM 4.11

General Remarks

This example problem investigates three states of pure shear.

Problem Statement

Draw the Mohr's circle associated with a state of pure shear with (a) a positive shear stress applied along the face having an outward normal in the x-direction, (b) a negative shear stress applied along the face having an outward normal in the x-direction, and (c) normal stresses of opposite sign applied in the x and y coordinate directions.

Solution

a) Figure Ex. 4.11A shows an example of pure shear that is imposed by a positive shear stress applied along the face having an outward normal in the x-direction.

The state of plane stress with respect to the face having an outward normal in the x-direction is thus

$$\sigma_x = \sigma_y = 0; \quad \tau_{xy} = \tau^* \qquad (4.11.1)$$

The points on the circle labeled x and y correspond to the (σ,τ) stress states associated with the x and y faces of the material element, respectively. Using Eq. (4.32), the coordinates of the center of the circle are (0,0). The radius of the circle is computed from Eq. (4.38), giving $R = \tau^*$. The magnitude of the maximum in-plane shear stress associated with this state of stress is thus $| \tau_{\max} | = R = | \tau^* |$.

b) Figure Ex. 4.11B shows an example of pure shear that is imposed by a negative shear stress applied along the face having an outward normal in the x-direction.

The state of plane stress with respect to the face having an outward normal in the x-direction is thus

$$\sigma_x = \sigma_y = 0; \quad \tau_{xy} = -\tau^* \qquad (4.11.2)$$

The points on the circle labeled x and y are simply interchanged as compared to those shown in Figure Ex. 4.11A. The coordinates of the center

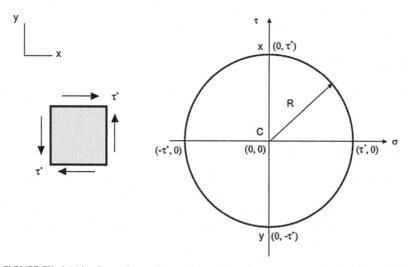

FIGURE EX. 4.11A State of pure shear with a positive shear stress on x-direction face and the associated Mohr's circle.

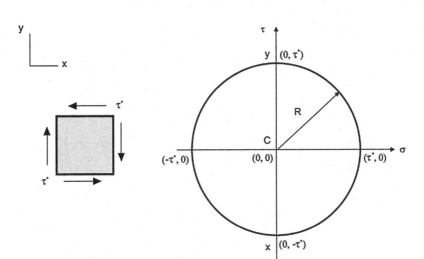

FIGURE EX. 4.11B State of pure shear with a negative shear stress on x-direction face and the associated Mohr's circle.

of the circle are again found to be (0,0) and the radius of the circle is again $R = \tau^*$. The magnitude of the maximum in-plane shear stress associated with this state of pure shear stress is thus again $|\tau_{max}| = R = |\tau^*|$.

c) Figure Ex. 4.11C shows an example of pure shear that is created by normal stresses applied in the x and y directions that are equal in magnitude but opposite in sign.

The state of plane stress with respect to the face having an outward normal in the x-direction is thus

$$\sigma_x = -\sigma_y = \sigma^*; \quad \tau_{xy} = 0 \qquad (4.11.3)$$

The points on the circle labeled x and y correspond to the (σ, τ) stress states associated with the x and y faces of the material element, respectively. Using Eq. (4.32), the coordinates of the center of the circle are again found to be (0,0). The radius of the circle is again computed from Eq. (4.38), giving $R = \sigma^*$. The magnitude of the maximum in-plane shear stress associated with this state of stress is thus $|\tau_{max}| = R = |\sigma^*|$.

EXAMPLE PROBLEM 4.12

General Remarks

This example problem investigates the use of the stress transformation equations presented in Section 4.2.5.

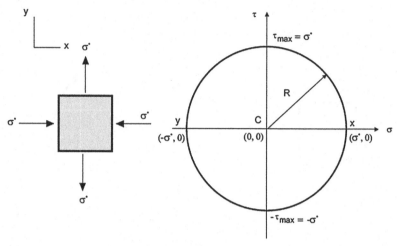

FIGURE EX. 4.11C State of pure shear created by normal stresses and the associated Mohr's circle.

Problem Statement

Given a major principal stress of 7600 psi, determine the minor principal stress such that the shear stress is limited to 3150 psi.

Solution

The maximum in-plane shear stress is given by Eq. (4.40), i.e.,

$$\tau_{max} = R = \sqrt{\left[\frac{1}{2}(\sigma_x - \sigma_y)\right]^2 + (\tau_{xy})^2} \qquad (4.12.1)$$

Recalling that the shear stress associated with a state of principal stress is zero, Eq. (4.12.1) gives

$$\tau_{max} = \sqrt{\left[\frac{1}{2}(7600 - \sigma_y)\right]^2 + (0.0)^2} = 3150 \text{ psi} \qquad (4.12.2)$$

Solving for the minor principal stress gives the desired result

$$3150 = \frac{1}{2}(7600 - \sigma_y) \Rightarrow \sigma_y = 7600 - 2(3150) = \mathbf{1300 \text{ psi}} \qquad (4.12.3)$$

EXAMPLE PROBLEM 4.13

General Remarks

This example problem illustrates the manner in which major and minor principal stresses are computed.

Problem Statement

Figure Ex. 4.13A shows the state of plane stress at a point in a body. Determine the principal stresses and the orientation of the planes on which they act.

Solution

In light of the sign convention used in analyzing soil mechanics problems, the state of stress shown in Figure Ex. 4.13A is thus

$$\sigma_x = 0.0; \quad \sigma_y = 80.0 \text{ MPa}; \quad \tau_{xy} = 20.0 \text{ MPa} \tag{4.13.1}$$

The σ-coordinate of the center of the Mohr's circle of stress associated with this problem is computed using Eq. (4.32), i.e.,

$$C = \frac{1}{2}(\sigma_x + \sigma_y) = \frac{1}{2}(0.0 + 80.0) = 40.0 \text{ MPa} \tag{4.13.2}$$

The radius of the Mohr's circle is next computed using Eq. (4.38)

$$R = \sqrt{\left[\frac{1}{2}(\sigma_x - \sigma_y)\right]^2 + (\tau_{xy})^2} = \sqrt{\left[\frac{1}{2}(0.0 - 80.0)\right]^2 + (20.0)^2} = 44.72 \text{ MPa} \tag{4.13.3}$$

The major and minor principal stresses are computed using Eq. (4.39), i.e.,

$$\sigma_1 = C + R = 40.0 + 44.72 = \textbf{84.72 MPa} \tag{4.13.4}$$

$$\sigma_2 = C - R = 40.0 - 44.72 = \textbf{-4.72 MPa} \tag{4.13.5}$$

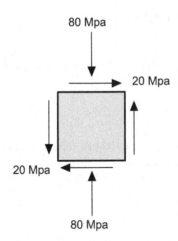

FIGURE EX. 4.13A State of plane stress considered in this example problem.

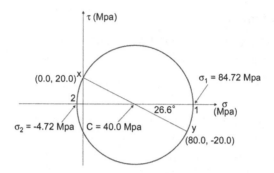

FIGURE EX. 4.13B Mohr's circle associated with this example problem (not to scale).

Figure Ex. 4.13B shows the Mohr's circle associated with this problem. The points on the circle labeled x and y correspond to the (σ, τ) stress states associated with the x and y faces of the material element, respectively. The points on the circle labeled 1 and 2 correspond to the major and minor principal stresses.

The angle between the original y-direction and the principal 1-direction is computed as follows:

$$\tan 2\theta_p = \frac{20}{(80.0 - 40.0)} = 0.50 \Rightarrow 2\theta_p = 26.6° \Rightarrow \theta_p = 13.3° \qquad (4.13.6)$$

Figure Ex. 4.13B shows the aforesaid double angle. Figure Ex. 4.13C shows the major and minor principal stresses and the orientation of the principal directions relative to the original $(x\text{-}y)$ directions.

EXAMPLE PROBLEM 4.14

General Remarks

This example problem computes the maximum in-plane shear stress from the major and minor principal stresses.

Problem Statement

Given a major principal stress of 400 kPa (compressive) and a minor principal stress of 100 kPa (compressive), determine the maximum in-plane shear stress and the orientation of the plane on which it acts.

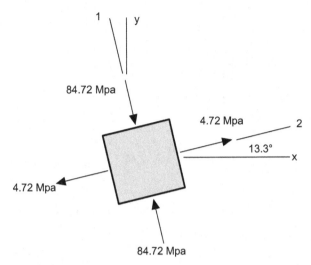

FIGURE EX. 4.13C   State of plane stress acting on principal planes.

Solution

Recalling Eq. (4.40), the maximum in-plane shear stress is equal to the radius of the Mohr's circle. Using Eq. (4.38) in conjunction with Eq. (4.40) gives

$$\tau_{max} = R = \sqrt{\left[\frac{1}{2}(\sigma_x - \sigma_y)\right]^2 + (\tau_{xy})^2} \qquad (4.14.1)$$

Substituting the given values of $\sigma_x = \sigma_1 = 400.0$ kPa, $\sigma_y = \sigma_2 = 100.0$ kPa, and $\tau_{xy} = 0.0$ into Eq. (4.12.1) gives

$$\tau_{max} = \sqrt{\left[\frac{1}{2}(400.0 - 100.0)\right]^2 + (0.0)^2} = \mathbf{150.0\ kPa} \qquad (4.14.2)$$

As shown in Figure 4.16, the maximum in-plane shear stress is oriented $2\theta = 90$ degrees from the major principal stress direction. The orientation at the point in the actual material is thus $\theta = 45$ degrees from the major principal stress direction.

The magnitude of the normal stress acting on the x and y planes associated with the maximum in-plane shear stress is

$$\sigma_x = \sigma_y = \frac{1}{2}(400.0 + 100.0) = \mathbf{250.0\ kPa} \qquad (4.14.3)$$

which is the normal stress corresponding to the center of the circle.

Figure Ex. 4.14A shows the original orientation and the one associated with the maximum in-plane shear stress.

Figure Ex. 4.14B shows the Mohr's circle associated with the problem.

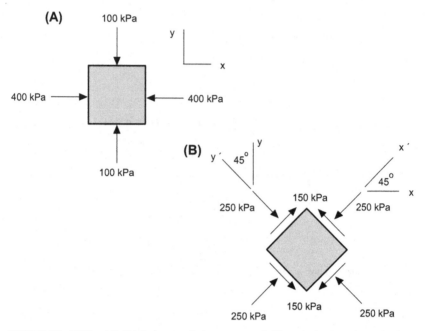

FIGURE EX. 4.14A (A) Original state of plane stress and (B) equivalent maximum in-plane shear stress.

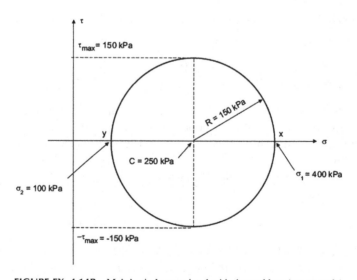

FIGURE EX. 4.14B Mohr's circle associated with the problem (not to scale).

EXAMPLE PROBLEM 4.15

General Remarks

This example problem illustrates the manner in which major and minor principal stresses are computed.

Problem Statement

Determine the principal stresses and the orientation of the principal axes of stress for the state of plane stress shown in Figure Ex. 4.15A.

Solution

In light of the sign convention used in analyzing soil mechanics problems, the state of stress shown in Figure Ex. 4.15A is

$$\sigma_x = 10,000 \text{ psi}; \quad \sigma_y = -20,000 \text{ psi}; \quad \tau_{xy} = -6000 \text{ psi} \qquad (4.15.1)$$

The center and radius of the Mohr's circle are computed using Eqs. (4.32) and (4.38), respectively, giving

$$C = \frac{1}{2}(10,000.0 - 20,000.0) = -5000 \text{ psi} \qquad (4.15.2)$$

$$
\begin{aligned}
R &= \sqrt{\left[\frac{1}{2}(\sigma_x - \sigma_y)\right]^2 + (\tau_{xy})^2} \\
&= \sqrt{\left[\frac{1}{2}(10,000 - (-20,000))\right]^2 + (-6000)^2} = 16,155 \text{ psi}
\end{aligned}
\qquad (4.15.3)
$$

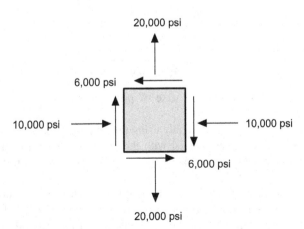

20,000 psi

6,000 psi

10,000 psi

10,000 psi

6,000 psi

20,000 psi

FIGURE EX. 4.15A   State of plane stress considered in this example problem.

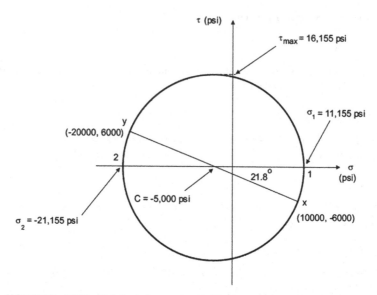

FIGURE EX. 4.15B   Mohr's circle associated with this example problem (not to scale).

The major and minor principal stresses are thus

$$\sigma_1 = C + R = -5000 + 16,155 = \mathbf{11,155 \text{ psi}} \qquad (4.15.4)$$

$$\sigma_2 = C - R = -5000 - 16,155 = \mathbf{-21,155 \text{ psi}} \qquad (4.15.5)$$

Figure Ex. 4.15B shows the Mohr's circle associated with the plane stress state shown in Figure Ex. 4.15A.

The orientation of the principal axes of stress is determined from the geometry shown in Figure Ex. 4.15B. In particular,

$$\tan 2\theta_p = \frac{6000}{10,000 - (-5000)} = \frac{6}{15} \Rightarrow 2\theta_p = 21.8° \qquad (4.15.6)$$

Thus,

$$\theta_p = \mathbf{10.9°} \qquad (4.15.7)$$

Figure Ex. 4.15C shows the state of principal stress and the associated orientation.

The solution is next repeated only using the pole method. Figure Ex. 4.15D shows the Mohr's circle associated with the problem. Beginning at the point labeled x, a line parallel to this plane (i.e., vertical) is drawn until it intersects the circle at the pole P.

The orientation associated with the major principal stress is obtained by drawing a line from point P through the point on the circle labeled 1. This line is oriented 10.9°, measured counterclockwise, from the vertical. The minor principal stress is oriented 90° from the major one (Figure Ex. 4.15C).

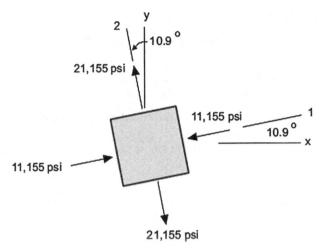

FIGURE EX. 4.15C State of plane stress acting on principal planes.

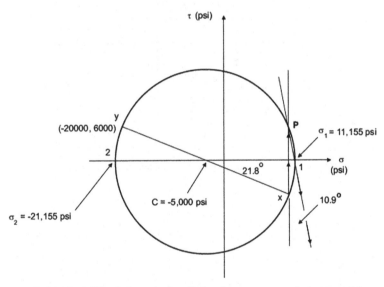

FIGURE EX. 4.15D Pole method applied to the present example (not to scale).

EXAMPLE PROBLEM 4.16

General Remarks

This example problem illustrates the manner in which stress components acting on an inclined plane, as well as the principal stresses and the maximum in-plane shear stress are computed.

Problem Statement

Given the state of plane stress shown in Figure Ex. 4.16A, compute the following:

a) The stresses acting on an element that is rotated 20° counterclockwise from the x-y axes as shown in Figure Ex. 4.16B.

b) The major and minor principal stresses and the orientation of the principal stress directions, and

c) The stresses acting on an element whose faces are aligned with the planes of maximum shear stress.

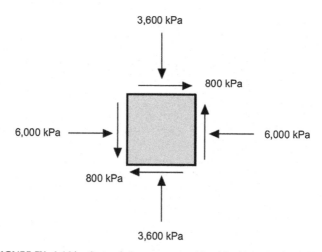

FIGURE EX. 4.16A State of plane stress considered in this example problem.

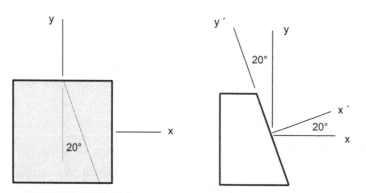

FIGURE EX. 4.16B Orientation of element face for stress transformation.

Solution

In light of the sign convention used in analyzing soil mechanics problems, the state of stress shown in Figure Ex. 4.16A is

$$\sigma_x = 6000 \text{ kPa}; \quad \sigma_y = 3600 \text{ kPa}; \quad \tau_{xy} = 800 \text{ kPa} \qquad (4.16.1)$$

a) Using the stress transformation Eqs. (4.10), (4.16), and (4.13) for the counterclockwise double angle of $2\theta = 40$ degrees gives

$$\sigma_{x'} = \frac{1}{2}(\sigma_x + \sigma_y) + \frac{1}{2}(\sigma_x - \sigma_y)\cos 2\theta - \tau_{xy} \sin 2\theta$$

$$= \frac{1}{2}(6000 + 3600) + \frac{1}{2}(6000 - 3600)\cos 40° - (800)\sin 40°$$

$$= \textbf{5205 kPa}$$

$$(4.16.2)$$

$$\sigma_{y'} = \frac{1}{2}(\sigma_x + \sigma_y) - \frac{1}{2}(\sigma_x - \sigma_y)\cos 2\theta + \tau_{xy} \sin 2\theta$$

$$= \frac{1}{2}(6000 + 3600) - \frac{1}{2}(6000 - 3600)\cos 40° + (800)\sin 40°$$

$$= \textbf{4395 kPa}$$

$$(4.16.3)$$

$$\tau_{x'y'} = \frac{1}{2}(\sigma_x - \sigma_y)\sin 2\theta + \tau_{xy} \cos 2\theta$$

$$= \frac{1}{2}(6000 - 3600)\sin 40° + (800)\cos 40° \qquad (4.16.4)$$

$$= \textbf{1384 kPa}$$

Figure Ex. 4.16C shows the stresses acting on an element that is rotated 20° counterclockwise from the x-y axes.

b) The center of Mohr's circle is computed using Eq. (4.32), i.e.,

$$C = \frac{1}{2}(\sigma_x + \sigma_y) = \frac{1}{2}(6000 + 3600) = 4800 \text{ kPa} \qquad (4.16.5)$$

Next, the radius of Mohr's circle is computed using Eq. (4.38), i.e.,

$$R = \sqrt{\left(\frac{\sigma_x - \sigma_y}{2}\right)^2 + (\tau_{xy})^2} = \sqrt{\left(\frac{6000 - 3600}{2}\right)^2 + (800)^2} = 1442 \text{ kPa}$$

$$(4.16.6)$$

The major and minor principal stresses are then computed using Eq. (4.39)

$$\sigma_1 = C + R = 4800 + 1442 = \textbf{6242 kPa} \qquad (4.16.7)$$

$$\sigma_2 = C - R = 4800 - 1442 = \textbf{3358 kPa} \qquad (4.16.8)$$

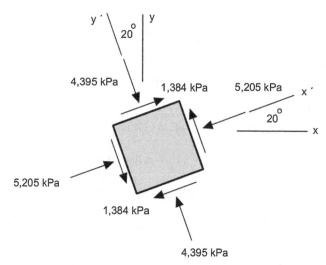

FIGURE EX. 4.16C   State of plane stress on an inclined plane.

Figure Ex. 4.16D shows the Mohr's circle associated with this problem. The points on the circle associated with the original state of plane stress are denoted by x and y. The points x' and y' are associated with the rotated stress state considered in part (a).

The major principal stress direction, denoted by the symbol "1", is located where the circle intersects the σ-axis. The *clockwise* angle between

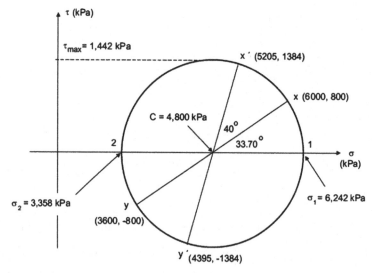

FIGURE EX. 4.16D   Mohr's circle associated with this example problem (not to scale).

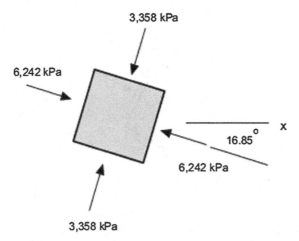

3,358 kPa

6,242 kPa

6,242 kPa

16.85°

x

3,358 kPa

FIGURE EX. 4.16E Principal stresses acting on a material element.

the original stress state (point x) and the major principal direction (point 1) is computed as follows:

$$\tan 2\theta_{p_1} = \frac{800}{(6000 - 4800)} = \frac{2}{3} \Rightarrow 2\theta_{p_1} = 33.70° \Rightarrow \theta_{p_1} = 16.85° \quad (4.16.9)$$

Figure Ex. 4.16E shows the principal stresses acting on a material element and the orientation of the major principal direction relative to the positive x-axis.

c) The maximum in-plane shear stress is computed from Eq. (4.40), i.e.,

$$\tau_{max} = R = \textbf{1442 kPa} \quad (4.16.10)$$

The associated normal stresses are

$$\sigma_x = \sigma_y = C = \textbf{4800 kPa} \quad (4.16.11)$$

The state of maximum in-plane shear stress is oriented 45° (counterclockwise) from the major principal direction (Figure Ex. 4.16D). This stress state is thus oriented $45 - 16.85 = 28.15°$ from the positive x-direction. Figure Ex. 4.16F shows this orientation, as well as the stresses acting on an element whose faces are aligned with the planes of maximum shear stress.

Parts (a)–(c) are next repeated only using the pole method. Figure Ex. 4.16G shows the Mohr's circle associated with the problem. Beginning at the point labeled x, a line parallel to this plane (i.e., vertically) is drawn until it intersects the circle at the pole P.

To determine the stresses on an element that is rotated 20° counterclockwise from the x-y axes, a line inclined 20° to the vertical is drawn. The point labeled x' denotes its intersection with the circle.

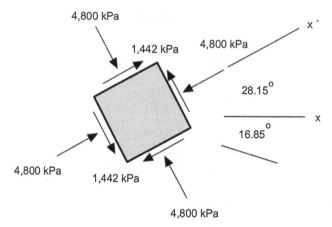

FIGURE EX. 4.16F Stresses acting on an element whose faces are aligned with the planes of maximum shear stress.

The orientation associated with the major principal stress is next obtained by drawing a line from point P through the point on the circle labeled 1. This line is oriented 16.85°, measured clockwise, from the vertical. The minor principal stress is oriented 90° from the major one (Figure Ex. 4.16E).

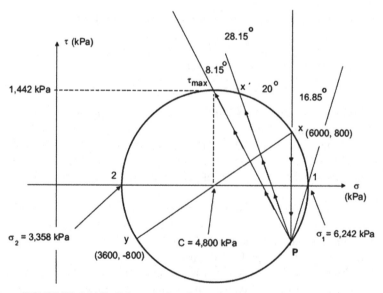

FIGURE EX. 4.16G Pole method applied to the present example (not to scale).

Finally, the orientation associated with the maximum in-plane shear stress is obtained by drawing a line from point P through the point on the circle labeled τ_{max}. Since the plane associated with the maximum shear stress is oriented 45° from the principal stress direction, this line is oriented 45°, measured counterclockwise, from the line through point 1 (recall Figure Ex. 4.16F). This line is thus oriented 28.15°, measured counterclockwise, from the vertical.

EXAMPLE PROBLEM 4.17

General Remarks

This example problem investigates the transformation of stresses using the equations presented in Section 4.2.5.

Problem Statement

Consider the state of stress shown in Figure Ex. 4.17.

The material will fail at a tensile stress of 12,000 psi acting in *any* direction.

a) Will the material fail in tension due to the stresses shown in Figure Ex. 4.17?

b) What is the maximum in-plane shear stress associated with the stress state shown in Figure Ex. 4.17?

Solution

For the stress state shown in Figure Ex. 4.17, $\sigma_x = -6450$ psi, $\sigma_y = 18,600$ psi, and $\tau_{xy} = -3850$ psi.

a) To answer the question posed, compute the principal stresses

$$C = \frac{1}{2}(\sigma_x + \sigma_y) = \frac{1}{2}(-6450 + 18,600) = 6075 \text{ psi} \qquad (4.17.1)$$

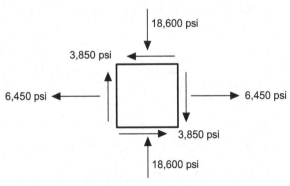

FIGURE EX. 4.17   State of plane stress at a point.

$$R = \sqrt{\left[\frac{1}{2}(\sigma_x - \sigma_y)\right]^2 + (\tau_{xy})^2} = \sqrt{\left[\frac{1}{2}(-6450 - 18{,}600)\right]^2 + (-3850)^2}$$

$$= 13{,}103 \text{ psi}$$

(4.17.2)

Thus, using Eq. (4.39) gives

$$\sigma_1 = C + R = 6075 + 13{,}103 = 19{,}178 \text{ psi} \qquad (4.17.3)$$

$$\sigma_2 = C - R = 6075 - 13{,}103 = -7028 \text{ psi} \qquad (4.17.4)$$

Since $\sigma_2 = -7028$ psi is the smallest tensile stress, and since -7028 psi $< -12{,}000$ psi, **the material will not fail**.

b) Recalling Eq. (4.40), the maximum in-plane shear stress is

$$\tau_{\max} = R = \textbf{13{,}103 psi} \qquad (4.17.5)$$

At this point

$$\sigma_x = \sigma_y = C = 6075 \text{ psi} \qquad (4.17.6)$$

EXAMPLE PROBLEM 4.18

General Remarks

This example problem investigates the general relationship between the stress states on two planes at a point using a Mohr's circle approach.

Problem Statement

Figure Ex. 4.18 shows the stresses states acting on two planes at a point in a body. The normal and shear stresses acting on these planes are given subscripts A and B. Relate the center (C) and radius (R) of the Mohr's circle, as well as the angles shown in Figure Ex. 4.18, to the stresses acting on the two planes.

Solution

To relate the given stresses to the center (C) of the Mohr's circle, use the general equation for the radius (R), i.e.,

$$R = \sqrt{(\sigma_A - C)^2 + (\tau_A)^2} = \sqrt{(C - \sigma_B)^2 + (\tau_B)^2} \qquad (4.18.1)$$

Squaring both sides of the equation and solving for C gives

$$C = \frac{(\sigma_A)^2 - (\sigma_B)^2 + (\tau_A)^2 - (\tau_B)^2}{2(\sigma_A - \sigma_B)} \qquad (4.18.2)$$

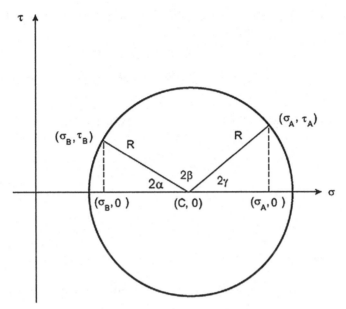

FIGURE EX. 4.18 Normal and shear stresses acting on two planes at a point.

Once C is known, R follows from Eq. (4.18.1). The major and minor principal stresses are then computed using Eq. (4.39), i.e.,

$$\sigma_1 = C + R; \quad \sigma_3 = C - R \qquad (4.18.3)$$

The angles shown in Figure Ex. 4.18 are computed as follows:

$$\tan 2\gamma = \frac{\tau_A}{\sigma_A - C}; \quad \tan 2\alpha = \frac{\tau_B}{C - \sigma_B} \qquad (4.18.4)$$

The angle between the two planes is thus

$$2\beta = 180° - 2\alpha - 2\gamma \Rightarrow \beta = 90° - (\alpha + \gamma) \qquad (4.18.5)$$

Next consider the following specific example: the normal and shear stresses on one plane are 16.0 and 4.0; on the second plane, the normal and shear stresses are 1.0 and 3.0. Thus, $\sigma_A = 16.0$, $\tau_A = 4.0$, $\sigma_B = 1.0$, and $\tau_B = 3.0$. The σ-coordinate of the center of the Mohr's circle is thus

$$C = \frac{(\sigma_A)^2 - (\sigma_B)^2 + (\tau_A)^2 - (\tau_B)^2}{2(\sigma_A - \sigma_B)} = \frac{(16.0)^2 - (1.0)^2 + (4.0)^2 - (3.0)^2}{2(16.0 - 1.0)} = 8.733$$

$$(4.18.6)$$

The radius of the Mohr's circle is next computed using Eq. (4.18.1), i.e.,

$$R = \sqrt{(\sigma_A - C)^2 + (\tau_A)^2} = \sqrt{(16.0 - 8.733)^2 + (4.0)^2} = 8.295 \quad (4.18.7)$$

Substituting Eqs. (4.18.6) and (4.18.7) into Eq. (4.18.3) gives the following the major and minor principal stresses:

$$\sigma_1 = C + R = 8.733 + 8.295 = 17.03 \qquad (4.18.8)$$

$$\sigma_3 = C - R = 8.733 - 8.295 = 0.44 \qquad (4.18.9)$$

The angles shown in Figure Ex. 4.18 are next computed

$$\tan 2\gamma = \frac{\tau_A}{\sigma_A - C} = \frac{4.0}{16.0 - 8.733} = 0.550 \Rightarrow 2\gamma = 28.83° \Rightarrow \gamma = 14.42 \qquad (4.18.10)$$

$$\tan 2\alpha = \frac{\tau_B}{C - \sigma_B} = \frac{3.0}{8.733 - 1.0} = 0.388 \Rightarrow 2\alpha = 21.20° \Rightarrow \alpha = 10.60° \qquad (4.18.11)$$

Thus,

$$\beta = 90° - (\alpha + \gamma) = 90° - (14.42° + 10.60°) = 64.98° \qquad (4.18.12)$$

EXAMPLE PROBLEM 4.19

General Remarks

This example problem illustrates the manner in which the angle associated with a stress transformation is computed from given stresses in the original and rotated configurations.

Problem Statement

Given the original state of plane stress shown in Figure Ex. 4.19A and the equivalent transformed stress state shown in Figure Ex. 4.19B, compute (1) the magnitude of the counterclockwise angle θ and (2) the value of σ_b.

Solution

a) Since σ_a is known, Eq. (4.10) is used to solve for θ. In particular,

$$\sigma_a = \frac{1}{2}(\sigma_x + \sigma_y) + \frac{1}{2}(\sigma_x - \sigma_y)\cos 2\theta - \tau_{xy} \sin 2\theta$$

$$-48.4 = \frac{1}{2}(-100.0 - 60.0) + \frac{1}{2}(-100.0 + 60.0)\cos 2\theta - (-48.0)\sin 2\theta \qquad (4.19.1)$$

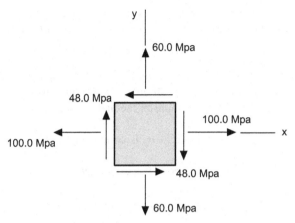

FIGURE EX. 4.19A   State of plane stress at a point.

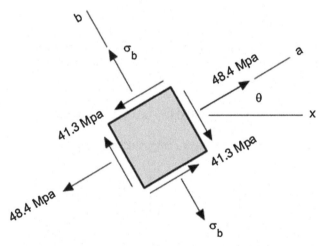

FIGURE EX. 4.19B   Equivalent state of plane stress on inclined planes.

This leads to the following homogeneous equation in 2θ:

$$48.0 \sin 2\theta - 20.0 \cos 2\theta - 31.60 = 0 \qquad (4.19.2)$$

A trial-and-error solution can be used to find the solution (root) of Eq. (4.19.2). Alternately, the solution can be obtained using an iterative root finding algorithm. For example, using the secant method[11] in conjunction with an error tolerance of 0.0001 gives, after five iterations,

$$2\theta = 1.0479 \text{ rad} = 60.0° \Rightarrow \theta = \mathbf{30.0°} \qquad (4.19.3)$$

This same result is obtained using other root finding approaches such as the bisection method (in 19 iterations), the Regula-Falsi method (in 8 iterations), and the Newton–Raphson method (in 4 iterations).

b) With θ known, the magnitude of σ_b is computed using Eq. (4.16), i.e.,

$$\sigma_b = \frac{1}{2}(\sigma_x + \sigma_y) - \frac{1}{2}(\sigma_x - \sigma_y)\cos 2\theta + \tau_{xy} \sin 2\theta$$

$$-48.4 = \frac{1}{2}(-100.0 - 60.0) - \frac{1}{2}(-100.0 + 60.0)\cos 60° + (-48.0)\sin 60°$$

$$= \mathbf{-111.6 \text{ MPa}}$$

$$(4.19.4)$$

Alternate Solution

The angle computed in Eq. (4.19.3) can likewise be determined using Mohr's circle and a trigonometry-based solution. The σ-coordinate of the center of the circle is computed using Eq. (4.32), i.e.,

$$C = \frac{1}{2}(\sigma_x + \sigma_y) = \frac{1}{2}(-100.0 - 60.0) = -80.0 \text{ MPa} \qquad (4.19.5)$$

The radius of the circle is next computed using Eq. (4.38), i.e.,

$$R = \sqrt{\left[\left(\frac{\sigma_x - \sigma_y}{2}\right)\right]^2 + (\tau_{xy})^2} = \sqrt{\left[\left(\frac{-100.0 + 60.0}{2}\right)\right]^2 + (-48.0)^2}$$

$$= -132.0 \text{ MPa}$$

$$(4.19.6)$$

Figure Ex. 4.19C shows the Mohr's circle associated with this problem. An equation for the unknown angle θ is now developed from the trigonometry of the circle.

That is, the three double angles shown in Figure Ex. 4.19C are related as follows:

$$2\alpha + 2\beta + 2\theta = 180° \Rightarrow \theta = 90° - \alpha - \beta \qquad (4.19.7)$$

---

11. Burden, R.L., Faires, J.D., Reynolds, A.C., 1980. Numerical Analysis. PWS Publishing, Boston, MA.

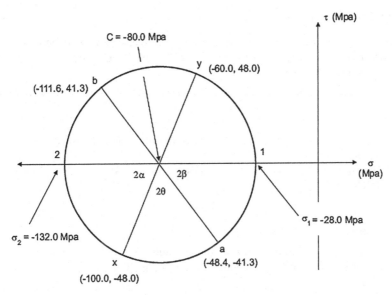

FIGURE EX. 4.19C Mohr's circle associated with this example problem (not to scale).

From the geometry of the Mohr's circle,

$$\tan 2\alpha = \frac{-48.0}{(-100.0 + 80.0)} = 2.400 \Rightarrow 2\alpha = 67.38° \Rightarrow \alpha = 33.69° \quad (4.19.8)$$

$$\tan 2\beta = \frac{-41.3}{(-80.0 + 48.4)} = 1.307 \Rightarrow 2\beta = 52.58° \Rightarrow \beta = 26.29° \quad (4.19.9)$$

Substituting for α and β into Eq. (4.19.7) gives

$$\theta = 90° - \alpha - \beta = 90° - 33.69° - 26.29° = \mathbf{30.0°} \quad (4.19.10)$$

which is *identical* to the solution obtained in Eq. (4.19.3).

Remark: The stress transformation equations are quite useful in many instances, but not all. All stress transformation problems can be solved using relations developed from the trigonometry of a *specific* Mohr's circle. As this example problem has shown, such a trigonometric-based approach is sometimes preferred.

EXAMPLE PROBLEM 4.20

This example problem illustrates the manner in which strain components acting on an inclined plane, as well as the principal strains and the maximum in-plane shear strain are computed.

Problem Statement

An element of a material is subjected to a state of plane strain with the following components: $\varepsilon_x = -220 \times 10^{-6}$, $\varepsilon_y = -480 \times 10^{-6}$, and $\gamma_{xy} = 180 \times 10^{-6}$. Compute the following:

a) The strains acting on an element that is rotated 55° counterclockwise from the x-y axes.

b) The major and minor principal strains and the orientation of the principal strain directions, and

c) The strains acting on an element whose faces are aligned with the planes of maximum shear strain.

Solution

a) Using Eqs. (4.49)–(4.51) for a double angle of $2\theta = 110$ degrees gives the desired strain components acting on the inclined element:

$$
\begin{aligned}
\varepsilon_{x'} &= \frac{1}{2}(\varepsilon_x + \varepsilon_y) + \frac{1}{2}(\varepsilon_x - \varepsilon_y)\cos 2\theta - \frac{\gamma_{xy}}{2}\sin 2\theta \\
&= \frac{1}{2}[(-220 - 480) + (-220 + 480)\cos 110° - (180)\sin 110°] \times 10^{-6} \\
&= -479.0 \times 10^{-6}
\end{aligned}
$$

$$(4.20.1)$$

$$
\begin{aligned}
\varepsilon_{y'} &= \frac{1}{2}(\varepsilon_x + \varepsilon_y) - \frac{1}{2}(\varepsilon_x - \varepsilon_y)\cos 2\theta + \frac{\gamma_{xy}}{2}\sin 2\theta \\
&= \frac{1}{2}[(-220 - 480) - (-220 + 480)\cos 110° + (180)\sin 110°] \times 10^{-6} \\
&= -221.0 \times 10^{-6}
\end{aligned}
$$

$$(4.20.2)$$

$$
\begin{aligned}
\gamma_{x'y'} &= (\varepsilon_x - \varepsilon_y)\sin 2\theta + \gamma_{xy}\cos 2\theta \\
&= [(-220 + 480)\sin 110° + (180)\cos 110°] \times 10^{-6} = 182.8 \times 10^{-6}
\end{aligned}
$$

$$(4.20.3)$$

b) The ε-coordinate of the center of the Mohr's circle of strain is computed using Eq. (4.53), i.e.,

$$
C = \frac{1}{2}(\varepsilon_x + \varepsilon_y) = \frac{1}{2}(-220 - 480) \times 10^{-6} = -350.0 \times 10^{-6} \text{ mm/mm}
$$

$$(4.20.4)$$

Using Eq. (4.54), the radius of the Mohr's circle of strain is next computed

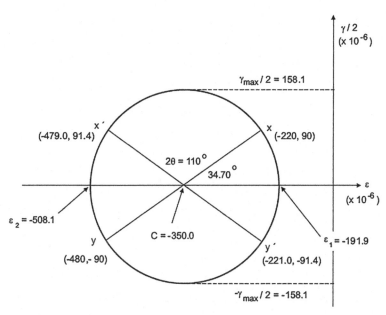

FIGURE EX. 4.20   Mohr's circle associated with this example problem (not to scale).

$$R = \sqrt{\left(\frac{\varepsilon_x - \varepsilon_y}{2}\right)^2 + \left(\frac{\gamma_{xy}}{2}\right)^2} = \sqrt{\left(\frac{-220 + 480}{2}\right)^2 + \left(\frac{180}{2}\right)^2} \times 10^{-6}$$

$$= 158.1 \times 10^{-6} \text{ mm/mm}$$

(4.20.5)

The principal strains are then computed using Eq. (4.52), i.e.,

$$\varepsilon_1 = C + R = (-350.0 + 158.1) \times 10^{-6} = -191.9 \times 10^{-6} \text{ mm/mm}$$

(4.20.6)

$$\varepsilon_2 = C - R = (-350.0 - 158.1) \times 10^{-6} = -508.1 \times 10^{-6} \text{ mm/mm}$$

(4.20.7)

Figure Ex. 4.20 shows the Mohr's circle associated with this problem. The principal direction is next computed as follows:

$$\tan 2\theta_p = \frac{90}{350 - 220} = 0.692 \Rightarrow 2\theta_p = 34.70° \Rightarrow \theta_p = 17.75° \quad (4.20.8)$$

c) The strains acting on an element whose faces are aligned with the planes of maximum shear strain consist of the maximum in-plane shear strain, which is computed from Eq. (4.56) as

$$\frac{1}{2}\gamma_{max} = R \Rightarrow \gamma_{max} = 2R = 2(158.1 \times 10^{-6}) = 316.3 \times 10^{-6} \text{ rad} \quad (4.20.9)$$

Since the ordinate in Mohr's circle is one half of the engineering shear strain, it follows that $\gamma_{max}/2 = 158.1$ rad is shown in Figure Ex. 4.20.

Chapter 5

Example Problems Involving In Situ Stresses Under Hydrostatic Conditions

5.0 GENERAL COMMENTS

The fluid in soils and porous rocks can be present in two forms, namely, as follows:

- Free water occupying part or all of the voids between the particles, and
- Adsorbed water films surrounding clay particles.

If the voids are completely filled with water, the material is *saturated* (i.e., $S = 100\%$) and the moisture is said to be continuous. If, on the other hand, the voids are only partially filled with water, the soil is *unsaturated*. The moisture is discontinuous and forms "wedges" of water between adjacent particles and moisture films around them.

Water is an important factor in most geotechnical engineering design and construction activities. The presence of water strongly affects the engineering behavior of most soils, especially fine-grained ones (e.g., silts and clays—recall the discussion of Atterberg limits in Chapter 2).

5.1 SURFACE TENSION

The boundary or interface between air and fluid in the voids is of particular importance. At liquid—air interfaces, the greater attraction of water molecules to each other (due to cohesion) rather than to molecules in air (due to adhesion) creates an unbalanced molecular attraction of the water. This in turn gives rise to *surface tension*, a force that acts parallel to the surface of the water in all directions[1] and causes water to behave as if its surface was covered with a stretched elastic membrane.

---

1. Sowers, G.B., Sowers, G.F., 1970. Introductory Soil Mechanics and Foundations. Macmillan Publishing Co., Inc., New York, NY.

Soil Mechanics. http://dx.doi.org/10.1016/B978-0-12-804491-9.00005-7

Surface tension is a contractive tendency of the surface of a liquid that allows it to resist an external force. Surface tension is evident, for example, any time an object that is denser than water is able to float or run along the water surface.

Because of the relatively high attraction of water molecules for each other, water has a high surface tension compared to that of most other liquids. Surface tension can be visualized as a tensile force per unit length (T_s) along the interface between air and water, acting parallel to the water surface. Surface tension thus has the dimension of *force* per *unit length* (FL^{-1}) or of energy per unit area. The magnitude of this force is approximately $T_s = -72.8$ mN/m $= -7.426 \times 10^{-2}$ g/cm at $20°C = -4.988 \times 10^{-3}$ lb/ft (recall that in soil mechanics tensile forces and stresses are denoted by a negative sign).

5.1.1 Surface Tension Phenomena

Surface tension manifests itself in several aspects of soil behavior, namely,

- In a hole dug in the ground, soil is found to be saturated long before the groundwater table is reached. This results from the *capillary rise* of water in the voids, a phenomenon that is discussed in Section 5.2.
- If a sample of saturated clay is dried, it decreases in volume in the process. Surface tension acting in the soil voids serves to compress the soil microfabric and decreases the volume of the sample.
- Dry sand cannot be molded into a ball. However, if the sand moistened, it can be packed and easily shaped. This moist strength is attributed to the tension in the interparticle moisture films. If the moist sand is immersed in water, the moisture films disappear and the sand will again lose its ability to be molded.

5.2 CAPILLARY PHENOMENA IN TUBES

Capillary rise or *capillarity* is a phenomenon in which liquid spontaneously rises or falls in a narrow space such as a thin tube or in the voids of a porous material. Surface tension is an important factor in the phenomenon of capillarity. The surface adhesion forces or internal cohesion present at the interface between a liquid and a solid stretch the liquid and form a curved surface called a *meniscus* (Figure 5.1A). The meniscus is the curve in the upper surface of a liquid close to the surface of the container or another object, caused by surface tension. It can be either concave or convex, depending on the liquid and the surface. Adhesion forces between water and a solid form a concave meniscus. Internal cohesion in mercury, on the other hand, pulls down the liquid to form a convex meniscus. Menisci are thus a manifestation of capillary action.

The stress associated with menisci is known as the *capillary tension*. It is computed for a cylindrical tube of diameter d (Figure 5.1A) by considering the

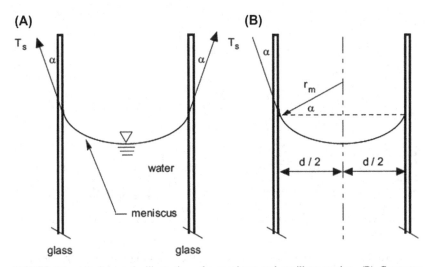

FIGURE 5.1 (A) Schematic illustration of a meniscus and capillary tension. (B) Geometry associated with the meniscus.

force developed by the stretched meniscus. If α is the angle of contact between the meniscus and the solid material (such as glass), then the total unbalanced upward force (F_u) developed along the perimeter of the meniscus is

$$F_u = (T_s \cos \alpha)(\pi d) \tag{5.1}$$

where a zero air pressure has been assumed. For an interface consisting of water and air-dried glass, $\alpha = 0$ degree. For an interface consisting of water and oven-dried glass, $\alpha = 45$ degrees.

The downward force (F_d) is equal to the weight of the water, which is computed from the product of the unit weight of water and the volume of the water that has risen to a capillary height of h_c in the tube (Figure 5.2A). Thus,

$$F_d = \gamma_w(h_c)\left(\frac{\pi d^2}{4}\right) \tag{5.2}$$

For force equilibrium, $F_u = F_d$, which leads to the following result:

$$h_c = \frac{4T_s \cos \alpha}{\gamma_w d} \tag{5.3}$$

For example, in the case of an interface consisting of water in contact with air-dried glass, the surface tension $\alpha = 0$, giving,

$$h_c = \frac{4T_s}{\gamma_w d} \Rightarrow h_c \propto \frac{1}{d} \tag{5.4}$$

(A)   **(B)**

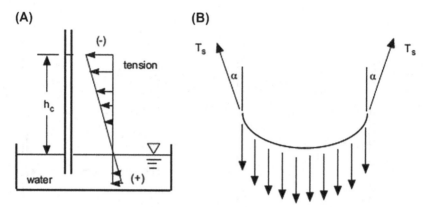

FIGURE 5.2 (A) Schematic illustration of capillary tension. (B) Water Hanging on a meniscus.

Furthermore, since $T_s = -72.8$ mN/m, the magnitude of the capillary rise as given by Eq. (5.3) is thus

$$h_c = \frac{4\left(72.8\dfrac{\text{mN}}{\text{m}}\right)\left(\dfrac{\text{N}}{1000 \text{ mN}}\right)\left(\dfrac{\text{kN}}{1000 \text{ N}}\right)}{(9.81 \text{ kN/m}^3)d} = \frac{2.968 \times 10^{-5}}{d} \approx \frac{0.00003}{d}$$

(5.5)

where d and h_c have units of *meters*.

If $h < h_c$, the angle α adjusts so as to satisfy the following equation:

$$\cos \alpha = \frac{\gamma_w dh}{4T_s}$$

(5.6)

The capillary stress is obtained by dividing F_u from Eq. (5.1) by the cross-sectional area (A) of the tube, giving

$$\sigma_c = \frac{F_u}{A} = \frac{(T_s \cos \alpha)(\pi d)}{\dfrac{\pi d^2}{4}} = \frac{4T_s \cos \alpha}{d} = \gamma_w h_c$$

(5.7)

The capillary tension can be related to the radius of the meniscus (r_m) by considering the geometry of the meniscus (Figure 5.1B). In particular, $d/2 = r_m \cdot \cos \alpha$. Substituting for d into Eq. (5.7) gives

$$\sigma_c = \frac{2T_s}{r_m}$$

(5.8)

Thus for water in contact with air, the capillary tension stress is dependent only on r_m and varies inversely with it.

The maximum capillary tension occurs when the meniscus radius is smallest, which corresponds to the case where the meniscus is tangent to the tube, implying that $\alpha = 0$ degrees and thus $r_m = d/2$. The maximum capillary tension will thus be $\sigma_{c_{max}} = 4T_s/d$, where it is understood that $\sigma_{c_{max}}$ will be negative.

5.3 CAPILLARY PHENOMENA IN SOILS

If pore water were subject only to the force of gravity, the soil above the groundwater table would be perfectly dry. In reality, however, every soil in the field is completely saturated for a certain distance above the groundwater table[2]; above this level, it is only partially saturated. Figure 5.3 shows the relation between a hypothetical saturated aquifer[3], the capillary zone, and the zone of varying degree of saturation.

If a soil is saturated, the air—water interfaces disappear and the capillary tension becomes zero. When a saturated soil is exposed to open air, capillary tension develops as soon as evaporation creates menisci at the surface[4]. Since the moisture in a saturated soil is continuous, the water tension stress developed at the air—water interfaces is felt throughout the mass.

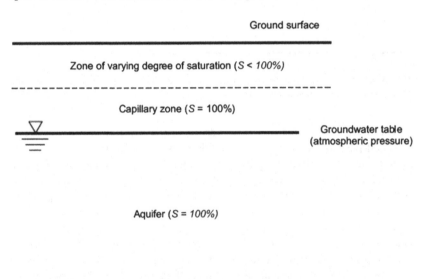

FIGURE 5.3 Schematic illustration of a saturated aquifer, capillary zone, and zone of varying saturation.

2. The *water table* or *phreatic surface* refers to the locus of the levels to which water would rise in observation wells, i.e., where the water pressure head is equal to the atmospheric pressure (where gauge pressure = 0). It is commonly visualized as the "surface" of the geomaterials that are saturated with groundwater in a given vicinity.
3. An aquifer is an underground layer of water-bearing permeable rock or unconsolidated materials (gravel, sand, or silt) from which groundwater can be extracted using a water well.
4. Capillarity is the reason that soils shrink as they dry out. In particular, capillary menisci pull the particles together.

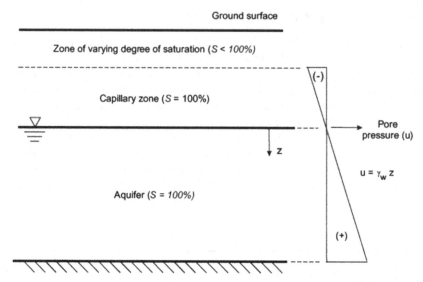

FIGURE 5.4 Schematic illustration of pore pressure distribution with depth in a soil deposit with capillary rise above the groundwater table.

The water obeys the law of hydrostatics; thus, $u = \gamma_w z$, where z is measured positive downward. The capillary rise of water in a soil above the groundwater table illustrates the combined effect of capillary tension and hydrostatic pressure. Referring to Figure 5.2A, at the groundwater elevation (free surface) the water pressure is zero. Below the free surface the pressure increases according to the aforesaid expression for u. In the capillary zone above the free surface, the water pressure decreases linearly, again in accordance with this expression (only with z being negative). Figure 5.4 shows the distribution of both positive (compressive) and negative (tensile) pore pressure with depth for a hypothetical soil with capillary rise above the groundwater table. Above the capillary zone the pore pressure will be a nonlinear function of the degree of saturation.

In contrast to the capillary phenomena discussed in Section 5.2, the continuous voids in soils have a variable width and are by no means straight. Indeed, the interconnected voids in a soil form a collection of irregular but definite capillary tubes. The maximum capillary tension that can develop will vary from point to point, depending on the pore diameter and degree of saturation[5]. Consequently, the capillary tube analogy is not directly applicable.

The thickness of the capillary zone in a soil thus depends on the size and shape of the pores. This, in turn, is a function of the particles sizes and shapes,

5. Sowers, G.B., Sowers, G.F., 1970. Introductory Soil Mechanics and Foundations. Macmillan Publishing Co., Inc., New York, NY.

as well as the void ratio. As the particle size decreases, the size of the voids likewise decreases, and the height of capillary rise in the soil (h_c) increases. Thus, in general,

$$h_c \propto \frac{1}{d_{eff}} \Rightarrow h_c = \frac{c}{d_{eff}} \tag{5.9}$$

where d_{eff} is some effective pore size and c is a constant of proportionality. When h_c is expressed in centimeters, c is typically taken to equal 0.3.

Along these same lines, Terzaghi and Peck[6] proposed the following empirical expression:

$$h_c = \frac{C}{eD_{10}} \tag{5.10}$$

where h_c is in units of *centimeters*, e is the void ratio, D_{10} is the effective grain size (in *centimeters*) as determined from a sieve analysis (recall Chapter 2), and C (units of cm^2) is an empirical constant that depends on the shape of particles and on surface impurities. It varies between 0.1 and 0.5 cm^2 for loose and dense sands, respectively.

5.4 IN SITU STRESSES IN SOILS UNDER HYDROSTATIC CONDITIONS

This section briefly reviews key aspects related to total stress, pore fluid pressure, and effective stress concept in soils. In all cases, the pore fluid is assumed to be at rest; i.e., under *hydrostatic*[7] conditions.

5.4.1 Total Stress

Consider the saturated soil deposit without seepage shown in Figure 5.5. The vertical total stress (σ_v) is obtained by summing up the densities of the solid and fluid phase above some point, multiplied by the gravitational acceleration (g). Mathematically this is written as

$$\sigma_v = \int_0^h \rho g \, dz \tag{5.11}$$

If ρg remains constant throughout the soil, then

$$\sigma_v = \rho g h = \gamma h \tag{5.12}$$

where $\rho g = \gamma$ is the moist unit weight of the soil and h is the depth below the ground surface (the origin of the coordinate system used).

---

6. Terzaghi, K., Peck, R.B., 1967. Soil Mechanics in Engineering Practice, second ed. John Wiley and Sons, New York, NY.
7. Hydrostatics or *fluid statics* is the branch of fluid mechanics that studies incompressible fluids at rest.

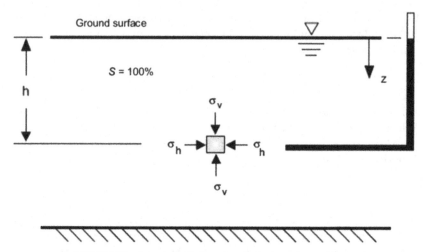

FIGURE 5.5 Saturated soil deposit under hydrostatic conditions.

5.4.2 Pore Fluid Pressure

From fluid mechanics, it is known that under hydrostatic (no seepage) conditions the pore fluid pressure (u) at some depth h is simply

$$u = \rho_w g h = \gamma_w h \tag{5.13}$$

Remark: The pore fluid pressure is also called the "neutral stress" because it has no shear stress components.[8]

5.4.3 Effective (Intergranular) Stress

The effective stress is defined as follows:

$$\sigma' = \sigma - u \tag{5.14}$$

where σ and u are again the total stress and pore fluid pressure, respectively. The *vertical* effective stress is thus

$$\sigma'_v = \sigma_v - u \tag{5.15}$$

The effective stress is approximately the force per unit area carried by the solid phase (i.e., the soil skeleton); it controls a soil's volume change and strength. For example, increases in σ' lead to a denser state of packing in cohesionless soils.

---

8. By definition, a liquid cannot support static shear stresses; it only has *normal* stress components that act *equally* in all directions.

Thus, the vertical total stress for the soil element shown in Figure 5.5 is

$$\sigma_v = \gamma_{sat} h \tag{5.16}$$

where γ_{sat} is the saturated unit weight of the soil. The pore fluid pressure at this point is

$$u = \rho_w g h = \gamma_w h \tag{5.17}$$

Finally, the vertical effective stress at this point is thus

$$\sigma_v' = \sigma_v - u = \gamma_{sat} h - \gamma_w h = (\gamma_{sat} - \gamma_w) h = \gamma_b h \tag{5.18}$$

where γ_b is the buoyant or submerged unit weight of the soil.

5.5 RELATIONSHIP BETWEEN HORIZONTAL AND VERTICAL STRESSES

From hydrostatics the pressure in a liquid is the *same* in all directions. This is not, however, true for soils, as the state of stress in situ is not necessarily hydrostatic.

The determination of the magnitude of the horizontal stress is not as straightforward as the vertical total stress (recall the discussion of Section 5.1.1). As such, the general relationship between horizontal and vertical total stresses is

$$\sigma_h = K\sigma_v \tag{5.19}$$

where K is a positive earth pressure coefficient.

Remark: Since the groundwater table can fluctuate, the total stress will change with such fluctuations. Thus, K is *not* a constant.

To remove the effect of a variable groundwater table on the determination of the horizontal stress, it is expedient to work in terms of *effective* stresses, i.e.,

$$\sigma_h' = K_0 \sigma_v' \tag{5.20}$$

where K_0 is the *coefficient of lateral earth pressure at rest*; it is *independent* of the location of the groundwater table. Even if the groundwater table fluctuates, K_0 will remain unchanged so long as the *same* soil layer is considered and its density remains *unchanged*.

The magnitude of K_0 is very sensitive to the geologic and engineering stress history that a soil has been subjected to in the past. In natural soil deposits,

- $K_0 = 0.4$–0.5 for sedimentary soils.
- K_0 may be as large as 3.0 for very heavily preloaded soils.

EXAMPLE PROBLEM 5.1

General Remarks

This example problem illustrates the manner in which to compute the capillary rise in a glass tube.

Problem Statement

Compute (a) the capillary tension (in g/cm) in a 0.002 mm diameter oven-dried glass tube and (b) the height of capillary rise (in *feet*) in the tube.

Solution

a) The capillary tension is computed from Eq. (5.7), i.e.,

$$\sigma_c = \frac{4T_s \cos \alpha}{d} \tag{5.1.1}$$

Since $\alpha = 45$ degrees for an interface consisting of water and oven-dried glass tube, Eq. (5.1.1) gives

$$\sigma_c = \frac{4T_s \cos \alpha}{d} = \frac{4\left(-7.426 \times 10^{-2} \ \frac{g}{cm}\right)(\cos 45°)}{(0.002 \ \text{mm})\left(\frac{cm}{10 \ \text{mm}}\right)} = -1.050 \times 10^3 \, g/cm^2 \tag{5.1.2}$$

b) The height of capillary rise in the tube is given by Eq. (5.7), i.e.,

$$\sigma_c = \gamma_w h_c \Rightarrow h_c = \frac{\sigma_c}{\gamma_w} = \frac{(1.050 \times 10^3 \, g/cm^3)}{1.0 \, g/cm^3} = 1.050 \times 10^3 \ \text{cm} \tag{5.1.3}$$

Converting to units of feet gives

$$h_c = (1.050 \times 10^3 \ \text{cm})\left(\frac{in}{2.54 \ \text{cm}}\right)\left(\frac{ft}{12 \ \text{in}}\right) = 34.5 \ \textbf{ft} \tag{5.1.4}$$

EXAMPLE PROBLEM 5.2

General Remarks

This example problem illustrates the manner in which to estimate capillary rise above the groundwater table in a sandy soil.

Problem Statement

The effective grain size (D_{10}) of a medium sand is 0.15 mm. The void ratio of the sand in a dense configuration is 0.45; in a loose configuration it is 0.81. What is the estimated capillary rise for this sand?

Solution

In the loose configuration, $C = 0.1 \text{ cm}^2$. Thus,

$$h_c = \frac{0.1 \text{ cm}^2}{(0.81)\left(0.15 \text{ mm} * \dfrac{\text{cm}}{10 \text{ mm}}\right)} = \textbf{8.2 cm} \qquad (5.2.1)$$

In the loose configuration, $C = 0.5 \text{ cm}^2$. Thus,

$$h_c = \frac{0.5 \text{ cm}^2}{(0.45)\left(0.15 \text{ mm} * \dfrac{\text{cm}}{10 \text{ mm}}\right)} = \textbf{74 cm} \qquad (5.2.2)$$

EXAMPLE PROBLEM 5.3

General Remarks

This example problem relates the effective pore size d_{eff} to the effective grain size D_{10}.

Problem Statement

Compute (a) the *maximum* capillary tension and (b) the theoretical height of capillary rise in a soil whose effective grain size (D_{10}) is 0.016 mm if the effective pore size (d_{eff}) is estimated to be $(D_{10})/5$.

Solution

a) Recalling that the capillary tension in a glass tube is given by Eq. (5.7), i.e.,

$$\sigma_c = \frac{4T_s \cos \alpha}{d} \qquad (5.3.1)$$

The maximum value will be realized for $\alpha = 0$ degree. Replacing d by d_{eff}, the estimated maximum capillary tension in the sand is thus

$$\sigma_{c_{max}} = \frac{4T_s}{d_{eff}} = \frac{4\left(-72.8 \dfrac{\text{mN}}{\text{m}}\right)\left(\dfrac{\text{N}}{1000 \text{ mN}}\right)}{\dfrac{1}{5}(0.016 \text{ mm})\left(\dfrac{\text{m}}{1000 \text{ mm}}\right)} = -9.100 \times 10^4 \text{ N/m}^2 \qquad (5.3.2)$$

$$= \textbf{−9.100} \times \textbf{10}^{\textbf{1}} \textbf{ kN/m}^2$$

where d_{eff} is some effective pore size.

b) The capillary rise in a glass tube is given by the second part of Eq. (5.7), i.e.,

$$\sigma_c = \frac{4T_s \cos \alpha}{d} = \gamma_w h_c \Rightarrow h_c = \frac{\sigma_c}{\gamma_w} \tag{5.3.3}$$

The magnitude of the estimated maximum capillary rise in the sand is thus

$$h_c = \frac{\sigma_{c_{max}}}{\gamma_w} = \frac{9.100 \times 10^1 \text{ kN/m}^2}{9.81 \text{ kN/m}^3} = \textbf{9.3 m} \tag{5.3.4}$$

EXAMPLE PROBLEM 5.4

General Remarks

This example problem illustrates how total stress, pore pressure, and effective stress are computed in the case where the groundwater table lies above the ground surface.

Problem Statement

Consider a case where the groundwater table is located above the surface of a saturated soil deposit (Figure Ex. 5.4A). Such conditions are typical of soils in lakes and in oceans. Determine the variation with depth below the groundwater table of the total stress, pore fluid pressure, and effective stress.

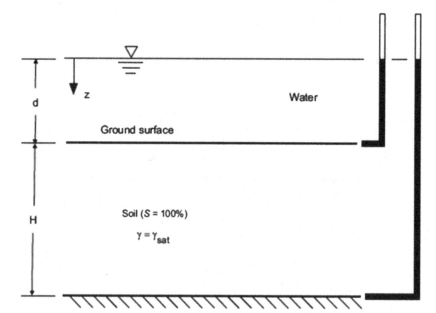

FIGURE EX. 5.4A Soil deposit with groundwater table above ground surface under hydrostatic conditions.

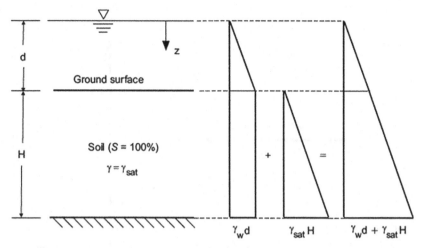

FIGURE EX. 5.4B   Schematic illustration of variation with depth of vertical total stress.

The vertical total stress distribution with depth has *two* contributions. The first ($\gamma_w d$) is from the layer of water that is located over the soil layer; the second ($\gamma_{sat} \cdot H$) is due to the saturated unit weight of the soil. Figure Ex. 5.4B shows the variation of vertical total stress with depth below the groundwater table.

The pore fluid pressure varies with depth in the usual linear fashion, i.e., $u = \gamma_w z$. Figure Ex. 5.4C shows the variation with depth of the pore fluid pressure.

Finally, the vertical effective stress is the difference between the total stress and the pore pressure. The maximum value is thus

$$\sigma'_v = \sigma_v - u = (\gamma_w d + \gamma_{sat} H) - \gamma_w (d + H) = (\gamma_{sat} - \gamma_w) H = \gamma_b H \quad (5.4.1)$$

where γ_b is the buoyant unit weight[9] of the soil. Figure Ex. 5.4D shows the variation of the effective stress with depth below the groundwater table.

EXAMPLE PROBLEM 5.5

General Remarks

This example problem illustrates how pore pressures, and thus effective stresses, are computed when capillary rise is present above the groundwater table.

Problem Statement

Given the soil profiles shown in Figures Ex. 5.5A and Ex. 5.5B, compute the total stress, pore fluid pressure, and effective stress at (a) points A and B in Figure Ex. 5.5A and (b) points C and D in Figure Ex. 5.5B.

---

9. Recall the discussion of saturated and buoyant (submerged) unit weights given in Chapter 1.

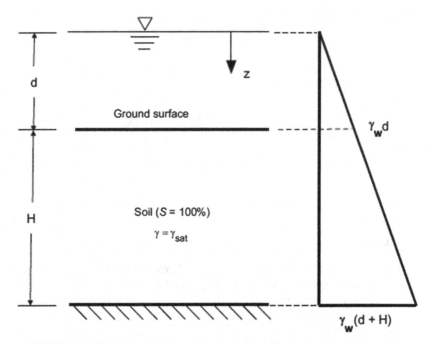

FIGURE EX. 5.4C Schematic illustration of variation with depth of pore fluid pressure.

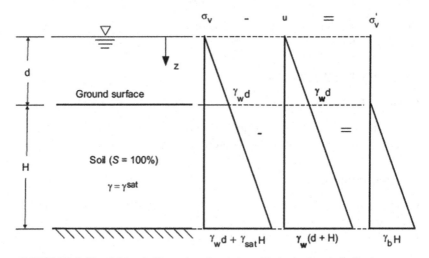

FIGURE EX. 5.4D Schematic illustration of variation with depth of vertical effective stress.

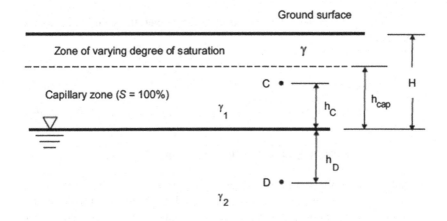

FIGURE EX. 5.5A   Soil deposit with capillary zone extending to ground surface.

FIGURE EX. 5.5B   Soil deposit with capillary zone not extending to ground surface.

Solution

a) In the case of the soil deposit shown in Figure Ex. 5.5A, the capillary rise above the groundwater table extends to the ground surface. In this figure, γ_1 and γ_2 are the saturated unit weights of the upper and lower soil layer, respectively.

At point A:

The vertical total stress is

$$\sigma_v = \gamma_1(H - h_A) \qquad (5.5.1)$$

Since point A lies *above* the groundwater table, the pore fluid pressure is *negative*, i.e.,

$$u = -\gamma_w h_A \qquad (5.5.2)$$

The vertical effective stress is thus

$$\sigma_v' = \sigma_v - u = \gamma_1(H - h_A) + \gamma_w h_A = \gamma_1 H - (\gamma_1 - \gamma_w)h_A = \gamma_1 H - \gamma_{b_1} h_A \qquad (5.5.3)$$

where γ_{b_1} is the buoyant unit weight of the upper soil layer.

At point B:

The vertical total stress is

$$\sigma_v = \gamma_1 H + \gamma_2 h_B \qquad (5.5.4)$$

Since point B lies *below* the groundwater table, the pore fluid pressure is *positive*, i.e.,

$$u = \gamma_w h_B \qquad (5.5.5)$$

The vertical effective stress is thus

$$\sigma_v' = \sigma_v - u = \gamma_1 H + \gamma_2 h_B - \gamma_w h_B = \gamma_1 H + (\gamma_2 - \gamma_w)h_B = \gamma_1 H - \gamma_{b_2} h_B \qquad (5.5.6)$$

where γ_{b_2} is the buoyant unit weight of the lower soil layer.

b) In the case of the soil deposit shown in Figure Ex. 5.5B, the capillary rise above the groundwater table does not extend to the ground surface. In this figure, g is the moist unit weight of the soil in the zone of varying degree of saturation, and γ_1 and γ_2 are again the saturated unit weights of the upper and lower soil layer, respectively.

At point C:

The vertical total stress is

$$\sigma_v = \gamma(H - h_{cap}) + \gamma_1(h_{cap} - h_C) \qquad (5.5.7)$$

Since point C lies *above* the groundwater table, the pore fluid pressure is *negative*, i.e.,

$$u = -\gamma_w h_C \qquad (5.5.8)$$

The vertical effective stress is thus

$$\sigma'_v = \sigma_v - u = \gamma(H - h_{cap}) + \gamma_1(h_{cap} - h_C) + \gamma_w h_C$$
$$= \gamma(H - h_{cap}) + \gamma_1 h_{cap} - (\gamma_1 - \gamma_w)h_C = \gamma(H - h_{cap}) + \gamma_1 h_{cap} - \gamma_{b_1} h_C \qquad (5.5.9)$$

where γ_{b_1} is again the buoyant unit weight of the upper soil layer.
At point D:
The vertical total stress is

$$\sigma_v = \gamma(H - h_{cap}) + \gamma_1 h_{cap} + \gamma_2 h_D \qquad (5.5.10)$$

Since point D lies *below* the groundwater table, the pore fluid pressure is *positive*, i.e.,

$$u = \gamma_w h_D \qquad (5.5.11)$$

The vertical effective stress is thus

$$\sigma'_v = \sigma_v - u = \gamma(H - h_{cap}) + \gamma_1 h_{cap} + \gamma_2 h_D - \gamma_w h_D$$
$$= \gamma(H - h_{cap}) + \gamma_1 h_{cap} + (\gamma_2 - \gamma_w)h_D = \gamma(H - h_{cap}) + \gamma_1 h_{cap} + \gamma_{b_2} h_D \qquad (5.5.12)$$

where γ_{b_2} is again the buoyant unit weight of the lower soil layer.

EXAMPLE PROBLEM 5.6

General Remarks

This example problem illustrates the manner in which in situ stresses are computed under hydrostatic conditions in the presence of capillary rise above the groundwater table.

Problem Statement

A 10 m thick soil deposit overlies a layer of soft rock. The groundwater table is approximately 5 m above the surface of the rock and the height of capillary rise is approximately 3.5 m. The soil has an average void ratio (e) of 0.36 and a specific gravity of solids (G_s) equal to 2.68. No seepage is present at the site. Determine the variation with depth of the vertical total stress, the pore fluid pressure, and the vertical effective stress in the deposit at depths of 2.5, 5.0, 7.5, and 10.0 m below the ground surface assuming (a) capillary rise above the groundwater table as stated above and (b) no capillary rise. Where not saturated, the soil has a moisture content (w) of 6% and a degree of saturation (S) equal to 45%.

Solution

Figure Ex. 5.6A shows the single soil profile and the extent of the capillary rise.

First, all of the necessary unit weights are determined. For the unsaturated portion of the soil deposit,

$$\gamma = \frac{\gamma_w(G_s + Se)}{1 + e} = \frac{(9.81\ \text{kN/m}^3)[2.68 + (0.45)(0.36)]}{1 + 0.36} = 20.50\ \text{kN/m}^3$$

$$(5.6.1)$$

Similarly, for the unsaturated portion of the soil deposit,

$$\gamma = \frac{\gamma_w(G_s + e)}{1 + e} = \frac{(9.81\ \text{kN/m}^3)[2.68 + 0.36]}{1 + 0.36} = 21.93\ \text{kN/m}^3 \qquad (5.6.2)$$

a) For capillary rise above the groundwater table as shown in Figure Ex. 5.6A, the vertical total stress, pore fluid pressure, and vertical effective stresses are next computed at the requested depths.
At a depth of 2.5 m:
    The vertical total stress is

$$\sigma_v = (20.50\ \text{kN/m}^3)(1.5\ \text{m}) + (21.93\ \text{kN/m}^3)(1.0\ \text{m}) = \textbf{52.68 kN/m}^2$$

$$(5.6.3)$$

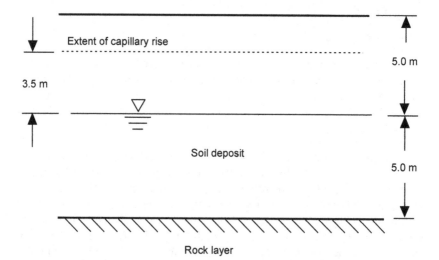

FIGURE EX. 5.6A   Profile consisting of a single soil layer.

The pore fluid pressure is

$$u = -(9.81 \text{ kN/m}^3)(1.0 \text{ m}) = \textbf{-9.81 kN/m}^2 \qquad (5.6.4)$$

The vertical effective stress is thus

$$\sigma'_v = \sigma_v - u = 52.68 - (-9.81) = \textbf{62.49 kN/m}^2 \qquad (5.6.5)$$

At a depth of 5.0 m:
The vertical total stress is

$$\sigma_v = (20.50 \text{ kN/m}^3)(1.5 \text{ m}) + (21.93 \text{ kN/m}^3)(3.5 \text{ m}) = \textbf{107.5 kN/m}^2$$
$$(5.6.6)$$

The pore fluid pressure is

$$u = \textbf{0.0} \qquad (5.6.7)$$

The vertical effective stress is thus

$$\sigma'_v = \sigma_v - u = 107.5 - 0 = \textbf{107.5 kN/m}^2 \qquad (5.6.8)$$

At a depth of 7.5 m:
The vertical total stress is

$$\sigma_v = (20.50 \text{ kN/m}^3)(1.5 \text{ m}) + (21.93 \text{ kN/m}^3)(6.0 \text{ m}) = \textbf{162.3 kN/m}^2$$
$$(5.6.9)$$

The pore fluid pressure is

$$u = (9.81 \text{ kN/m}^3)(2.5 \text{ m}) = \textbf{24.53 kN/m}^2 \qquad (5.6.10)$$

The vertical effective stress is thus

$$\sigma'_v = \sigma_v - u = 162.3 - 24.53 = \textbf{137.8 kN/m}^2 \qquad (5.6.11)$$

At a depth of 10.0 m:
The vertical total stress is

$$\sigma_v = (20.50 \text{ kN/m}^3)(1.5 \text{ m}) + (21.93 \text{ kN/m}^3)(8.5 \text{ m}) = \textbf{217.2 kN/m}^2$$
$$(5.6.12)$$

The pore fluid pressure is

$$u = (9.81 \text{ kN/m}^3)(5.0 \text{ m}) = \textbf{49.05 kN/m}^2 \qquad (5.6.13)$$

The vertical effective stress is thus

$$\sigma'_v = \sigma_v - u = 217.2 - 49.05 = \textbf{168.1 kN/m}^2 \qquad (5.6.14)$$

Figure Ex. 5.6B shows the variation with depth of the vertical total stress, pore pressure, and vertical effective stress.

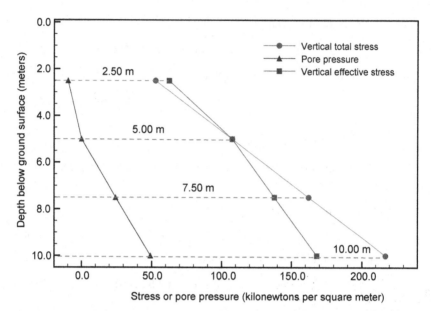

FIGURE EX. 5.6B Variation with depth of vertical total stress, pore fluid pressure, and vertical effective stress when considering capillary rise.

b) If capillary rise above the groundwater table is *ignored*, the vertical total stress, pore fluid pressure, and vertical effective stresses are next computed at the requested depths. The degree of saturation above the groundwater table is still 45%.

At a depth of 2.5 m:

The vertical total stress is

$$\sigma_v = (20.50 \text{ kN/m}^3)(2.5 \text{ m}) = \textbf{51.25 kN/m}^2 \qquad (5.6.15)$$

The pore fluid pressure is

$$u = \textbf{0.0} \qquad (5.6.16)$$

The vertical effective stress is thus

$$\sigma_v' = \sigma_v - u = 51.25 - 0 = \textbf{51.25 kN/m}^2 \qquad (5.6.17)$$

At a depth of 5.0 m:

The vertical total stress is

$$\sigma_v = (20.50 \text{ kN/m}^3)(5.0 \text{ m}) = \textbf{102.5 kN/m}^2 \qquad (5.6.18)$$

The pore fluid pressure is

$$u = \textbf{0.0} \qquad (5.6.19)$$

The vertical effective stress is thus

$$\sigma'_v = \sigma_v - u = 102.5 - 0 = \textbf{102.5 kN/m}^2 \qquad (5.6.20)$$

At a depth of 7.5 m:
The vertical total stress is

$$\sigma_v = (20.50 \text{ kN/m}^3)(5.0 \text{ m}) + (21.93 \text{ kN/m}^3)(2.5 \text{ m}) = \textbf{157.3 kN/m}^2$$
$$(5.6.21)$$

The pore fluid pressure is

$$u = (9.81 \text{ kN/m}^3)(2.5 \text{ m}) = \textbf{24.53 kN/m}^2 \qquad (5.6.22)$$

The vertical effective stress is thus

$$\sigma'_v = \sigma_v - u = 157.3 - 24.53 = \textbf{132.8 kN/m}^2 \qquad (5.6.23)$$

At a depth of 10.0 m:
The vertical total stress is

$$\sigma_v = (20.50 \text{ kN/m}^3)(5.0 \text{ m}) + (21.93 \text{ kN/m}^3)(5.0 \text{ m}) = \textbf{212.2 kN/m}^2$$
$$(5.6.24)$$

The pore fluid pressure is

$$u = (9.81 \text{ kN/m}^3)(5.0 \text{ m}) = \textbf{49.05 kN/m}^2 \qquad (5.6.25)$$

The vertical effective stress is thus

$$\sigma'_v = \sigma_v - u = 212.2 - 49.05 = \textbf{163.1 kN/m}^2 \qquad (5.6.26)$$

Figure Ex. 5.6C shows the variation with depth of the vertical total stress, pore fluid pressure, and vertical effective stress.

EXAMPLE PROBLEM 5.7

General Remarks

This example problem illustrates the manner in which in situ stresses are computed in the presence of capillary rise above the groundwater table.

Problem Statement

Figure Ex. 5.7A shows the soil profile at a specific site. No seepage is present at the site. The following properties are known for the respective soil layers:

- Sand layer: $G_s = 2.70$; moisture content of 30%.
- Silt layer: saturated unit weight of 127 lb/ft^3.
- Weald clay layer: buoyant unit weight 45 lb/ft^3.

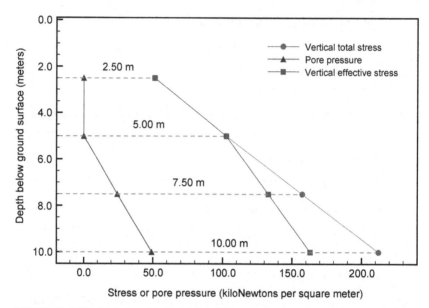

FIGURE EX. 5.6C Variation with depth of vertical total stress, pore fluid pressure, and vertical effective stress when ignoring capillary rise.

Assume the groundwater table to be 8 ft below the ground surface with capillary rise in the sand layer that extends to the ground surface. Determine the vertical total stress, the pore fluid pressure, and the vertical effective stress at depths of 0, 8, 20, 25, and 45 ft.

Solution

The correct unit weights to use for the respective soil layers are first determined. It is important to note that in this problem, all of the layers are *saturated*.

Using the expression developed in Case 1.3 of Chapter 1, for the sand layer

$$e = \frac{G_s w}{S} = \frac{(2.70)(0.30)}{1.0} = 0.810 \tag{5.7.1}$$

Thus, from Case 1.8 of Chapter 1,

$$\gamma_{sat\_sand} = \frac{(G_s + e)\gamma_w}{1 + e} = \frac{(2.70 + 0.810)(62.4 \text{ lb/ft}^3)}{1 + 0.810} = 121.0 \text{ lb/ft}^3 \tag{5.7.2}$$

For the silt layer the saturated unit weight is given, i.e., $\gamma_{sat\_silt} = 127 \text{lb/ft}^3$.

Finally, for the Weald clay,

$$\gamma_{sat\_clay} = \gamma' + \gamma_w = 45.0 + 62.4 = 107.4 \text{ lb/ft}^3 \tag{5.7.3}$$

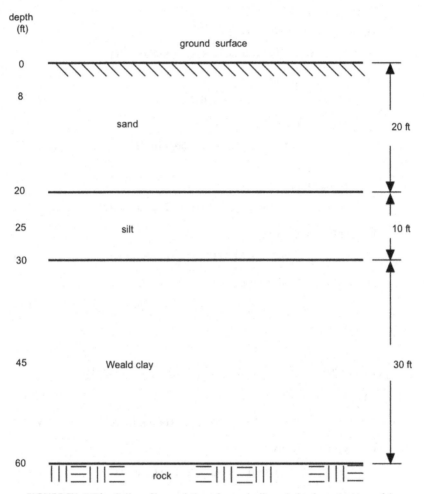

FIGURE EX. 5.7A Soil profile consisting of a sand, silt, and clay layer (not to scale).

At a depth of 0 ft:
The vertical total stress is

$$\sigma_v = \mathbf{0.0} \tag{5.7.4}$$

The pore fluid pressure is

$$u = -(62.4 \text{ lb/ft}^3)(8 \text{ ft}) = \mathbf{-499.2 \text{ lb/ft}^2} \tag{5.7.5}$$

The vertical effective stress is thus

$$\sigma'_v = 0 - (-499.2) = \mathbf{499.2 \text{ lb/ft}^2} \tag{5.7.6}$$

At a depth of 8 ft:
The vertical total stress is

$$\sigma_v = \left(121.0 \text{ lb/ft}^3\right)(8 \text{ ft}) = \textbf{968.0 lb/ft}^2 \qquad (5.7.7)$$

The pore fluid pressure is

$$u = \textbf{0.0} \qquad (5.7.8)$$

The vertical effective stress is thus

$$\sigma'_v = 968.0 - 0 = \textbf{968.0 lb/ft}^2 \qquad (5.7.9)$$

At a depth of 20 ft:
The vertical total stress is

$$\sigma_v = \left(121.0 \text{ lb/ft}^3\right)(20 \text{ ft}) = \textbf{2420.0 lb/ft}^2 \qquad (5.7.10)$$

The pore fluid pressure is

$$u = \left(62.4 \text{ lb/ft}^3\right)(12 \text{ ft}) = \textbf{748.8 lb/ft}^2 \qquad (5.7.11)$$

The vertical effective stress is thus

$$\sigma'_v = 2420.0 - 748.8 = \textbf{1671.2 lb/ft}^2 \qquad (5.7.12)$$

At a depth of 25 ft:
The vertical total stress is

$$\sigma_v = 2420.0 \text{ lb/ft}^2 + \left(127.0 \text{ lb/ft}^3\right)(5 \text{ ft}) = \textbf{3055.0 lb/ft}^2 \qquad (5.7.13)$$

The pore fluid pressure is

$$u = \left(62.4 \text{ lb/ft}^3\right)(17 \text{ ft}) = \textbf{1060.8 lb/ft}^2 \qquad (5.7.14)$$

The vertical effective stress is thus

$$\sigma'_v = 3055.0 - 1060.8 = \textbf{1994.2 lb/ft}^2 \qquad (5.7.15)$$

At a depth of 45 ft:
The vertical total stress is

$$\sigma_v = 3055.0 \text{ lb/ft}^2 + \left(127.0 \text{ lb/ft}^3\right)(5 \text{ ft}) + \left(107.4 \text{ lb/ft}^3\right)(15 \text{ ft})$$
$$= \textbf{5301.0 lb/ft}^2 \qquad (5.7.16)$$

The pore fluid pressure is

$$u = \left(62.4 \text{ lb/ft}^3\right)(12 + 10 + 15 \text{ ft}) = \textbf{2308.8 lb/ft}^2 \qquad (5.7.17)$$

The vertical effective stress is thus

$$\sigma'_v = 5301.0 - 2308.8 = \textbf{2992.2 lb/ft}^2 \qquad (5.7.18)$$

Figure Ex. 5.7B shows the variation with depth of the vertical total stress, pore fluid pressure, and vertical effective stress.

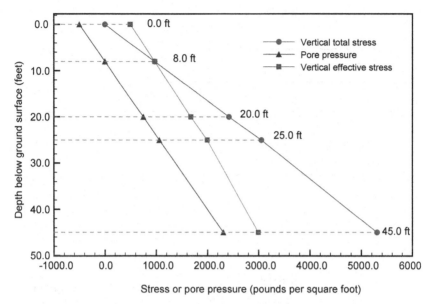

FIGURE EX. 5.7B Variation with depth of vertical total stress, pore fluid pressure, and effective stress in soil profile consisting of a sand, silt, and clay layer.

EXAMPLE PROBLEM 5.8

General Remarks

This example problem illustrates the manner in which in situ stresses are computed under hydrostatic conditions. The second part of the problem considers the rapid rise of the groundwater table and its affect on the pore fluid pressure and effective stress state.

Problem Statement

Borehole data at a site reveals the soil profile shown in Figure Ex. 5.8A. The groundwater table is found at a depth of 3.6 m. No seepage is present at the site.

Some details pertaining to the soil profile are given as follows:

- The top 2.0 m consists of very fine, wet sand with silt. Laboratory tests indicate that for this soil the moisture content (w) is 5%, the degree of saturation (S) is 40%, and the specific gravity of solids (G_s) equals 2.69.
- The next 3.4 m consists of fine sand. Laboratory tests indicate that for this soil, $G_s = 2.68$. Above the groundwater table, $w = 8\%$ and $S = 78\%$. Below the groundwater table, $w = 12\%$.
- The final 15.2 m consists of soft blue clay. Laboratory tests indicate that for this soil, $w = 32\%$ and $G_s = 2.71$.

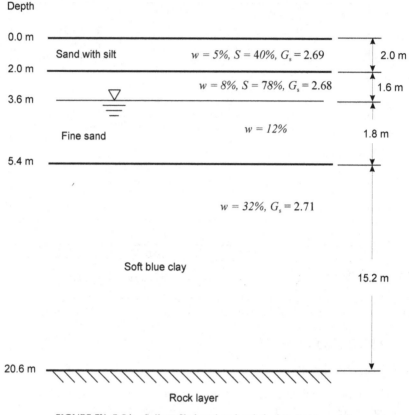

a) Determine the vertical total stress, pore pressure, and vertical effective stress at depths of 0.0, 2.0, 3.6, 5.4, and 20.6 m. If the coefficient of lateral earth pressure at rest (K_0) is equal to 0.55, determine the lateral effective stress at the same depths.

b) If the groundwater table were to rise rapidly to the ground surface, determine the vertical total stress, pore pressure, and vertical effective stress at depths of 2.0, 5.4, and 20.6 m. Assume that the void ratio in the sand layers remains *unchanged* during the rise in groundwater.

Solution

First, all of the necessary unit weights are determined. Since the moisture content, degree of saturation, and the specific gravity of solids are known for each of the soils, it is timely to substitute the relation $e = G_s w/S$ into the general expression for moist unit weight to give the following relation:

$$\gamma = \frac{\gamma_w G_s(1+w)}{1+e} = \frac{\gamma_w G_s(1+w)}{1 + \left(\dfrac{G_s w}{S}\right)} \tag{5.8.1}$$

For the very fine wet sand with silt, $w = 5\%$, $S = 40\%$, and $G_s = 2.69$. The moist unit weight is thus

$$\gamma = \frac{(9.81 \text{ kN/m}^3)(2.69)(1+0.05)}{1 + \left[\dfrac{(2.69)(0.05)}{0.40}\right]} = 20.74 \text{ kN/m}^3 \tag{5.8.2}$$

For the fine sand *above* the groundwater table, $w = 8\%$, $S = 78\%$, and $G_s = 2.68$. The moist unit weight is thus

$$\gamma = \frac{(9.81 \text{ kN/m}^3)(2.68)(1+0.08)}{1 + \left[\dfrac{(2.68)(0.08)}{0.78}\right]} = 22.27 \text{ kN/m}^3 \tag{5.8.3}$$

For the fine sand *below* the groundwater table, $w = 12\%$, $S = 100\%$, and $G_s = 2.68$. The moist unit weight of the fine sand is thus

$$\gamma = \frac{(9.81 \text{ kN/m}^3)(2.68)(1+0.12)}{1 + \left[\dfrac{(2.68)(0.12)}{1.00}\right]} = 22.28 \text{ kN/m}^3 \tag{5.8.4}$$

Finally, for the soft blue clay, $w = 32\%$, $S = 100\%$, and $G_s = 2.71$. The moist unit weight of this soil is thus

$$\gamma = \frac{(9.81 \text{ kN/m}^3)(2.71)(1+0.32)}{1 + \left[\dfrac{(2.71)(0.32)}{1.00}\right]} = 18.79 \text{ kN/m}^3 \tag{5.8.5}$$

a) The vertical total stress, pore fluid pressure, and vertical and horizontal (lateral) effective stresses are next computed at the requested depths. At a depth of 0.0 m:

$$\sigma_v = \textbf{0.0} \quad u = \textbf{0.0}, \quad \sigma_v' = \sigma_h' = \textbf{0.0} \tag{5.8.6}$$

At a depth of 2.0 m:
The vertical total stress is

$$\sigma_v = (20.74 \text{ kN/m}^3)(2.0 \text{ m}) = \textbf{41.48 kN/m}^2 \tag{5.8.7}$$

Since the soil is not saturated, capillary rise in the very fine wet sand with silt is ignored. The pore fluid pressure is thus

$$u = \mathbf{0.0} \tag{5.8.8}$$

The vertical effective stress is thus equal to the vertical total stress, i.e.,

$$\sigma'_v = \sigma_v - u = \mathbf{41.48 \ kN/m^2} \tag{5.8.9}$$

The horizontal (lateral) effective stress is thus

$$\sigma'_h = K_0\sigma'_v = 0.55\left(41.48 \ kN/m^2\right) = \mathbf{22.81 \ kN/m^2} \tag{5.8.10}$$

At a depth of 3.6 m:
The vertical total stress is

$$\sigma_v = 41.48 \ kN/m^2 + \left(22.27 \ kN/m^3\right)(1.6 \ m) = \mathbf{77.11 \ kN/m^2} \tag{5.8.11}$$

Since the soil is not saturated, capillary rise in the fine sand is ignored. The pore fluid pressure is thus

$$u = \mathbf{0.0} \tag{5.8.12}$$

The vertical effective stress is thus again equal to the vertical total stress, i.e.,

$$\sigma'_v = \sigma_v - u = \mathbf{77.11 \ kN/m^2} \tag{5.8.13}$$

The horizontal (lateral) effective stress is thus

$$\sigma'_h = K_0\sigma'_v = 0.55\left(77.11 \ kN/m^2\right) = \mathbf{42.41 \ kN/m^2} \tag{5.8.14}$$

At a depth of 5.4 m:
The vertical total stress is

$$\sigma_v = 77.11 \ kN/m^2 + \left(22.28 \ kN/m^3\right)(1.8 \ m) = \mathbf{117.2 \ kN/m^2} \tag{5.8.15}$$

The pore fluid pressure is

$$u = \left(9.81 \ kN/m^3\right)(1.8 \ m) = \mathbf{17.66 \ kN/m^2} \tag{5.8.16}$$

The vertical effective stress is thus

$$\sigma'_v = \sigma_v - u = 117.2 - 17.66 = \mathbf{99.54 \ kN/m^2} \tag{5.8.17}$$

Finally, the horizontal (lateral) effective stress is

$$\sigma'_h = K_0\sigma'_v = 0.55\left(99.54 \ kN/m^2\right) = \mathbf{54.75 \ kN/m^2} \tag{5.8.18}$$

At a depth of 20.6 m:
The vertical total stress is

$$\sigma_v = 117.2 \ kN/m^2 + \left(18.79 \ kN/m^3\right)(15.2 \ m) = \mathbf{402.8 \ kN/m^2} \tag{5.8.19}$$

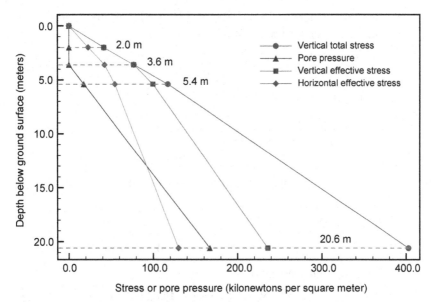

FIGURE EX. 5.8B   Variation with depth of vertical total stress, pore fluid pressure, and vertical and horizontal effective stress.

The pore fluid pressure is

$$u = (9.81 \text{ kN/m}^3)(1.8 + 15.2 \text{ m}) = \mathbf{166.8 \text{ kN/m}^2} \quad (5.8.20)$$

The vertical effective stress is thus

$$\sigma_v' = \sigma_v - u = 402.8 - 166.8 = \mathbf{236.0 \text{ kN/m}^2} \quad (5.8.21)$$

Finally, the horizontal (lateral) effective stress is

$$\sigma_h' = K_0 \sigma_v' = 0.55(236.0 \text{ kN/m}^2) = \mathbf{129.8 \text{ kN/m}^2} \quad (5.8.22)$$

Figure Ex. 5.8B shows the variation with depth of the vertical total stress, pore fluid pressure, and vertical and horizontal effective stress.

b) If the groundwater table rises rapidly to the ground surface, the very fine, wet sand with silt, as well as the fine sand will be *saturated*. For both soils, the rapid rise in groundwater is assumed to take place *without* change in the void ratio. The initial void ratio in the wet sand with silt layer before the rise in groundwater table is

$$e = \frac{G_s w}{S} = \frac{(2.69)(0.05)}{0.40} = 0.336 \quad (5.8.23)$$

Since this void ratio is assumed to be unchanged, the saturated unit weight in the wet sand with silt layer is thus

$$\gamma_{sat} = \frac{\gamma_w(G_s + e)}{1 + e} = \frac{(9.81 \text{ kN/m}^3)(2.69 + 0.336)}{1 + 0.336} = 22.22 \text{ kN/m}^3 \quad (5.8.24)$$

For the entire fine sand layer,

$$\gamma_{sat} = \frac{(9.81 \text{ kN/m}^3)(2.68)(1 + 0.12)}{1 + (2.68)(0.12)} = 22.28 \text{ kN/m}^3 \quad (5.8.25)$$

At a depth of 0.0 m:

$$\sigma_v = \textbf{0.0}, \quad u = \textbf{0.0}, \quad \sigma'_v = \sigma'_h = \textbf{0.0} \quad (5.8.26)$$

At a depth of 2.0 m:
The vertical total stress is

$$\sigma_v = (22.22 \text{ kN/m}^3)(2.0 \text{ m}) = \textbf{44.44 kN/m}^2 \quad (5.8.27)$$

The pore fluid pressure is

$$u = (9.81 \text{ kN/m}^3)(2.0 \text{ m}) = \textbf{19.62 kN/m}^2 \quad (5.8.28)$$

The vertical effective stress is thus

$$\sigma'_v = \sigma_v - u = 44.44 - 19.62 = \textbf{24.82 kN/m}^2 \quad (5.8.29)$$

Finally, the horizontal (lateral) effective stress is

$$\sigma'_h = K_0\sigma'_v = 0.55(24.82 \text{ kN/m}^2) = \textbf{13.65 kN/m}^2 \quad (5.8.30)$$

At a depth of 3.6 m:
The vertical total stress is

$$\sigma_v = 44.44 \text{ kN/m}^2 + (22.28 \text{ kN/m}^3)(1.6 \text{ m}) = \textbf{80.09 kN/m}^2 \quad (5.8.31)$$

The pore fluid pressure is

$$u = (9.81 \text{ kN/m}^3)(3.6 \text{ m}) = \textbf{35.32 kN/m}^2 \quad (5.8.32)$$

The vertical effective stress is thus

$$\sigma'_v = \sigma_v - u = 80.09 - 35.32 = \textbf{44.77 kN/m}^2 \quad (5.8.33)$$

The horizontal (lateral) effective stress is thus

$$\sigma'_h = K_0\sigma'_v = 0.55(44.77 \text{ kN/m}^2) = \textbf{24.63 kN/m}^2 \quad (5.8.34)$$

At a depth of 5.4 m:
The vertical total stress is

$$\sigma_v = 80.09 \text{ kN/m}^2 + (22.28 \text{ kN/m}^3)(1.8 \text{ m}) = \textbf{120.2 kN/m}^2 \quad (5.8.35)$$

The pore fluid pressure is

$$u = (9.81 \text{ kN/m}^3)(5.4 \text{ m}) = \textbf{52.97 kN/m}^2 \qquad (5.8.36)$$

The vertical effective stress is thus

$$\sigma'_v = \sigma_v - u = 120.2 - 52.97 = \textbf{67.23 kN/m}^2 \qquad (5.8.37)$$

The horizontal (lateral) effective stress is thus

$$\sigma'_h = K_0\sigma'_v = 0.55(67.23 \text{ kN/m}^2) = \textbf{37.00 kN/m}^2 \qquad (5.8.38)$$

At a depth of 20.6 m:
The vertical total stress is

$$\sigma_v = 120.2 \text{ kN/m}^2 + (18.80 \text{ kN/m}^3)(15.2 \text{ m}) = \textbf{406.0 kN/m}^2 \quad (5.8.39)$$

The pore fluid pressure is

$$u = (9.81 \text{ kN/m}^3)(20.6 \text{ m}) = \textbf{202.1 kN/m}^2 \qquad (5.8.40)$$

The vertical effective stress is thus

$$\sigma'_v = \sigma_v - u = 406.0 - 202.1 = \textbf{203.9 kN/m}^2 \qquad (5.8.41)$$

The horizontal (lateral) effective stress is thus

$$\sigma'_h = K_0\sigma'_v = 0.55(203.9 \text{ kN/m}^2) = \textbf{112.1 kN/m}^2 \qquad (5.8.42)$$

Figure Ex. 5.8C shows the variation with depth of the vertical total stress, pore fluid pressure, and vertical and horizontal effective stress.

EXAMPLE PROBLEM 5.9

General Remarks

This example problem illustrates the manner in which in situ stresses are computed under hydrostatic conditions. The second part of the problem considers a lowering of the groundwater table (e.g., by the process of dewatering) and its effect on the pore fluid pressure and effective stress state.

Problem Statement

Figure Ex. 5.9A shows a soil profile consisting of a 3 m thick gravel fill, an 8 m thick layer of sand, a 10 m thick layer of soft silty clay, and an 8 m thick stiff clay layer. The 2 m portion of the sand layer above the ground water layer is saturated by capillary rise, while the gravel fill is unaffected by the capillarity. No seepage is present at the site. The following properties are known for the respective soil layers:

- Gravel fill: $G_s = 2.75$, $w = 12.5\%$, $S = 67\%$, $\gamma_d = 17.8 \text{ kN/m}^3$, $K_0 = 1.20$.
- Sand layer: $G_s = 2.69$, $w = 16.0\%$, $K_0 = 0.470$.

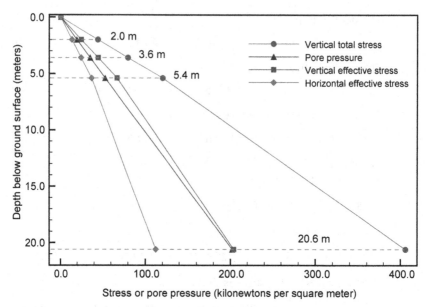

FIGURE EX. 5.8C Variation with depth of vertical total stress, pore fluid pressure, and vertical and horizontal effective stress.

- Soft silty clay: $G_s = 2.65$, $w = 65.0\%$, $K_0 = 0.658$.
- Stiff clay layer: $G_s = 2.68$, $w = 20.0\%$, $K_0 = 1.00$.

where K_0 is the coefficient of lateral earth pressure at rest.

a) Determine the vertical total stress (σ_v), pore pressure (u), vertical effective stress (σ_v'), and horizontal effective stress (σ_h') at depths of 1.5, 4, 8, 16, and 25 m.

b) If the groundwater table is next *lowered* by 6 m, what is the vertical total stress, the pore pressure, vertical effective stress, and horizontal effective stress at a depth of 16 m? Assume that the sand layer remains saturated.

Solution

The correct unit weights to use for the respective soil layers are first determined. For the gravel fill:

$$\gamma = \gamma_d(1 + w) = (17.8 \text{ kN/m}^3)(1 + 0.125) = 20.03 \text{ kN/m}^3 \qquad (5.9.1)$$

Depth

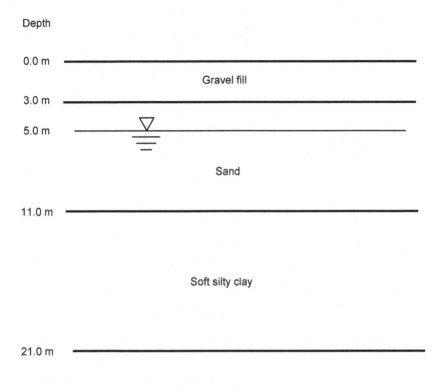

FIGURE EX. 5.9A   Soil profile consisting of gravel fill, and sand, silty clay, and stiff clay layers.

For the sand layer:

$$\gamma_{sat} = \frac{\gamma_w G_s(1+w)}{1+G_s w} = \frac{(9.81\ \text{kN/m}^3)(2.69)(1+0.16)}{1+(2.69)(0.16)} = 21.40\ \text{kN/m}^3$$

(5.9.2)

For the soft silty clay layer:

$$\gamma_{sat} = \frac{\gamma_w G_s(1+w)}{1+G_s w} = \frac{(9.81\ \text{kN/m}^3)(2.65)(1+0.65)}{1+(2.65)(0.65)} = 15.76\ \text{kN/m}^3$$

(5.9.3)

For the stiff clay layer:

$$\gamma_{sat} = \frac{\gamma_w G_s (1+w)}{1 + G_s w} = \frac{(9.81 \text{ kN/m}^3)(2.68)(1+0.20)}{1 + (2.68)(0.20)} = 20.54 \text{ kN/m}^3$$

$$(5.9.4)$$

a) The hydrostatic stresses are determined for the initial location of the groundwater table as shown in Figure Ex. 5.4A.
At a depth of 1.5 m:
   The vertical total stress is

$$\sigma_v = (20.03 \text{ kN/m}^3)(1.5 \text{ m}) = \mathbf{30.04 \text{ kN/m}^2} \qquad (5.9.5)$$

The pore fluid pressure is

$$u = \mathbf{0.0} \qquad (5.9.6)$$

The vertical effective stress is thus

$$\sigma'_v = \sigma_v - u = \mathbf{30.04 \text{ kN/m}^2} \qquad (5.9.7)$$

The horizontal effective stress is thus

$$\sigma'_h = K_0 \sigma'_v = (1.20)(30.04 \text{ kN/m}^2) = \mathbf{36.05 \text{ kN/m}^2} \qquad (5.9.8)$$

At a depth of 4.0 m:
   The vertical total stress is

$$\sigma_v = (20.03 \text{ kN/m}^3)(3.0 \text{ m}) + (21.40 \text{ kN/m}^3)(1.0 \text{ m}) = \mathbf{81.49 \text{ kN/m}^2}$$

$$(5.9.9)$$

The pore fluid pressure is

$$u = -(9.81 \text{ kN/m}^3)(1.0 \text{ m}) = \mathbf{-9.81 \text{ kN/m}^2} \qquad (5.9.10)$$

The vertical effective stress is thus

$$\sigma'_v = 81.49 - (-9.81) = \mathbf{91.30 \text{ kN/m}^2} \qquad (5.9.11)$$

The horizontal effective stress is thus

$$\sigma'_h = K_0 \sigma'_v = (0.470)(91.30 \text{ kN/m}^2) = \mathbf{42.91 \text{ kN/m}^2} \qquad (5.9.12)$$

At a depth of 8.0 m:
   The vertical total stress is

$$\sigma_v = (20.03 \text{ kN/m}^3)(3.0 \text{ m}) + (21.40 \text{ kN/m}^3)(5.0 \text{ m}) = \mathbf{167.1 \text{ kN/m}^2}$$

$$(5.9.13)$$

The pore fluid pressure is

$$u = \left(9.81 \text{ kN/m}^3\right)(3.0 \text{ m}) = \mathbf{29.43 \text{ kN/m}^2} \qquad (5.9.14)$$

The vertical effective stress is thus

$$\sigma'_v = 167.1 - 29.43 = \mathbf{137.7 \text{ kN/m}^2} \qquad (5.9.15)$$

The horizontal effective stress is thus

$$\sigma'_h = K_0\sigma'_v = (0.470)\left(137.7 \text{ kN/m}^2\right) = \mathbf{64.70 \text{ kN/m}^2} \qquad (5.9.16)$$

At a depth of 16.0 m:
The vertical total stress is

$$\sigma_v = \left(20.03 \text{ kN/m}^3\right)(3.0 \text{ m}) + \left(21.40 \text{ kN/m}^3\right)(8.0 \text{ m})$$
$$+ \left(15.76 \text{ kN/m}^3\right)(5.0 \text{ m}) = \mathbf{310.1 \text{ kN/m}^2} \qquad (5.9.17)$$

The pore fluid pressure is

$$u = \left(9.81 \text{ kN/m}^3\right)(11.0 \text{ m}) = \mathbf{107.9 \text{ kN/m}^2} \qquad (5.9.18)$$

The vertical effective stress is thus

$$\sigma'_v = 310.1 - 107.9 = \mathbf{202.2 \text{ kN/m}^2} \qquad (5.9.19)$$

The horizontal effective stress is thus

$$\sigma'_h = K_0\sigma'_v = (0.658)\left(202.2 \text{ kN/m}^2\right) = \mathbf{133.0 \text{ kN/m}^2} \qquad (5.9.20)$$

At a depth of 25.0 m:
The vertical total stress is

$$\sigma_v = \left(20.03 \text{ kN/m}^3\right)(3.0 \text{ m}) + \left(21.40 \text{ kN/m}^3\right)(8.0 \text{ m})$$
$$+ \left(15.76 \text{ kN/m}^3\right)(10.0 \text{ m}) + \left(20.54 \text{ kN/m}^3\right)(4.0 \text{ m}) = \mathbf{471.1 \text{ kN/m}^2}$$
$$(5.9.21)$$

The pore fluid pressure is

$$u = \left(9.81 \text{ kN/m}^3\right)(20.0 \text{ m}) = \mathbf{196.2 \text{ kN/m}^2} \qquad (5.9.22)$$

The vertical effective stress is thus

$$\sigma'_v = 471.1 - 196.2 = \mathbf{274.9 \text{ kN/m}^2} \qquad (5.9.23)$$

The horizontal effective stress is thus

$$\sigma'_h = K_0\sigma'_v = (1.0)\left(274.9 \text{ kN/m}^2\right) = \mathbf{274.9 \text{ kN/m}^2} \qquad (5.9.24)$$

Figure Ex. 5.9B shows the variation with depth of the vertical total stress, pore fluid pressure, and vertical effective stress.

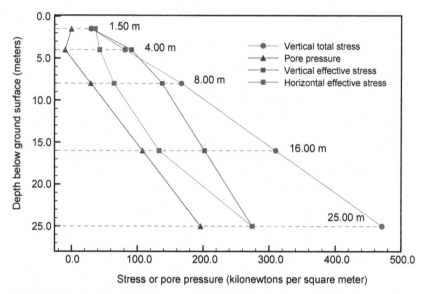

FIGURE EX. 5.9B Variation with depth of vertical total stress, pore fluid pressure, and vertical and horizontal effective stress when ignoring capillary rise.

b) The hydrostatic stresses are next determined for the case where the groundwater table is *lowered* by 6 m.

At a depth of 16.0 m:

The vertical total stress is unchanged, i.e.,

$$\sigma_v = (20.03 \text{ kN/m}^3)(3.0 \text{ m}) + (21.40 \text{ kN/m}^3)(8.0 \text{ m}) \\ + (15.76 \text{ kN/m}^3)(5.0 \text{ m}) = \textbf{310.1 kN/m}^2 \tag{5.9.25}$$

The pore fluid pressure is

$$u = (9.81 \text{ kN/m}^3)(5.0 \text{ m}) = \textbf{49.05 kN/m}^2 \tag{5.9.26}$$

The vertical effective stress is thus

$$\sigma_v' = 310.1 - 49.05 = \textbf{261.1 kN/m}^2 \tag{5.9.27}$$

Finally, the horizontal effective stress is thus

$$\sigma_h' = K_0\sigma_v' = (0.658)(261.1 \text{ kN/m}^2) = \textbf{171.8kN/m}^2 \tag{5.9.28}$$

EXAMPLE PROBLEM 5.10

General Remarks

This example problem illustrates how information is obtained from in situ stresses computed under hydrostatic conditions.

Problem Statement

At a given site the soil stratum consists of a thick clay layer. The groundwater table is located 1.5 m below the ground surface. Above the groundwater table the degree of saturation (S) is 92.5%. If the specific gravity of solids (G_s) is equal to 2.71 and the void ratio in the clay is 1.21, determine (a) at which depth (d) below the ground surface the vertical effective stress will be equal to 120 kPa and (b) the vertical effective stress at this depth immediately after the groundwater table is lowered by 2.0 m.

Solution

a) Above the groundwater table, the moist unit weight is computed using the expression determined in Case 1.7 of Chapter 1, i.e.,

$$\gamma = \frac{\gamma_w(G_s + Se)}{1+e} = \frac{(9.81 \text{ kN/m}^3)[2.71 + (0.925)(1.21)]}{1 + 1.21} = 17.0 \text{ kN/m}^3$$

$$(5.10.1)$$

Below the groundwater table, the moist unit weight is computed using the expression determined in Case 1.8 of Chapter 1, i.e.,

$$\gamma_{sat} = \frac{\gamma_w(G_s + e)}{1+e} = \frac{(9.81 \text{ kN/m}^3)[2.71 + 1.21]}{1 + 1.21} = 17.4 \text{ kN/m}^3 \quad (5.10.2)$$

At a depth (d) below the ground surface the vertical total stress is

$$\sigma_v = \gamma(1.5 \text{ m}) + \gamma_{sat}(d - 1.5 \text{ m}) \quad (5.10.3)$$

The pore pressure at the same depth is

$$u = \gamma_w(d - 1.5 \text{ m}) \quad (5.10.4)$$

Finally, the vertical effective stress at a depth (d) below the ground surface is

$$\sigma'_v = \sigma_v - u = \gamma(1.5 \text{ m}) + (\gamma_{sat} - \gamma_w)(d - 1.5 \text{ m}) \quad (5.10.5)$$

Solving Eq. (5.10.5) for d gives

$$d = \frac{\sigma'_v - \gamma(1.5 \text{ m})}{(\gamma_{sat} - \gamma_w)} + 1.5 \text{ m} \quad (5.10.6)$$

Substituting all known values into Eq. (5.10.6) gives the desired result

$$d = \frac{(120 \text{ kPa}) - (17.0 \text{ kN/m}^3)(1.5 \text{ m})}{(17.4 - 9.81) \text{ kN/m}^3} + 1.5 \text{ m} = \mathbf{14.0 \text{ m}} \quad (5.10.7)$$

b) Since the vertical effective stress is computed immediately following the lowering of the groundwater table, the unit weights of the partially and fully saturated portions of the clay layer remain unchanged. The vertical total stress at a depth $d = 14.0$ m below the ground surface is

$$\sigma_v = \gamma(1.5 \text{ m}) + \gamma_{sat}(14.0 - 1.5 \text{ m})$$
$$= (17.0 \text{ kN/m}^3)(1.5 \text{ m}) + (17.4 \text{ kN/m}^3)(14.0 - 1.5 \text{ m}) = 243.0 \text{ kPa}$$
$$(5.10.8)$$

The pore pressure at the same depth is now

$$u = \gamma_w(14.0 - 1.5 - 2.0 \text{ m}) = (9.81 \text{ kN/m}^3)(10.5 \text{ m}) = 103.0 \text{ kPa}$$
$$(5.10.9)$$

The vertical effective stress at a depth $d = 14.0$ m below the ground surface is thus

$$\sigma'_v = \sigma_v - u = 243.0 - 103.0 = \mathbf{140.0 \text{ kPa}} \qquad (5.10.10)$$

Chapter 6

Example Problems Involving One-Dimensional Fluid Flow in Soils

6.0 GENERAL COMMENTS

In soil mechanics there are *three* general problem types that require a clear understanding of fluid flow through the soil, namely those seeking to determine the following:

1. The *rate* at which fluid flows through a soil (e.g., leakage through an earth dam),
2. The *rate* of settlement due to consolidation (i.e., the expulsion of excess pore fluid from the pores), and
3. The *strength* of slopes, embankments, etc.

The flow of fluid through a soil can be either *steady state* or *transient*. In the case of *steady-state* flow,

- Pore fluid pressures remain constant.
- The rate of flow through the soil is a constant.
- The effective stresses remain constant.
- The soil does not deform.

In the case of *transient* flow of fluid through a soil,

- Pore fluid pressures and thus effective stresses vary with time.
- The soil deforms.
- There exists a complex interrelationship between pore fluid pressure, flow, and deformation.

Soil Mechanics. http://dx.doi.org/10.1016/B978-0-12-804491-9.00006-9

6.1 CONSERVATION OF MASS

If the mass of a system remains unchanged, then

$$\frac{dm}{dt} = \frac{d}{dt} \int_V \rho dV = 0 \tag{6.1}$$

where m is the mass of the system, ρ is the density of the material, V is the volume of the system, and t is time.

For *incompressible* flow the density of the fluid remains constant. The law of conservation of mass then reduces to the *equation of continuity of flow*, i.e.,

$$\frac{dm}{dt} = \rho \frac{d}{dt} \int_V dV = \rho \frac{dV}{dt} = 0 \Rightarrow \frac{dV}{dt} = 0 \tag{6.2}$$

Consider two sections along the saturated aquifer shown in Figure 6.1. Define the following quantities:

A_1 = cross-sectional area perpendicular to the direction of flow at Section 1 [units of L^2].
A_2 = cross-sectional area perpendicular to the direction of flow at Section 2 [units of L^2].
v_1 = velocity of flow at Section 1 [units of Lt^{-1}].
v_2 = velocity of flow at Section 2 [units of Lt^{-1}].
Q = rate of discharge [units of $L^3 t^{-1}$].

To satisfy the equation of continuity of flow it follows that

$$\frac{dV}{dt} = 0 \Rightarrow Q = v_1 A_1 = v_2 A_2 \tag{6.3}$$

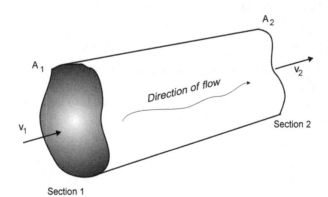

FIGURE 6.1 Schematic illustration of flow through a saturated aquifer.

6.2 BERNOULLI'S ENERGY EQUATION

Applying Bernoulli's[1] principle to an incompressible, steady flow of a fluid gives Bernoulli's energy equation for incompressible flow. In particular, at some point along a streamline,

$$\frac{p}{\rho_w} + gz + \frac{v^2}{2} = C \tag{6.4}$$

where C is a constant and p = pressure at a point [units of FL^{-2}]; ρ_w = density of the fluid [units of $FL^{-4}t$]; g = gravitational acceleration [units of Lt^{-2}]; z = elevation of the point above a datum [units of L]; v = velocity of flow at the point [units of Lt^{-1}].

This expression represents the steady-flow energy equation in terms of energy per unit of mass of fluid [units of $L^2 t^{-2}$]. The term p/ρ_w is the *pressure energy* per unit mass. The term gz is the *potential energy* per unit mass.[2] Finally, the term $v^2/2$ is the *kinetic energy* per unit mass.[3] Thus, in words, "the energy per unit mass is conserved along a streamline."

Diving Bernoulli's energy equation for incompressible flow by g gives

$$\frac{p}{\gamma_w} + z + \frac{v^2}{2g} = C^* \tag{6.5}$$

where $C^* = C/g$ is a constant and $\rho_w g = \gamma_w$. The equation is now expressed in terms of energy per unit weight [units of L]. The respective terms in the equation are defined as follows:

- The term p/γ_w is the pressure or piezometric head.
- The term z represents the elevation head.
- The term $v^2/2g$ is the velocity head.

Remark: In most soils the voids (pores) are so small that the flow is laminar (i.e., nonturbulent).

Remark: Since the velocity of flow is typically small in soils and rocks, the velocity head is typically *negligible*.

---

1. Daniel Bernoulli (1700–1782).
2. The potential energy is equal to *mgz*.
3. The kinetic energy is equal to *mv²/2*.

Bernoulli's equation thus reduces to

$$\frac{p}{\gamma_w} + z = C^* \tag{6.6}$$

where C^* is a constant total head.

Inserting imaginary standpipes into a soil (Figure 6.2) is a convenient method of visualizing the total head.

For some point A located in the saturated soil, the total head (h) is simply the sum of the elevation head (z_A) of the standpipe with respect to the selected datum and the pressure head p_A/γ_w. The latter represents the elevation that the pore fluid raises in the tube.

To gain additional insight into the pressure head, note that the pore fluid pressure at point A, at a depth d below the groundwater table, is

$$u_A = \rho_w g d = \gamma_w d \tag{6.7}$$

But d is also the pressure head at point A, i.e., $d = p_A/\gamma_w$. Thus,

$$u_A = \gamma_w d = \gamma_w \left(\frac{p_A}{\gamma_w}\right) = p_A \tag{6.8}$$

indicating that p_A is exactly equal to the pore pressure at point A.

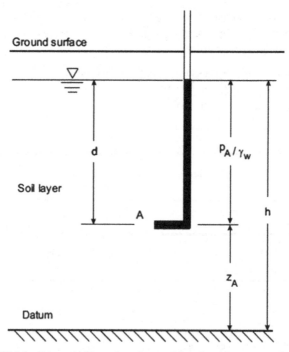

FIGURE 6.2 Schematic illustration of a standpipe placed in a saturated aquifer.

Remark: When fluid flows through soils and rock, energy, or head is lost through friction, much as in flow through pipes and in open channels.

6.3 HEAD LOSS

For flow to occur between two points a and b, a loss in head must occur between these points. Let h_a and h_b be the total heads at points a and b, respectively; thus,

$$h_a = \frac{p_a}{\gamma_w} + z_a; \ h_b = \frac{p_b}{\gamma_w} + z_b \qquad (6.9)$$

If $\Delta h = h_a - h_b \neq 0$, fluid flow will take place between points a and b. In particular,

- If $\Delta h > 0$ (i.e., $h_a > h_b$), the flow will be from point a to point b.
- If $\Delta h < 0$ (i.e., $h_a < h_b$), the flow will be from point b to point a.

6.4 HYDRAULIC GRADIENT

Writing the head loss in nondimensional form gives

$$\frac{\Delta h}{L} \equiv i \qquad (6.10)$$

where i is the *hydraulic gradient*, and L is the distance between points a and b, measured *along the direction of flow*.

To better understand the proper definition of L, consider the one-dimensional flow between points 1 and 2 shown in Figure 6.3. Since $h_1 > h_2$, the fluid will flow from point 1 to point 2. The distance between points 1 and 2 (L) is measured along the flow tube (as opposed to using the horizontal distance between these two points).

6.5 SEEPAGE VELOCITY

In general, all voids (pores) in soils are assumed to be connected to neighboring voids. The voids of soils are small in diameter and irregular. As fluid flows through a soil, it follows a *tortuous* path. Any flow calculated using the theory of pipe flow will thus be in error!

Rather than computing specific velocities through particular voids, it is more realistic to instead compute an *average* or *superficial* velocity through a given area of soil. Consider the one-dimensional flow through the saturated soil sample shown in Figure 6.4. The cross-sectional area of the tube containing the sample is A.

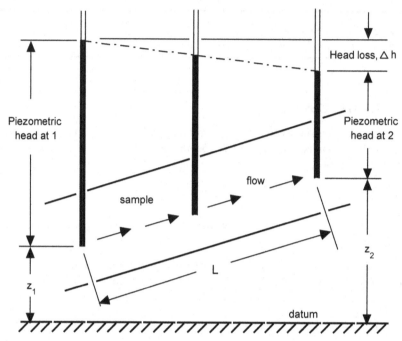

FIGURE 6.3 One-dimensional inclined flow example.

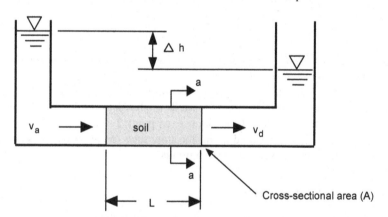

FIGURE 6.4 One-dimensional flow example.

Let v_a and v_d be the approach and discharge velocities [units of Lt^{-1}], respectively, and Q be the discharge [units of L^3t^{-1}]. From the continuity of flow,

$$Q = v_a A = v_d A \qquad (6.11)$$

or

$$v_a = v_d = \frac{Q}{A} \equiv v \qquad (6.12)$$

where v is the *average* or *superficial* velocity. It is this velocity that would be macroscopically measured.

Next consider section $a-a$ through the soil sample, taken at right angles to the direction of flow. Figure 6.5 shows the actual soil at section $a-a$, as well as the associated phase diagram. Assume a *unit* thickness of the sample in the direction of flow (i.e., into the page).

Denote the *actual* or *seepage*[4] velocity of flow through the voids by v_s. From the continuity of flow,

$$Q = v_a A = v_d A = vA = v_s A_v \qquad (6.13)$$

where A_v is the area of the voids (Figure 6.5). Thus,

$$vA = v(A_v + A_s) = v_s A_v \Rightarrow v_s = \left(\frac{A_v + A_s}{A_v}\right)v \qquad (6.14)$$

Multiplying numerator and denominator by the length along the flow path (L) gives

$$v_s = \left(\frac{A_v + A_s}{A_v}\right)v\left(\frac{L}{L}\right) = \left(\frac{V}{V_v}\right)v = \frac{v}{n} \qquad (6.15)$$

where n is the porosity. Since $n < 1$, it follows that the seepage velocity (v_s) will always be greater than the superficial velocity (v).

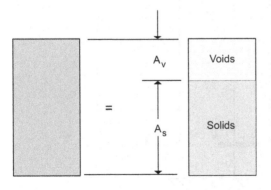

FIGURE 6.5 Schematic illustration of section a—a.

Actual soil   Phase diagram

---

4. Seepage is defined as the slow escape of fluid through a porous material.

6.6 DARCY'S LAW

The French waterworks engineer Henri Darcy[5] conducted vertical column flow experiments (Figure 6.6) on homogeneous sand filters in connection with the foundation of the city of Dijon, France.

Darcy concluded that the rate of flow Q (volume per unit time) was

- proportional to the constant cross-sectional area A of the sand filter;
- proportional to the difference in total head $(h_1 - h_2)$, where h_1 and h_2 are the total heads at points 1 and 2, respectively;
- inversely proportional to the length L of the sand filter.

When combined, the aforesaid observations give the famous *Darcy formula* (or *Darcy's law*), i.e.,

$$Q = kA \left(\frac{h_1 - h_2}{L} \right) \tag{6.16}$$

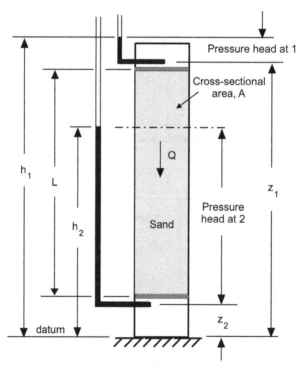

FIGURE 6.6 Schematic illustration of Darcy's experiment.

---

5. Henry Darcy (1803–1858).

where the coefficient of proportionality k is commonly referred to as the *permeability* or *hydraulic conductivity* (units of Lt^{-1}). The quantity $(h_1 - h_2)/L$ is seen to be the hydraulic gradient i. Thus, Darcy's formula is typically written as

$$Q = kiA \qquad (6.17)$$

Darcy's formula is thus a phenomenologically derived constitutive law that relates the rate of flow (per unit area) to the hydraulic gradient. It is analogous to Fick's law in diffusion theory, Fourier's law in heat conduction, and Ohm's law in the field of electrical circuits.

The rather simplified configuration shown in Figure 6.6 likewise holds for the more general inclined case shown in Figure 6.7.

Remark: The coefficient of permeability (k) expresses the ease with which fluid passes through a soil.

Remark: When a soil is said to have a certain coefficient of permeability, this value assumes the pore fluid to be *water*, typically at 20°C.

Darcy's law is also written as

$$Q = \frac{\overline{k}}{\eta} \gamma_w \, iA \qquad (6.18)$$

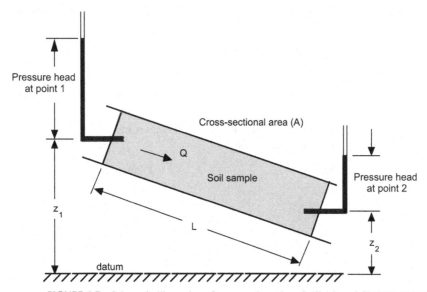

FIGURE 6.7 Schematic illustration of seepage through an inclined sand filter.

where $\bar{k}$ is the *intrinsic permeability* [units of L^2] and η is the *dynamic viscosity* of water [units of $ML^{-1}t^{-1}$]. To relate the intrinsic permeability to k, equate Eqs. (6.17) and (6.18), giving

$$kiA = \frac{\bar{k}}{\eta}\gamma_w iA \implies k = \frac{\bar{k}}{\eta}\gamma_w \quad or \quad \frac{k}{\gamma_w} = \frac{\bar{k}}{\eta} \tag{6.19}$$

The permeability of a soil depends on the characteristics of the pore fluid, as well as the solid phase. The following factors thus affect the permeability:

For the fluid phase,

- the fluid density (γ_w);
- the magnitude of the dynamic viscosity (η), which depends on temperature.

For the solid phase,

- the particle size distribution in the microfabric;
- the shape of particles (or pores);
- the tortuosity (i.e., the degree of "crookedness" of the pore space);
- the specific surface (i.e., the magnitude of the surface area/unit mass);
- the porosity or void ratio;
- the degree of saturation.

Table 6.1 lists some typical values for the coefficient of permeability. It is evident that the values of k differ significantly (e.g., by six orders of magnitude in going from gravels to clays).

6.7 EXPERIMENTAL DETERMINATION OF PERMEABILITY

Values for the coefficient of permeability are typically determined from the following:

1. relatively simple laboratory tests;
2. field pumping tests;
3. empirical correlations.

TABLE 6.1 Typical Permeability Values for Common Soil Types

Soil Type	k (m/s)
Gravel	$>10^{-2}$
Sand	10^{-2} to 10^{-5}
Silt	10^{-5} to 10^{-8}
Clay	$<10^{-8}$

Additional details pertaining to field pumping tests[6] and empirical correlations[7] are given elsewhere. Attention is instead turned to the constant-head and falling-head permeability tests.

6.7.1 Constant-Head Permeability Test

The constant-head method is typically used on relatively permeable soils such as gravels and clean sands.[8] In this test water is allowed to flow through the soil under a steady-state head condition, with the quantity (volume) of water flowing through the soil specimen being measured over a period of time. Figure 6.8 gives a schematic illustration of the constant-head permeability test.

By knowing the quantity $\overline{Q}$ of water measured (units of L^3), length L of specimen, the cross-sectional area A of the specimen, the time t required for the quantity of water $\overline{Q}$ to be discharged, and the head Δh, from Darcy's law it follows that

$$\overline{Q} = k\left(\frac{\Delta h}{L}\right)A \tag{6.20}$$

The permeability is then calculated from

$$k = \frac{\overline{Q}L}{At\Delta h} = \frac{\overline{Q}}{Ait} \tag{6.21}$$

where i is the hydraulic gradient as defined in Eq. (6.10).

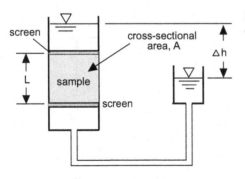

FIGURE 6.8 Schematic illustration of a constant-head permeability test.

6. Lambe, T.W., Whitman, R.V., 1979. Soil Mechanics, SI Version. John Wiley and Sons, New York, NY.
7. Das, B.M., 2010. Principles of Geotechnical Engineering, seventh ed. Cengage Learning, Stamford, CT.
8. Kezdi, A., 1974. Handbook of Soil Mechanics. Elsevier, Amsterdam, The Netherlands.

6.7.2 Falling-Head Permeability Test

The falling-head test is quite similar to the constant-head test in its initial configuration. The soil sample is first saturated under a specific head condition. However, unlike the case of the constant-head test, the water is then allowed to flow through the soil *without* maintaining a constant pressure head. This gives the falling-head test the advantage of being used for both fine-grained and coarse-grained soils. Figure 6.9 gives a schematic illustration of the falling-head permeability test.

During a given increment of time dt the incremental change in volume of fluid flowing through the sample is

$$dV = -a\,dh \qquad (6.22)$$

where a is the cross-sectional area of the standpipe and dh represents an increment in hydraulic head. From Darcy's law

$$dV = k\left(\frac{h}{L}\right)A\,dt \qquad (6.23)$$

Equating the two aforesaid expressions for dV gives

$$-a\,dh = k\left(\frac{h}{L}\right)A\,dt \;\Rightarrow\; -\frac{dh}{h} = \frac{k}{L}\frac{A}{a}dt \qquad (6.24)$$

Integrating this equation gives

$$-\ln h = \frac{k}{L}\frac{A}{a}t + C \qquad (6.25)$$

where C is a constant. Noting that at $t = 0$, $h = h_0$ (Figure 6.9) it follows that $C = -\ln h_0$.

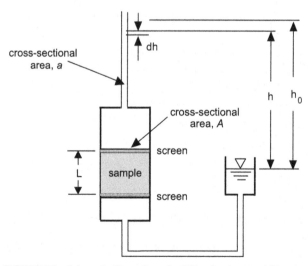

FIGURE 6.9 Schematic illustration of a falling-head permeability test.

Eq. (6.25) thus becomes

$$\ln\left(\frac{h_0}{h}\right) = \frac{k}{L}\frac{A}{a}t \tag{6.26}$$

The permeability is then calculated from

$$k = \left(\frac{aL}{At}\right)\ln\left(\frac{h_0}{h}\right) = 2.303\left(\frac{aL}{At}\right)\log_{10}\left(\frac{h_0}{h}\right) \tag{6.27}$$

The determination of values for the coefficient of permeability is also illustrated in several example problems appearing later in this chapter.

6.8 HYDROSTATIC CONDITIONS COMPARED TO UPWARD AND DOWNWARD SEEPAGE

Having discussed elevation, pressure, and total heads, it is timely to look at three specific cases associated with one-dimensional flow through a saturated soil.

6.8.1 No Seepage (Hydrostatic Conditions)

Consider the saturated soil sample shown in Figure 6.10.

At point a:
The elevation head, $z_a = e + H$
The pressure head, $p_a/\gamma_w = d$
The total head, $h_a = (e + H) + d$.

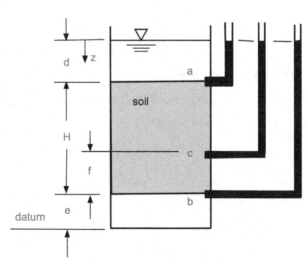

FIGURE 6.10 Saturated soil sample without seepage.

At point b:

The elevation head, $z_b = e$

The pressure head, $p_b/\gamma_w = H + d$

The total head, $h_b = e + (H + d)$.

Since $h_a = h_b$, there is no head loss along the sample (i.e., $\Delta h = 0$) and thus no flow. To verify this finding, consider point c.

The elevation head, $z_c = e + f$

The pressure head, $p_c/\gamma_w = (H - f) + d$

The total head, $h_c = (e + f) + [(H - f) + d] = e + H + d$.

Thus, $h_c = h_a = h_b$, which further confirms the hydrostatic conditions. To relate these results to the problems presented in Chapter 5, compute the pore pressures at points a, b, and c.

At point a:

$$\frac{p_a}{\gamma_w} = d \Rightarrow p_a = \gamma_w d \tag{6.28}$$

At point b:

$$\frac{p_b}{\gamma_w} = H + d \Rightarrow p_b = \gamma_w(H + d) \tag{6.29}$$

At point c:

$$\frac{p_c}{\gamma_w} = (H - f) + d \Rightarrow p_a = \gamma_w\big[(H - f) + d\big] \tag{6.30}$$

The aforesaid results are precisely the pore pressures associated with hydrostatic conditions. Next, compute the vertical total and effective stress at points b and c.

At point b:

The vertical total stress is

$$\sigma_v = \gamma_w d + \gamma_{sat} H \tag{6.31}$$

The vertical effective stress is

$$\sigma'_v = \sigma_v - p_b = (\gamma_w d + \gamma_{sat} H) - \gamma_w(H + d) = (\gamma_{sat} - \gamma_w)H = \gamma_b H \tag{6.32}$$

where γ_b is the buoyant unit weight.

At point c:

The vertical total stress is

$$\sigma_v = \gamma_w d + \gamma_{sat}(H - f) \tag{6.33}$$

The vertical effective stress is

$$\sigma'_v = \sigma_v - p_c = \gamma_w d + \gamma_{sat}(H - f) - \gamma_w\big[(H - f) + d\big] = (\gamma_{sat} - \gamma_w)(H - f)$$
$$= \gamma_b(H - f) \tag{6.34}$$

The aforesaid results are consistent with those presented in Chapter 5 for hydrostatic conditions.

6.8.2 Downward Seepage

Next consider the saturated soil sample shown in Figure 6.11.

At point a:
  The elevation head, $z_a = e + H$
  The pressure head, $p_a/\gamma_w = d$
  The total head, $h_a = (e + H) + d$.
At point b:
  The elevation head, $z_b = e$
  The pressure head, $p_b/\gamma_w = H + d - \Delta h$
  The total head, $h_b = e + (H + d - \Delta h)$.
At point c:
  The elevation head, $z_c = e + f$
  The pressure head, $p_c/\gamma_w = (H - f + d) - \Delta h(H - f)/H$
  The total head, $h_c = e + f + p_c/\gamma_w$.

Next, compute the vertical total and effective stress at points b and c.

At point b:
  The vertical total stress is unchanged from the case with no seepage (Figure 6.8), i.e.,

$$\sigma_v = \gamma_w d + \gamma_{sat} H \qquad (6.35)$$

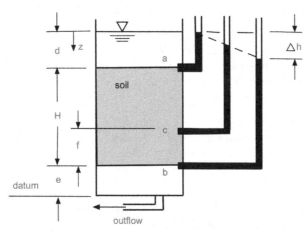

FIGURE 6.11 Saturated soil sample with downward seepage.

The vertical effective stress is

$$\sigma'_v = \sigma_v - p_b = (\gamma_w d + \gamma_{sat} H) - \gamma_w (H + d - \Delta h)$$
$$= (\gamma_{sat} - \gamma_w)H + \gamma_w \Delta h = \gamma_b H + \gamma_w \Delta h \tag{6.36}$$

Thus, for the case of downward seepage, the effective stress at the bottom of the sample is *increased* by the amount $\gamma_w \Delta h$.

At point c:

The vertical total stress is again unchanged from the case with no seepage, i.e.,

$$\sigma_v = \gamma_w d + \gamma_{sat}(H - f) \tag{6.37}$$

The vertical effective stress is

$$\sigma'_v = \gamma_w d + \gamma_{sat}(H - f) - \gamma_w \left[H - f + d - \frac{\Delta h(H - f)}{H} \right]$$
$$= (\gamma_{sat} - \gamma_w)(H - f) + \gamma_w \left[\frac{\Delta h(H - f)}{H} \right] \tag{6.38}$$
$$= \gamma_b(H - f) + \gamma_w \, i \, (H - f)$$

where i is the hydraulic gradient. To make sense of the aforesaid expression for vertical effective stress, let $f = H/2$. Then,

$$\sigma'_v = \gamma_b \frac{H}{2} + \gamma_w \frac{\Delta h}{2} \tag{6.39}$$

If $f = 0$, the aforesaid expression reduces to that determined earlier for point b. If $f = H$, the effective stress is zero, which is correct for point a.

6.8.3 Upward Seepage

Finally, consider the saturated soil sample shown in Figure 6.12.

At point a:
  The elevation head, $z_a = e + H$
  The pressure head, $p_a/\gamma_w = d$
  The total head, $h_a = (e + H) + d$.

At point b:
  The elevation head, $z_b = e$
  The pressure head, $p_b/\gamma_w = H + d + \Delta h$
  The total head, $h_b = e + (H + d + \Delta h)$.

At point c:
  The elevation head, $z_c = e + f$
  The pressure head, $p_c/\gamma_w = (H - f + d) + \Delta h(H - f)/H$
  The total head, $h_c = e + f + p_c/\gamma_w$.

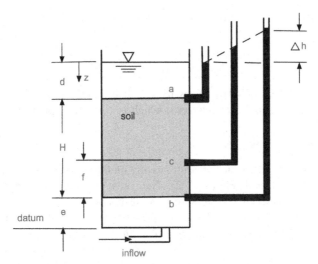

FIGURE 6.12 Saturated soil sample with upward seepage.

Next, compute the vertical total and effective stress at points b and c.

At point b:

The vertical total stress is unchanged from the case with no seepage (Figure 6.8), i.e.,

$$\sigma_v = \gamma_w d + \gamma_{sat} H \tag{6.40}$$

The vertical effective stress is

$$
\begin{aligned}
\sigma'_v = \sigma_v - p_b &= (\gamma_w d + \gamma_{sat} H) - \gamma_w (H + d + \Delta h) \\
&= (\gamma_{sat} - \gamma_w)H - \gamma_w \Delta h = \gamma_b H - \gamma_w \Delta h
\end{aligned} \tag{6.41}
$$

Thus, for the case of upward seepage, the effective stress at the bottom of the sample is *decreased* by the amount $\gamma_w \Delta h$.

At point c:

The vertical total stress is again unchanged from the case with no seepage, i.e.,

$$\sigma_v = \gamma_w d + \gamma_{sat}(H - f) \tag{6.42}$$

The vertical effective stress is

$$
\begin{aligned}
\sigma'_v &= \gamma_w d + \gamma_{sat}(H - f) - \gamma_w \left[H - f + d + \frac{\Delta h(H - f)}{H}\right] \\
&= (\gamma_{sat} - \gamma_w)(H - f) - \gamma_w \left[\frac{\Delta h(H - f)}{H}\right] \\
&= \gamma_b(H - f) - \gamma_w \, i \, (H - f)
\end{aligned} \tag{6.43}
$$

where i is again the hydraulic gradient.

To make sense of the aforesaid expression for vertical effective stress, let $f = H/2$. Then,

$$\sigma'_v = \gamma_b\frac{H}{2} - \gamma_w\frac{\Delta h}{2} \tag{6.44}$$

If $f = 0$, Eq. (6.44) reduces to that determined earlier for point b. If $f = H$, the effective stress is zero, which is correct for point a.

6.9 SEEPAGE FORCES

Figure 6.13 shows the water pressures acting on the saturated soil sample with upward seepage (recall Figure 6.12). The water pressure acting at point a is equal to the product of the pressure head at this point and the unit weight of water, i.e., $\gamma_w d$. Similarly, the water pressure acting at the bottom of the sample (point b) is equal to $\gamma_w(H + d + \Delta h)$. Figure 6.13A shows both of these pressures; since they act at the sample boundaries, they are sometimes referred to as the *boundary water pressures*.[9]

Figure 6.13B shows the water pressures that would exist under hydrostatic conditions, i.e., if there was no flow. The pressure acting at point a is again $\gamma_w d$. At point b it is equal to $\gamma_w(H + d)$. These values are sometimes referred to as the *buoyancy water pressures*.

The difference between the boundary water pressures and the hydrostatic pressures is equal to the *seepage pressure* $\gamma_w\Delta h$ that is shown in Figure 6.13C.

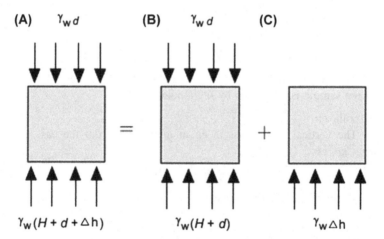

FIGURE 6.13 Water pressure acting on soil sample with upward seepage. (A) Boundary water pressures, (B) hydrostatic water pressures, and (C) pressure associated with upward seepage.

9. Lambe, T.W., Whitman, R.V., 1979. Soil Mechanics, SI Version. John Wiley and Sons, New York, NY.

The upwardly flowing pore fluid exerts the seepage pressure; it is uniformly and completely dissipated in the course of upward flow through the soil.

To obtain the seepage force, the seepage stress must be multiplied by the cross-sectional area (A) of the sample, giving

$$F_{seepage} = \gamma_w \Delta h A \qquad (6.45)$$

A convenient way in which to express the seepage force is the force per unit of volume;

$$j = \frac{F_{seepage}}{V} = \frac{\Delta h A \gamma_w}{AH} = i\gamma_w \qquad (6.46)$$

where $i = \Delta h/L$ is again the hydraulic gradient.

6.10 CRITICAL HYDRAULIC GRADIENT FOR UPWARD SEEPAGE

If a cohesionless soil is subjected to a pore fluid flow condition that results in zero (or near zero) effective stress, the strength of the soil goes to zero. This is often referred to as a "quick" condition.[10] Under such conditions, the seepage forces overcome the gravitational forces and the pore pressure equals the total stress. The excess pore pressure then forces the overlying soil mass to rise and *heave*. In cohesionless soils, the soil bubbles in a "boil"; since the soil has no strength, it often washes out.

There are two common situations in which the pore pressure equals the total stress, namely as follows:

1. In the case of upward seepage (Figure 6.12), where the seepage force equals the submerged weight of the soil.
2. When a sudden loading is applied to a loose saturated soil; this causes a volume decrease in the soil and results in the effective stress being transferred to the pore pressure.

The hydraulic gradient associated with a "quick" condition near an unrestricted soil surface is called the critical gradient (i_c). For the case of upward seepage, the vertical effective stress at the bottom of the sample (i.e., at point b in Figure 6.12) is

$$\sigma'_v = \gamma_b H - \gamma_w \Delta h \qquad (6.47)$$

where γ_b is the buoyant unit weight. Setting $\sigma'_v = 0$ gives $\Delta h = \gamma_b/\gamma_w H$. The critical hydraulic gradient is thus

---

10. Since cohesive soils can have shear strength even at zero effective stress, they do not necessarily exhibit "quick" conditions.

$$i_c = \frac{\Delta h}{H} = \frac{\gamma_b}{\gamma_w} \tag{6.48}$$

Since $\gamma_b = \gamma_{sat} - \gamma_w$,

$$i_c = \frac{\gamma_{sat} - \gamma_w}{\gamma_w} = \frac{\gamma_{sat}}{\gamma_w} - 1 \tag{6.49}$$

From Case 1.8 in Chapter 1,

$$\gamma_{sat} = \frac{\gamma_w(G_s + e)}{1 + e} \tag{6.50}$$

Thus,

$$i_c = \frac{G_s + e}{1 + e} - 1 = \frac{G_s - 1}{1 + e} \tag{6.51}$$

6.11 ONE-DIMENSIONAL SEEPAGE THROUGH ANISOTROPIC SOIL STRATA

Due to their mode of deposition, natural soils are often nonhomogeneous. Consequently, they possess a permeability anisotropy, i.e., they have different permeabilities in different coordinate directions.

6.11.1 Equivalent Horizontal Permeability

Figure 6.14 shows a hypothetical nonhomogeneous soil stratum of thickness H that consists of n layers. The direction of flow is assumed to be parallel to the global x-axis, i.e., in the horizontal direction.

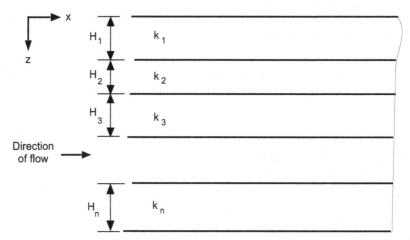

FIGURE 6.14 Schematic illustration of nonhomogeneous layered soil stratum: flow in horizontal direction.

The head loss along each layer will be the same, i.e.,

$$i_1 = i_2 = \cdots = i_n \equiv i \tag{6.52}$$

Let Q_x be the total discharge in the horizontal direction. From the continuity of flow equation,

$$Q_x = \sum_{j=1}^{n} Q_{x_j} = \sum_{j=1}^{n} k_{x_j} i_j A_j \tag{6.53}$$

where k_{x_j} is the permeability in the horizontal (x) direction for layer j, and A_j is the cross-sectional area of the layer j. Assuming a unit width into the page gives $A_j = (1)H_j$.

Thus,

$$Q_x = \tilde{k}_x iH = \sum_{j=1}^{n} k_{x_j} i_j H_j \tag{6.54}$$

where $\tilde{k}_x$ is the *equivalent* horizontal permeability. Since the head loss along each layer is the same,

$$\tilde{k}_x = \frac{1}{H} \sum_{j=1}^{n} k_{x_j} H_j = \frac{\sum_{j=1}^{n} k_{x_j} H_j}{\sum_{j=1}^{n} H_j} \tag{6.55}$$

6.11.2 Equivalent Vertical Permeability

Next assume the direction of flow to parallel to the global z-axis, i.e., in the vertical direction (Figure 6.15).

From the continuity of flow equation, the discharge across each layer must be the same, i.e.,

$$Q_z = Q_{z_1} = Q_{z_2} = \cdots = Q_{z_n} \tag{6.56}$$

Since the cross-sectional area for flow is the *same* for all the layers, the aforesaid expression is rewritten as follows:

$$v_z A = v_{z_1} A = v_{z_2} A = \cdots = v_{z_n} A \Rightarrow v_z = v_{z_1} = v_{z_2} = \cdots = v_{z_n} \tag{6.57}$$

Using Darcy's law,

$$v_z = \tilde{k}_z i = \tilde{k}_{z_1} i_1 = \tilde{k}_{z_2} i_2 = \cdots = \tilde{k}_{z_n} i_n \tag{6.58}$$

where $\tilde{k}_z$ is the *equivalent* vertical permeability and $i = \Delta h/H$ is the total hydraulic gradient. The total head loss (Δh) across the stratum in the vertical (z) direction will be the sum of the losses for each layer, i.e.,

$$\Delta h_1 + \Delta h_2 + \cdots + \Delta h_n = \Delta h \tag{6.59}$$

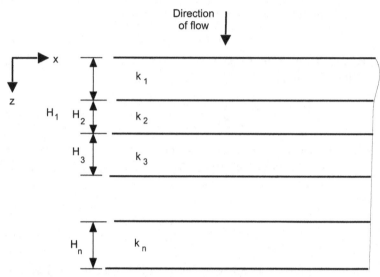

FIGURE 6.15 Schematic illustration of nonhomogeneous layered soil stratum: flow in vertical direction.

Thus,

$$v_z = \tilde{k}_z i = \tilde{k}_z \left(\frac{\Delta h}{H}\right) = \frac{\tilde{k}_z}{H} \sum_{j=1}^{n} \Delta h_j \Rightarrow \tilde{k}_z = \frac{v_z H}{\sum_{j=1}^{n} \Delta h_j} \tag{6.60}$$

But

$$v_{z_j} = k_{z_j} i_j = k_{z_j} \left(\frac{\Delta h_j}{H_j}\right) \Rightarrow \Delta h_j = \frac{v_{z_j} H_j}{k_{z_j}} \tag{6.61}$$

Since $v_z = v_{z1}$, it follows that

$$\tilde{k}_z = \frac{v_z H}{\left(\sum_{j=1}^{n} \Delta h_j\right)} = \frac{v_z H}{\left(\sum_{j=1}^{n} \frac{v_{z_j} H_j}{k_{z_j}}\right)} = \frac{H}{\left(\sum_{j=1}^{n} \frac{H_j}{k_{z_j}}\right)} \tag{6.62}$$

or

$$\tilde{k}_z = \frac{\sum_{j=1}^{n} H_j}{\left(\sum_{j=1}^{n} \frac{H_j}{k_{z_j}}\right)} \tag{6.63}$$

EXAMPLE PROBLEM 6.1

General Remarks

This example problem illustrates the manner in which the elevation head, pressure head, and total head are determined for the case of one-dimensional steady-state flow. In this problem the flow is directed toward decreasing elevation head.

Problem Statement

The soil shown in Figure Ex. 6.1A is supported on a porous stone and disk. Determine the elevation head, pressure head, and total head at points a, b, and c.

Solution

At point a:
Elevation head, $z_a = 0.60 + 0.80 + 1.60 = \textbf{3.00 m}$
Pressure head, $p_a/\gamma_w = \textbf{1.20 m}$
Total head, $h_a = 3.00 + 1.20 = \textbf{4.20 m}$.

At point c:
Elevation head, $z_c = \textbf{0.60 m}$
Pressure head, $p_c/\gamma_w = \textbf{-0.60 m}$
Total head, $h_c = 0.60 + (-0.60) = \textbf{0.0 m}$.

At point b:
Elevation head, $z_b = 0.60 + 0.80 = \textbf{1.40 m}$.

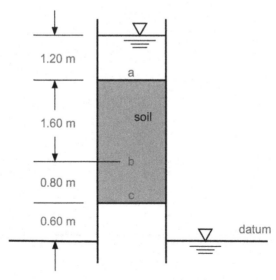

FIGURE EX. 6.1A Hypothetical apparatus for one-dimensional vertical flow (not to scale).

The total head is computed from the total heads at points a and c. Since the soil is homogeneous, the variation in total head along the soil is *linear*. As such, the total head at point b is thus

$$\frac{h_a - h_b}{h_a - h_c} = \frac{1.60 \text{ m}}{1.60 \text{ m} + 0.80 \text{ m}} \Rightarrow h_b = h_a - \left(\frac{1.60 \text{ m}}{2.40 \text{ m}}\right)(h_a - h_c) \quad (6.1.1)$$

Substituting the known total heads gives

$$h_b = 4.20 \text{ m} - \left(\frac{2}{3}\right)(4.20 - 0) = \mathbf{1.40 \text{ m}} \quad (6.1.2)$$

The pressure head is thus equal to the difference between the total and elevation head, i.e.,

Pressure head, $p_b/\gamma_w = h_b - z_b = 1.40 - 1.40 = \mathbf{0.0 \text{ m}}$.

The total head loss (Δh) across the sample is thus $h_a - h_c = 4.20 - 0.0 = 4.20$ m, which is consistent with Figure Ex. 6.1A.

Figure Ex. 6.1B graphically represents the variation of the elevation, pressure, and total heads with elevation above the datum.

Alternate Solution

To illustrate that the location of the datum is indeed arbitrary, the solution is repeated with the datum placed at point a.

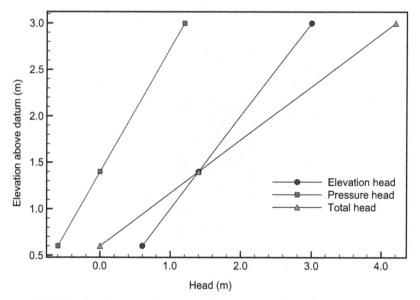

FIGURE EX. 6.1B Variation of elevation, pressure, and total head with elevation.

At point a:

Elevation head, $z_a = $ **0.00 m**

Pressure head, $p_a/\gamma_w = $ **1.20 m**

Total head, $h_a = 0.00 + 1.20 = $ **1.20 m**.

At point c:

Elevation head, $z_c = -1.60 - 0.80 = $ **−2.40 m**

Pressure head, $p_c/\gamma_w = $ **−0.60 m**

Total head, $h_c = -2.40 + (-0.60) = $ **−3.00 m**.

At point b:

Elevation head, $z_b = $ **−1.60 m**

The total head is again computed by linearly interpolating between the total heads at points a and c, i.e.,

$$h_b = h_a - \left(\frac{1.60 \text{ m}}{2.40 \text{ m}}\right)(h_a - h_c) = 1.20 \text{ m} - \left(\frac{1.60 \text{ m}}{2.40 \text{ m}}\right)(1.20 - (-3.00))$$

$$= -1.60 \text{ m}$$

$$(6.1.3)$$

Thus, the pressure head, $p_b/\gamma_w = h_b - z_b = -1.60 - (-1.60) = $ **0.0 m**.

Thus, changing the location of the datum affects the elevation and total heads. The pressure head, however, is unaffected.

The total head loss (Δh) across the sample is thus $h_a - h_c = 1.20 - (3.00) = 4.20$ m, which agrees with the earlier result.

EXAMPLE PROBLEM 6.2

General Remarks

This example problem illustrates the manner in which the elevation head, pressure head, and total head are determined for the case of one-dimensional steady-state flow. In this problem the flow is directed toward increasing elevation head.

Problem Statement

The soil shown in Figure Ex. 6.2A is supported on a porous stone and disk. Determine the elevation head, pressure head, and total head at points a, b, and c.

Solution

At point a:

Elevation head, $z_a = 0.60 + 0.80 + 1.60 = $ **3.00 m**

Pressure head, $p_a/\gamma_w = $ **1.20 m**

Total head, $h_a = 3.00 + 1.20 = $ **4.20 *m***.

At point c:

Elevation head, $z_c = $ **0.60 m**

Pressure head, $p_c/\gamma_w = 0.80 + 1.60 + 1.20 + 0.80 = $ **4.40 m**

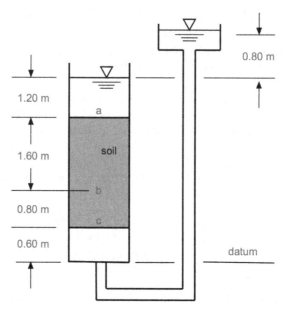

FIGURE EX. 6.2A Hypothetical apparatus for one-dimensional vertical flow (not to scale).

Total head, $h_c = 0.60 + 4.40 = \textbf{5.00 m}$.
At point b:
Elevation head, $z_b = 0.60 + 0.80 = \textbf{1.40 m}$.
The total head is computed from the total heads at points a and c. Since the soil is homogeneous, the variation in total head along the soil is *linear.* As such, the total head at point b is thus

$$\frac{h_c - h_b}{h_c - h_a} = \frac{0.80 \text{ m}}{1.60 \text{ m} + 0.80 \text{ m}} \Rightarrow h_b = h_c - \left(\frac{0.80 \text{ m}}{2.40 \text{ m}}\right)(h_c - h_a) \qquad (6.2.1)$$

Substituting the known total heads gives

$$h_b = 5.00 \text{ m} - \left(\frac{1}{3}\right)(5.00 - 4.20) = \textbf{6.40 m} \qquad (6.2.2)$$

The pressure head is thus equal to the difference between the total and & elevation head, i.e., pressure head, $p_b/\gamma_w = h_b - z_b = 6.40 - 1.40 = \textbf{5.00 m}$.
The total head loss (Δh) across the sample is thus $h_c - h_a = 5.00 - 4.20 = 0.80$ m, which is consistent with Figure Ex. 6.2A.
Figure Ex. 6.2B graphically represents the variation of the elevation, pressure, and total heads with elevation above the datum.

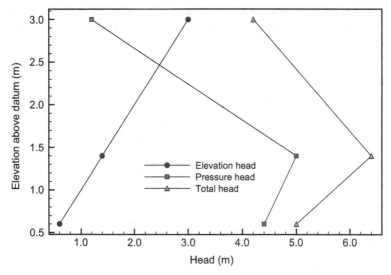

FIGURE EX. 6.2B   Variation of elevation, pressure, and total head with elevation.

EXAMPLE PROBLEM 6.3

General Remarks

This example problem illustrates the manner in which the average or superficial velocity and seepage velocity are computed for the case of one-dimensional steady-state flow at constant elevation head. The case of a nonhomogeneous soil sample is also addressed.

Problem Statement

A clean sand having a permeability of 0.055 cm/s and a void ratio of 0.62 is placed in a horizontal flow apparatus such as that shown in Figure Ex. 6.3A. Compute a)

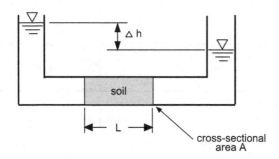

FIGURE EX. 6.3A   Hypothetical horizontal flow apparatus with a single soil.

the superficial velocity (v), b) the seepage velocity (v_s) when the change in head (Δh) equals 100 cm, and c) the discharge Q. The cross-sectional area of the horizontal pipe is 100 cm^2 and the length (L) of the soil sample is 0.50 m.

Solution

a) The hydraulic gradient for the problem is

$$i = \frac{\Delta h}{L} = \frac{100 \text{ cm}}{0.50 \text{ m}}\left(\frac{\text{m}}{100 \text{ cm}}\right) = \textbf{2.0} \tag{6.3.1}$$

The superficial velocity is then

$$v = ki = (0.055 \text{ cm/s})(2.0) = \textbf{0.11 cm/s} \tag{6.3.2}$$

b) The seepage velocity is next computed with the porosity (n) replaced by the void ratio (e), i.e.,

$$v_s = \frac{v}{n} = \left(\frac{1+e}{e}\right)v = \left(\frac{1+0.62}{0.62}\right)(0.11 \text{ cm/s}) = \textbf{0.29 cm/s} \tag{6.3.3}$$

c) Finally, the discharge is

$$Q = vA = (0.11 \text{ cm/s})(100 \text{ cm}^2) = \textbf{11.0 cm}^3\textbf{/s} \tag{6.3.4}$$

Next assume that the sample consists of *two* different soils, arranged in series in the manner shown in Figure Ex. 6.3B. The permeability coefficients for soils 1 and 2 are denoted by k_1 and k_2, respectively.

The head loss Δh and thus the hydraulic gradient i are assumed to be unchanged from the values used earlier. From Darcy's law,

$$v_1 = k_1 i_1 = k_1\left(\frac{\Delta h_1}{L_1}\right); \quad v_2 = k_2 i_2 = k_2\left(\frac{\Delta h_2}{L_2}\right) \tag{6.3.5}$$

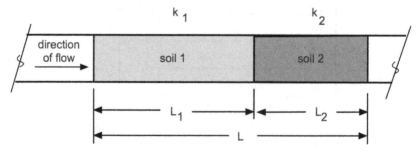

FIGURE EX. 6.3B Two soils arranged in series with the horizontal flow apparatus.

The discharge through the two materials is then

$$Q_1 = k_1 i_1 A_1 = k_1 \left(\frac{\Delta h_1}{L_1}\right) A_1; \quad Q_2 = k_2 i_2 A_2 = k_2 \left(\frac{\Delta h_2}{L_2}\right) A_2 \qquad (6.3.6)$$

From the continuity of flow equation, $Q_1 = Q_2$. In addition, since $A_1 = A_2 = A$, it follows that

$$k_1 \left(\frac{\Delta h_1}{L_1}\right) = k_2 \left(\frac{\Delta h_2}{L_2}\right) \qquad (6.3.7)$$

In addition, the head losses across the two soils must add up to the total head loss, i.e.,

$$\Delta h_1 + \Delta h_2 = \Delta h \qquad (6.3.8)$$

Substituting for $\Delta h_1 = \Delta h - \Delta h_2$ into the continuity of flow equation and solving for Δh_2 gives

$$\Delta h_2 = \left(\frac{k_1 L_2}{k_1 L_2 + k_2 L_1}\right) \Delta h \qquad (6.3.9)$$

Then,

$$\Delta h_1 = \Delta h - \Delta h_2 = \left(1 - \frac{k_1 L_2}{k_1 L_2 + k_2 L_1}\right) \Delta h = \left(\frac{k_2 L_1}{k_1 L_2 + k_2 L}\right) \Delta h \qquad (6.3.10)$$

The discharge in each of the two soils is thus

$$Q_1 = k_1 \left(\frac{\Delta h_1}{L_1}\right) A_1 = k_1 \left(\frac{1}{L_1}\right) \left(\frac{k_2 L_1}{k_1 L_2 + k_2 L_1}\right) \Delta h A = \left(\frac{k_1 k_2}{k_1 L_2 + k_2 L_1}\right) \Delta h A \qquad (6.3.11)$$

Similarly,

$$Q_2 = k_2 \left(\frac{\Delta h_2}{L_2}\right) A_2 = k_2 \left(\frac{1}{L_2}\right) \left(\frac{k_1 L_2}{k_1 L_2 + k_2 L_1}\right) \Delta h A = \left(\frac{k_1 k_2}{k_1 L_2 + k_2 L_1}\right) \Delta h A = Q_1 \qquad (6.3.12)$$

which verifies that the continuity of flow equation has indeed been satisfied.

EXAMPLE PROBLEM 6.4

General Remarks

This example problem illustrates the manner in which the superficial velocity is determined for the case of one-dimensional steady-state flow in an inclined aquifer.

Problem Statement

Consider the saturated, confined aquifer shown in Figure Ex. 6.4. The aquifer has a permeability of 0.022 cm/s. The total head loss across the aquifer is 4.2 m. If $H = 3.5$ m, $L = 75.0$ m, and the aquifer makes an angle $\beta = 12.0$ degrees with respect to the horizontal. Determine the flow rate (at right angles to the cross section) in m³/h per meter (into the page).

Solution

Using Darcy's law in conjunction with the continuity of flow, the flow rate is written as follows: $Q = kiA$. The determination of the hydraulic gradient is slightly complicated by the fact that the aquifer is inclined. In particular, the length in the (inclined) direction of flow is equal to $L/\cos \beta$. The hydraulic gradient is thus

$$i = \frac{\Delta h}{(L/\cos \beta)} = \frac{\Delta h}{L} \cos \beta \qquad (6.4.1)$$

The cross-sectional area for flow through the aquifer (for a unit thickness directed into the page) is

$$A = H \cos \beta (1) \qquad (6.4.2)$$

The discharge is thus

$$Q = kiA = k\left(\frac{\Delta h}{L} \cos \beta\right)(H \cos \beta) = k\left(\frac{\Delta h}{L}\right)H \cos^2 \beta \qquad (6.4.3)$$

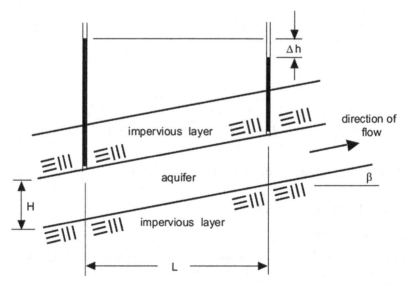

FIGURE EX. 6.4 Schematic Illustration of an inclined aquifer.

Substituting the given values of $k = 0.022$ cm/s $= 2.20 \times 10^{-4}$ m/s, $\Delta h = 4.2$ m, along with $L = 75$ m, $H = 3.5$ m, and $\beta = 12$ degrees gives

$$Q = \left(2.20 \times 10^{-4}\ \text{m/s}\right)\left(\frac{4.2\ \text{m}}{75\ \text{m}}\right)(3.5\ \text{m})(\cos 12°)^2 = 4.126 \times 10^{-5}\ \text{m}^3/\text{s/m}$$

$$(6.4.4)$$

Converting to the desired gives

$$Q = \left(4.126 \times 10^{-5}\ \text{m}^3/\text{s/m}\right)\left(\frac{60\ \text{s}}{\text{min}}\right)\left(\frac{60\ \text{min}}{\text{h}}\right) = \textbf{0.149 m}^3/\textbf{h/m}\quad (6.4.5)$$

EXAMPLE PROBLEM 6.5

General Remarks

This example problem illustrates the manner in which the elevation head, pressure head, and total head are determined for the case of one-dimensional steady-state flow.

Problem Statement

Consider the soil profile shown in Figure Ex. 6.5. Each soil layer is homogeneous. The groundwater table is at the ground surface. Piezometers a, b, and c are

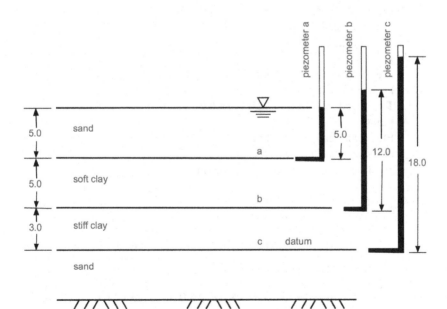

FIGURE EX. 6.5   Soil profile with three piezometers (not to scale—all distances in units of meters).

installed at interfaces between layers as shown in Figure Ex. 6.5. The measured elevation of fluid in each piezometer, relative to its tip, is also shown. All distances are units of *meters*. The datum is located at the elevation of point *c*.

The saturated unit weight for the upper sand layer, the soft clay, and the stiff clay are 20.5, 16.8, and 18.5 kN/m³, respectively.

a) First determine the elevation head, pressure head, and total head at points *a*, *b*, and *c*.
b) Next determine the vertical total stress, pore fluid pressure, and vertical effective stress at points *a*, *b*, and *c*.
c) Next determine the ratio between the coefficient of permeability for the soft clay and the stiff clay, i.e., k_{soft}/k_{stiff}.
d) Finally, assume that the location of the groundwater table drops to the elevation of point *a*. Assume that all the unit weights remain unchanged and neglect capillary rise in the upper sand layer. The height of water in piezometer *a* drops to zero, and that in piezometer *c* remains at 18 m. The reading of piezometer *b* is deemed unreliable. Determine the vertical total stress, pore fluid pressure, and vertical effective stress at points *a*, *b*, and *c* for the new location of the groundwater table.

Solution

a) At point *a*:
   Elevation head, $z_a = 3.0 + 5.0 = \textbf{8.0 m}$
   Pressure head, $p_a/\gamma_w = \textbf{5.0 m}$
   Total head, $h_a = 8.0 + 5.0 = \textbf{13.0 m}$.
   At point *b*:
   Elevation head, $z_b = \textbf{3.0 m}$
   Pressure head, $p_b/\gamma_w = \textbf{12.0 m}$
   Total head, $h_b = 3.0 + 12.0 = \textbf{15.0 m}$.
   At point *c*:
   Elevation head, $z_c = \textbf{0.0 m}$
   Pressure head, $p_c/\gamma_w = \textbf{18.0 m}$
   Total head, $h_c = 0.0 + 18.0 = \textbf{18.0 m}$.

b) At point *a*:
   The vertical total stress is

$$\sigma_{v_a} = (20.5 \text{ kN/m}^3)(5.0 \text{ m}) = \textbf{102.5 kPa} \tag{6.5.1}$$

   The pore fluid pressure is

$$\frac{p_a}{\gamma_w} = 5.0 \text{ m} \Rightarrow p_a = (9.81 \text{ kN/m}^3)(5.0 \text{ m}) = \textbf{49.1 kPa} \tag{6.5.2}$$

The vertical effective stress is

$$\sigma'_{v_a} = \sigma_{v_a} - p_a = 102.5 - 49.1 = \mathbf{53.4\ kPa} \tag{6.5.3}$$

At point b:
The vertical total stress is

$$\sigma_{v_b} = 102.5\ \text{kPa} + \left(16.8\ \text{kN/m}^3\right)(5.0\ \text{m}) = \mathbf{186.5\ kPa} \tag{6.5.4}$$

The pore fluid pressure is

$$\frac{p_b}{\gamma_w} = 12.0\ \text{m} \Rightarrow p_b = \left(9.81\ \text{kN/m}^3\right)(12.0\ \text{m}) = \mathbf{117.7\ kPa} \tag{6.5.5}$$

The vertical effective stress is

$$\sigma'_{v_b} = \sigma_{v_b} - p_b = 186.5 - 117.7 = \mathbf{68.8\ kPa} \tag{6.5.6}$$

At point c:
The vertical total stress is

$$\sigma_{v_c} = 186.5\ \text{kPa} + \left(18.5\ \text{kN/m}^3\right)(3.0\ \text{m}) = \mathbf{242.0\ kPa} \tag{6.5.7}$$

The pore fluid pressure is

$$\frac{p_c}{\gamma_w} = 18.0\ \text{m} \Rightarrow p_c = \left(9.81\ \text{kN/m}^3\right)(18.0\ \text{m}) = \mathbf{176.6\ kPa} \tag{6.5.8}$$

The vertical effective stress is

$$\sigma'_{v_c} = \sigma_{v_c} - p_c = 242.0 - 176.6 = \mathbf{65.4\ kPa} \tag{6.5.9}$$

c) The elevation of pore fluid in the piezometers indicates an *upward* flow. From the continuity of flow equation, $Q_{soft} = Q_{stiff}$, where

$$Q_{soft} = k_{soft} i_{soft} A_{soft} \tag{6.5.10}$$

$$Q_{stiff} = k_{stiff} i_{stiff} A_{stiff} \tag{6.5.11}$$

with $A_{soft} = A_{stiff}$. Thus,

$$k_{soft} i_{soft} = k_{stiff} i_{stiff} \quad \Rightarrow \quad \frac{k_{soft}}{k_{stiff}} = \frac{i_{stiff}}{i_{soft}} \tag{6.5.12}$$

Computing the necessary hydraulic gradients gives the desired result

$$\frac{k_{soft}}{k_{stiff}} = \frac{i_{stiff}}{i_{soft}} = \frac{\dfrac{(18.0\ \text{m} - 15.0\ \text{m})}{3.0\ \text{m}}}{\dfrac{(15.0\ \text{m} - 13.0\ \text{m})}{5.0\ \text{m}}} = \frac{5}{2} \tag{6.5.13}$$

d) At point a:

Elevation head, $z_a = 3.0 + 5.0 = \textbf{8.0 m}$

Pressure head, $p_a/\gamma_w = \textbf{0.0 m}$

Total head, $h_a = 8.0 + 0.0 = \textbf{8.0 m}$.

At point b:

Elevation head, $z_b = \textbf{3.0 m}$.

Since the reading of piezometer b was deemed unreliable, it is not possible to directly determine the pressure head. As such, for now, $p_b/\gamma_w = ?$

Total head, $h_b = 3.0 + p_b/\gamma_w$.

At point c:

Elevation head, $z_c = \textbf{0.0 m}$

Pressure head, $p_c/\gamma_w = \textbf{18.0 m}$

Total head, $h_c = 0.0 + 18.0 = \textbf{18.0 m}$.

Since the permeability of the two clay layers is unchanged, the pressure head at point b is computed using the results of part c. Since the flow is still upward, $h_c > h_b > h_a$. The hydraulic gradients associated with the soft and stiff clay layers are thus

$$i_{soft} = \frac{h_b - h_a}{5.0 \text{ m}} = \frac{\left(3.0 + \dfrac{p_b}{\gamma_w}\right) - 8.0}{5.0} = \frac{\dfrac{p_b}{\gamma_w} - 5.0}{5.0} \tag{6.5.14}$$

$$i_{stiff} = \frac{h_c - h_b}{3.0 \text{ m}} = \frac{18.0 - \left(3.0 + \dfrac{p_b}{\gamma_w}\right)}{3.0} = \frac{15.0 - \dfrac{p_b}{\gamma_w}}{3.0} \tag{6.5.15}$$

Substituting these hydraulic gradients into Eq. (6.5.13) gives

$$\frac{i_{stiff}}{i_{soft}} = \frac{5}{2} = \frac{\left(15.0 - \dfrac{p_b}{\gamma_w}\right) \Big/ 3.0}{\left(\dfrac{p_b}{\gamma_w} - 5.0\right) \Big/ 5.0} \Rightarrow \frac{15.0 - \dfrac{p_b}{\gamma_w}}{\dfrac{p_b}{\gamma_w} - 5.0} = \frac{3}{2} \tag{6.5.16}$$

Solving for the pressure at b head gives

$$\frac{p_b}{\gamma_w} = 9.0 \text{ m} \tag{6.5.17}$$

The total head at point b is thus

$$h_b = z_b + \frac{p_b}{\gamma_w} = 3.0 + 9.0 = \textbf{12.0 m} \tag{6.5.18}$$

The vertical total and effective stresses and the pore pressure are next computed. Since the unit weights have not changed, the total stresses are *unchanged* from part b.

At point a:
  The vertical total stress is

$$\sigma_{v_a} = \mathbf{102.5\ kPa} \tag{6.5.19}$$

  The pore fluid pressure is

$$\frac{p_a}{\gamma_w} = 0.0\ m \Rightarrow p_a = \mathbf{0.0\ kPa} \tag{6.5.20}$$

  The vertical effective stress is

$$\sigma'_{v_a} = \sigma_{v_a} - p_a = 102.5 - 0.0 = \mathbf{102.5\ kPa} \tag{6.5.21}$$

At point b:
  The vertical total stress is

$$\sigma_{v_b} = \mathbf{186.5\ kPa} \tag{6.5.22}$$

  The pore fluid pressure is

$$\frac{p_b}{\gamma_w} = 9.0\ m \Rightarrow p_b = \left(9.81\ kN/m^3\right)(9.0\ m) = \mathbf{88.3\ kPa} \tag{6.5.23}$$

  The vertical effective stress is

$$\sigma'_{v_b} = \sigma_{v_b} - p_b = 186.5 - 88.3 = \mathbf{98.2\ kPa} \tag{6.5.24}$$

At point c:
  The vertical total stress is

$$\sigma_{v_c} = \mathbf{242.0\ kPa} \tag{6.5.25}$$

  The pore fluid pressure is

$$\frac{p_c}{\gamma_w} = 18.0\ m \Rightarrow p_c = \left(9.81\ kN/m^3\right)(18.0\ m) = \mathbf{176.6\ kPa} \tag{6.5.26}$$

  The vertical effective stress is

$$\sigma'_{v_c} = \sigma_{v_c} - p_c = 242.0 - 176.6 = \mathbf{65.4\ kPa} \tag{6.5.27}$$

EXAMPLE PROBLEM 6.6

General Remarks

This example problem illustrates the manner in which the elevation head, pressure head, and total head are determined for the case of one-dimensional steady-state flow.

Problem Statement

a) Determine the pressure, elevation, and total head at points A–E for the apparatus shown in Figure Ex. 6.6. b) If the coefficient of permeability for the

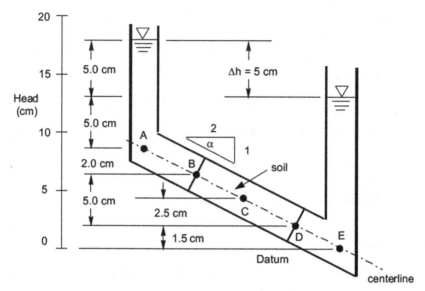

FIGURE EX. 6.6 Schematic illustration of flow apparatus.

soil is estimated to be $k = 0.05$ cm/s, what is the average or discharge (Darcy) velocity of flow through the soil (the soil is located between points B and D)?

Solution

a) Noting the assumed datum, the elevation head at a given point is first determined. Then, recalling that the pressure head is the height that the fluid will rise in a tube placed at the point in question, the pressure head at a given point is determined. Finally, the total head is the sum of the elevation head and pressure head. Following this solution strategy gives the following results:

At point A:

    Elevation head $= 1.5 + 5.0 + 2.0 = $ **8.5 cm**

    Pressure head $= 5.0 + 5.0 = $ **10.0 cm**

    Total head $= 8.5 + 10.0 = $ **18.5 cm**.

At point B:

    Elevation head $= 1.5 + 5.0 = $ **6.5 cm**

    Pressure head $= 2.0 + 5.0 + 5.0 = $ **12.0 cm**

    Total head $= 6.5 + 12.0 = $ **18.5 cm**.

At point D:

    Elevation head $= $ **1.5 cm**

    Pressure head $= 5.0 + 2.0 + 5.0 + 5.0 - 5.0 = $ **12.0 cm**

    Total head $= 1.5 + 12.0 = $ **13.5 cm**.

At point C:
Elevation head $= 1.5 + 2.5 = $ **4.0 cm**
Total head $= (18.5 + 13.5)/2 = $ **16.0 cm**
Pressure head $= 16.0 - 4.0 = $ **12.0 cm**.
At point E:
Elevation head $= $ **0.0 cm**
Pressure head $= 12.0 + 1.5 = $ **13.5 cm**
Total head $= 0.0 + 13.5 = $ **13.5 cm**.
Notes:
- Since point C is equidistant between points B and D, the total head at point C is computed as the *average* of the total heads at B and D.
- The pressure head at point C is then computed as the difference between the total head and elevation head at this point.

b) To determine the superficial velocity of flow through the soil requires the hydraulic gradient. To this end, the inclination of the apparatus shown in Figure Ex. 6.6, quantified by the angle α, is first computed

$$\alpha = \tan^{-1}\left(\frac{1}{2}\right) = 26.6 \text{ degrees} \tag{6.6.1}$$

The length of the flow is thus

$$L = \frac{5 \text{ cm}}{\sin \alpha} \tag{6.6.2}$$

The hydraulic gradient is thus

$$i = \frac{\Delta h}{L} = \frac{5 \text{ cm}}{5 \text{ cm}/\sin \alpha} \sin \alpha = 0.447 \tag{6.6.3}$$

Finally, the superficial velocity of flow through the soil is simply

$$v = ki = (0.05 \text{ cm/s})(0.447) = \textbf{2.236} \times \textbf{10}^{-2} \text{ cm/s} \tag{6.6.4}$$

EXAMPLE PROBLEM 6.7

General Remarks

This example problem illustrates the manner in which the elevation head, pressure head, and total head are determined for the case of one-dimensional steady-state flow. In addition, it shows how the average or superficial and seepage velocities are computed.

Problem Statement

A fully saturated soil sample is maintained between points a and b in the inclined tube shown in Figure Ex. 6.7A. In the vicinity of the soil sample, the

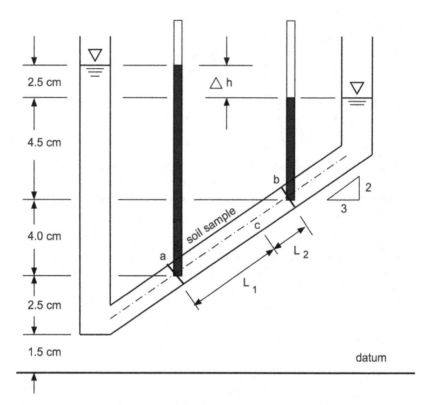

FIGURE EX. 6.7A One-dimensional flow example involving an inclined tube.

tube has a uniform cross-sectional area (A). Given that $L_1 = 2L_2$, please do the following:

a) Determine the elevation head (z), pressure head, total head (h) (in centimeters), and pore pressure (in kPa) at points a, b, and c.

b) The soil sample has permeability of 0.001 cm/s, a specific gravity of solids (G_s) of 2.68, and a moisture content (w) of 42%. Determine the discharge (Darcy) velocity v and the seepage velocity v_s (both in units of cm/s).

c) The soil between points a and c now has a permeability k_{ac}; the soil between points c and b now has a permeability k_{cb}, where $k_{ac} = mk_{cb}$, with $m \neq 0$. If $m \neq 1$, the total head loss over the full length of the soil sample will *no longer* be linear. Recalling that the fluid is incompressible and that the discharge Q must thus remain unchanged from a to b, determine the head loss Δh_{ac} and Δh_{cb} across each of the two soils. Express the answers in terms of m and the total head loss (Δh) across the entire soil sample (ab).

Solution

a) The datum is located as shown in Figure Ex. 6.7A.

At point a:

Elevation head, $z_a = 1.5 + 2.5 = $ **4.0 cm**
Pressure head, $p_a/\gamma_w = 4.0 + 4.5 + 2.5 = $ **11.0 cm**
Total head, $h_a = 4.0 + 11.0 = $ **15.0 cm**.
The pore pressure at this point is thus

$$p_a = \gamma_w(11.0 \text{ cm})\left(\frac{\text{m}}{100 \text{ cm}}\right) = (9.81 \text{ kN/m}^3)(11.0 \text{ m})\left(\frac{\text{m}}{100 \text{ cm}}\right) = \mathbf{1.079 \text{ kPa}}$$

(6.7.1)

At point b:

Elevation head, $z_b = 1.5 + 2.5 + 4.0 = $ **8.0 cm**
Pressure head, $p_b/\gamma_w = $ **4.5 cm**
Total head, $h_b = 8.0 + 4.5 = $ **12.5 cm**.
The pore pressure at this point is thus

$$p_b = \gamma_w(4.5 \text{ cm})\left(\frac{\text{m}}{100 \text{ cm}}\right) = (9.81 \text{ kN/m}^3)(4.5 \text{ m})\left(\frac{\text{m}}{100 \text{ cm}}\right) = \mathbf{0.441 \text{ kPa}}$$

(6.7.2)

At point c:

Figure Ex. 6.7B shows the geometry of the inclined tube between points a and b. Since the vertical distance between these two points is 4.0 cm, and since the tube is inclined two vertical on three horizontal, it follows that $e = 6.0$ cm. Since $L_1 = 2L_2$, the vertical distance to point c

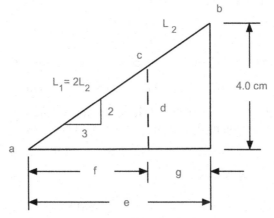

FIGURE EX. 6.7B   Geometry associated with a portion of the inclined tube.

from the datum is $d = 2(4.0 \text{ cm})/3 = 8/3$ cm. The elevation head at point c is thus

$$z_c = 4.0 + \frac{8}{3} = \frac{20}{3} \text{ cm} \tag{6.7.3}$$

Since the soil is homogeneous, the total head will vary *linearly* from point a to point b. Using the values of h_a and h_b determined earlier, the total head at point c is thus

$$h_c = h_a - \frac{2}{3}(h_a - h_b) = 15.0 - \frac{2}{3}(15.0 - 12.5) = \frac{40}{3} = \textbf{13.33 cm} \tag{6.7.4}$$

The pressure head is next computed

$$\frac{p_c}{\gamma_w} = h_c - z_c = \frac{40}{3} - \frac{20}{3} = \frac{20}{3} \text{ cm} \tag{6.7.5}$$

The pore pressure at this point is thus

$$p_c = \gamma_w \left(\frac{20}{3} \text{ cm}\right)\left(\frac{m}{100 \text{ cm}}\right) = (9.81 \text{ kN/m}^3)\left(\frac{20}{3} \text{ cm}\right)\left(\frac{m}{100 \text{ cm}}\right) = \textbf{0.65 kPa} \tag{6.7.6}$$

b) The head loss across the soil sample is $\Delta h = h_a - h_b = 15.0 - 12.5 = 2.5$ cm. To determine the hydraulic gradient, the length $(L = L_1 + L_2)$ along the direction of flow must be determined. This length is computed using the Pythagorean theorem applied to the right triangle shown in Figure Ex. 6.7B, i.e.,

$$L = \sqrt{(4.0 \text{ cm})^2 + (6.0 \text{ cm})^2} = 7.211 \text{ cm} \tag{6.7.7}$$

The hydraulic gradient is thus

$$i = \frac{\Delta h}{L} = \frac{2.5 \text{ cm}}{7.211 \text{ cm}} = 0.347 \tag{6.7.8}$$

The discharge velocity is next computed using Darcy's law

$$v = ki = (0.001 \text{ cm/s})(0.347) = \textbf{3.467} \times \textbf{10}^{-4} \text{ cm/s} \tag{6.7.9}$$

Since the soil is saturated and $G_s = 2.68$ and $w = 42\%$, the void ratio in the sample is computed using Case 1.3 of Chapter 1, i.e.,

$$e = wG_s = (0.42)(2.68) = 1.126 \tag{6.7.10}$$

The porosity is thus

$$n = \frac{e}{1+e} = \frac{1.126}{1+1.126} = 0.530 \tag{6.7.11}$$

The seepage velocity is then

$$v_s = \frac{v}{n} = \frac{3.467 \times 10^{-4} \text{ cm/s}}{0.530} = \textbf{6.541} \times \textbf{10}^{-4} \textbf{ cm/s} \qquad (6.7.12)$$

c) For an incompressible fluid, the continuity of flow equation gives

$$Q_{ac} = v_{ac}A = Q_{cb} = v_{cb}A \Rightarrow v_{ac} = v_{cb} \qquad (6.7.13)$$

From Darcy's law

$$k_{ac}i_{ac} = k_{cb}i_{cb} \Rightarrow k_{ac}\left(\frac{\Delta h_{ac}}{L_1}\right) = k_{cb}\left(\frac{\Delta h_{cb}}{L_2}\right) \qquad (6.7.14)$$

where $\Delta h_{ac} + \Delta h_{cb} = \Delta h$. Substituting for Δh_{ac} into the aforesaid expression gives

$$k_{ac}\left(\frac{\Delta h - \Delta h_{cb}}{L_1}\right) = k_{cb}\left(\frac{\Delta h_{cb}}{L_2}\right) \Rightarrow \Delta h_{cb} = \frac{k_{ac}L_2\Delta h}{k_{ac}L_2 + k_{cb}L_1} \qquad (6.7.15)$$

But $k_{ac} = mk_{cb}$, so

$$\Delta h_{cb} = \frac{(m\,k_{cb})L_2\Delta h}{(m\,k_{cb})L_2 + k_{cb}L_1} = \left(\frac{mL_2}{mL_2 + L_1}\right)\Delta h \qquad (6.7.16)$$

Next, since $L_1 = 2L_2$, the aforesaid expression becomes

$$\Delta h_{cb} = \left(\frac{mL_2}{mL_2 + 2L_2}\right)\Delta h = \left(\frac{m}{m+2}\right)\Delta h \qquad (6.7.17)$$

Finally,

$$\Delta h_{ac} = \Delta h - \Delta h_{cb} = \left[1 - \left(\frac{m}{m+2}\right)\right]\Delta h = \left(\frac{2}{m+2}\right)\Delta h \qquad (6.7.18)$$

As a check, let $m = 1$, which implies a *homogeneous* material with a constant permeability. Then,

$$\Delta h_{ac} = \left(\frac{2}{1+2}\right)\Delta h = \frac{2}{3}\Delta h \qquad (6.7.19)$$

$$\Delta h_{cb} = \left(\frac{1}{1+2}\right)\Delta h = \frac{1}{3}\Delta h \qquad (6.7.20)$$

which is consistent with the fact that for a homogeneous soil the head loss varies linearly.

EXAMPLE PROBLEM 6.8

General Remarks

This example problem investigates the effect of a sand seam on the flow of pore fluid in a soil profile.

Problem Statement

Consider the soil profile shown in Figure Ex. 6.8. The moist unit weight of the top sand layer is 20.0 kN/m³. Capillary rise is present above the groundwater table. The saturated unit weight of clay 1 and clay 2 is 17.8 and 18.5 kN/m³, respectively. Between these two clay layers is a sand seam of negligible thickness. The tip of piezometer 1 is within this sand seam. The tip of piezometer 2 is at the bottom of the clay 2 layer.

Determine the vertical total stress, the pore fluid pressure, and the vertical effective stress at points A, B, C, D, E, and F.

Solution

The datum is placed a distance 12.0 m below the ground surface. Based on the height of fluid in the two piezometers, it is evident that upward seepage is taking place in clay 1, while in clay 2 the seepage is downward.

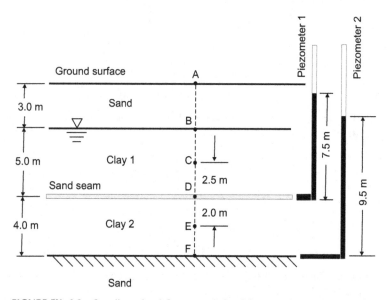

FIGURE EX. 6.8 One-dimensional flow example involving a sand seam (not to scale).

At point A:

The vertical total stress is

$$\sigma_{v_A} = \textbf{0.0 kPa} \tag{6.8.1}$$

The pore fluid pressure is

$$p_A = -\gamma_w(3.0 \text{ m}) = -(9.81 \text{ kN/m}^3)(3.0 \text{ m}) = \textbf{−29.43 kPa} \tag{6.8.2}$$

The vertical effective stress is thus

$$\sigma'_{v_A} = \sigma_{v_A} - p_A = 0.0 - (-29.43 \text{ kPa}) = \textbf{29.4 kPa} \tag{6.8.3}$$

At point B:

The elevation head, $z_B = 9.0$ m
The pressure head, $p_B/\gamma_w = 0.0$ m
The total head, $h_B = 9.0 + 0.0 = 9.0$ m.
The vertical total stress is

$$\sigma_{v_B} = (20.0 \text{ kN/m}^3)(3.0 \text{ m}) = \textbf{60.0 kPa} \tag{6.8.4}$$

The pore fluid pressure is

$$p_B = \textbf{0.0 kPa} \tag{6.8.5}$$

The vertical effective stress is thus

$$\sigma'_{v_B} = \sigma_{v_B} - p_B = 60.0 - 0.0 = \textbf{60.0 kPa} \tag{6.8.6}$$

At point D:

The elevation head, $z_D = 4.0$ m
The pressure head, $p_D/\gamma_w = 7.5$ m
The total head, $h_D = 4.0 + 7.5 = 11.5$ m.
The vertical total stress is

$$\sigma_{v_D} = 60.0 + (17.8 \text{ kN/m}^3)(5.0 \text{ m}) = \textbf{149.0 kPa} \tag{6.8.7}$$

The pore fluid pressure is

$$p_B = \gamma_w(7.5 \text{ m}) = (9.81 \text{ kN/m}^3)(7.5 \text{ m}) = \textbf{73.6 kPa} \tag{6.8.8}$$

The vertical effective stress is thus

$$\sigma'_{v_D} = \sigma_{v_D} - p_D = 149.0 - 73.6 = \textbf{75.4 kPa} \tag{6.8.9}$$

As a check, note that in the absence of upward seepage, the hydrostatic pore pressure at point D would be $\gamma_w(5.0 \text{ m}) = 49.1$ kPa; the vertical effective stress would thus be equal to $149.0 - 49.1 = 99.9$ kPa. Recalling the discussion of Section 6.8.3, in the presence of upward seepage, the hydrostatic effective stress is *reduced* by the amount $\gamma_w\Delta h = (9.81 \text{ kN/m}^3)$ $(11.5 - 9.0 \text{ m}) = 24.5$ kPa. Thus, $99.9 - 24.5 = 75.4$ kPa, which is identical to the aforesaid value for σ'_{v_D}.

At point C:
The elevation head, $z_C = 4.0 + 2.5 = 6.5$ m.
Since the material is homogeneous, the total head at point C is linearly interpolated from h_B and h_D, i.e.,

$$h_C = h_D - \frac{1}{2}(h_B - h_D) = \frac{1}{2}(h_B + h_D) = \frac{1}{2}(11.5 + 9.0) = 10.25 \text{ m} \quad (6.8.10)$$

The pressure head is thus $p_C/\gamma_w = h_C - z_C = 10.25 - 6.5 = 3.75$ m.
The vertical total stress is

$$\sigma_{vC} = 60.0 + (17.8 \text{ kN/m}^3)(2.5 \text{ m}) = \mathbf{104.5 \text{ kPa}} \quad (6.8.11)$$

The pressure head is

$$p_C = \gamma_w(3.75 \text{ m}) = (9.81 \text{ kN/m}^3)(3.75 \text{ m}) = \mathbf{36.8 \text{ kPa}} \quad (6.8.12)$$

The vertical effective stress is thus

$$\sigma'_{vC} = \sigma_{vC} - p_C = 104.5 - 36.8 = \mathbf{67.7 \text{ kPa}} \quad (6.8.13)$$

At point F:
The elevation head, $z_F = 0.0$ m
The pressure head, $p_F/\gamma_w = 9.5$ m
The total head, $h_F = 0.0 + 9.5 = 9.5$ m.
The vertical total stress is

$$\sigma_{vF} = 149.0 + (18.5 \text{ kN/m}^3)(4.0 \text{ m}) = \mathbf{223.0 \text{ kPa}} \quad (6.8.14)$$

The pore fluid pressure is

$$p_F = \gamma_w(9.5 \text{ m}) = (9.81 \text{ kN/m}^3)(9.5 \text{ m}) = \mathbf{93.2 \text{ kPa}} \quad (6.8.15)$$

The vertical effective stress is thus

$$\sigma'_{vF} = \sigma_{vF} - p_F = 223.0 - 93.2 = \mathbf{129.8 \text{ kPa}} \quad (6.8.16)$$

At point E:
The elevation head, $z_E = 2.0$ m.
Since the material is homogeneous, the total head at point E is linearly interpolated from h_D and h_F, i.e.,

$$h_E = h_D - \frac{1}{2}(h_D - h_F) = \frac{1}{2}(h_D + h_F) = \frac{1}{2}(11.5 + 9.5) = 10.50 \text{ m} \quad (6.8.17)$$

The pressure head is thus $p_E/\gamma_w = h_E - z_E = 10.50 - 2.0 = 8.5$ m.
The vertical total stress is

$$\sigma_{vE} = 149.0 + (18.5 \text{ kN/m}^3)(2.0 \text{ m}) = \mathbf{186.0 \text{ kPa}} \quad (6.8.18)$$

The pressure head is

$$p_E = \gamma_w(8.5 \text{ m}) = (9.81 \text{ m/s}^2)(8.5 \text{ m}) = \textbf{83.4 kPa} \qquad (6.8.19)$$

The vertical effective stress is thus

$$\sigma'_{v_E} = \sigma_{v_E} - p_E = 186.0 - 83.4 = \textbf{102.6 kPa} \qquad (6.8.20)$$

EXAMPLE PROBLEM 6.9

General Remarks

This example problem illustrates the manner in which the value of the coefficient of permeability is determined from a constant-head permeability test.

Problem Statement

Consider the simple water filter system shown in Figure Ex. 6.9. It consists of two soil filters, connected in series. Each filter traps different size contaminants and thus contains soil: the upper filter traps coarse particles; the lower

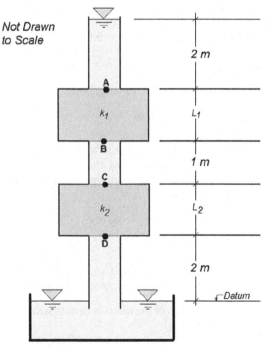

FIGURE EX. 6.9   Hypothetical two-level soil filter[11].

---

11. Leshchinsky, D., 2010. Personal communication.

one traps finer ones. The cross-sectional area of each filter is $1.2\ m^2$. The length of the upper filter (L_1) is 1.0 m; the length of the lower filter (L_2) is 2.0 m. The permeability of the upper filter (k_1) is 0.40 m/h; the permeability of the lower filter (k_2) is 0.18 m/h. Determine the following:

a) The total head at points A, B, C, and D.

b) The water pressure at points A, B, C, and D.

c) The discharge through each filter (i.e., Q_1 and Q_2) in units of m^3/h.

d) Is there potential for soil "heave" (boiling)? If so, where will this heave occur?

Solution

a) Begin the solution by determining the elevation head, pressure head, and then the total head at the four points in question.

At point A:

The elevation head, $z_A = 6.0$ m

The pressure head, $p_A/\gamma_w = 2.0$ m

The total head, $h_A = 6.0 + 2.0 = \mathbf{8.0\ m}$.

At point D:

The elevation head, $z_{ed} = 2.0$ m

The pressure head, $p_{ad}/\gamma_w = -2.0$ m

The total head, $h_D = 2.0 + (-2.0) = \mathbf{0.0\ m}$.

At point B:

The elevation head, $z_B = 5.0$ m

The pressure head, $p_{ub}/\gamma_w = ?$

At point C:

The elevation head, $z_C = 4.0$ m

The pressure head, $p_c/\gamma_w = ?$

The determination of the pressure head at points B and C first of all requires consideration of continuity of flow. In particular, since the two filters are arranged "in series," the discharge through both is the *same*, i.e., $Q_1 = Q_2$. Using Darcy's law,

$$k_1 i_1 A_1 = k_2 i_2 A_2 \qquad (6.9.1)$$

where $A_1 = A_2$. Thus,

$$k_1 i_1 = k_2 i_2 \Rightarrow k_1 \left(\frac{\Delta h_1}{L_1}\right) = k_2 \left(\frac{\Delta h_2}{L_2}\right) \qquad (6.9.2)$$

The head loss across each of the filters is determined from Figure Ex. 6.9, giving

$$k_1 \left(\frac{h_A - h_B}{L_1}\right) = k_2 \left(\frac{h_C - h_D}{L_2}\right) \qquad (6.9.3)$$

Since no soil is present between points B and C, there is no head loss between these two points, i.e., $h_B = h_C$. Thus,

$$k_1 \left(\frac{h_A - h_B}{L_1} \right) = k_2 \left(\frac{h_B - h_D}{L_2} \right) \tag{6.9.4}$$

Solving for h_B gives

$$h_B = \frac{k_1 h_A}{L_1} \frac{1}{\left(\dfrac{k_2}{L_2} + \dfrac{k_1}{L_1} \right)} \tag{6.9.5}$$

Substituting all known quantities gives

$$h_B = \frac{(0.40 \text{ m/h})(8.0 \text{ m})}{1.0 \text{ m}} \frac{1}{\dfrac{(0.18 \text{ m/h})}{2.0 \text{ m}} + \dfrac{(0.40 \text{ m/h})}{1.0 \text{ m}}} \tag{6.9.6}$$

$$= \textbf{6.531 m} = h_C$$

b) The water pressure at points $A-D$ requires the pressure head.
At point A:
The pressure head is $p_A/\gamma_w = 2.0$ m, thus

$$p_A = \gamma_w (2.0 \text{ m}) = (9.81 \text{ kN/m}^3)(2.0 \text{ m}) = \textbf{19.62 kPa} \tag{6.9.7}$$

At point B:
The pressure head is

$$\frac{p_B}{\gamma_w} = h_B - z_B = 6.531 - 5.0 = 1.531 \text{ m} \tag{6.9.8}$$

Thus,

$$p_B = \gamma_w (1.531 \text{ m}) = (9.81 \text{ kN/m}^3)(1.531 \text{ m}) = \textbf{15.02 kPa} \tag{6.9.9}$$

At point C:
The pressure head is

$$\frac{p_C}{\gamma_w} = h_C - z_C = 6.531 - 4.0 = 2.531 \text{ m} \tag{6.9.10}$$

Thus

$$p_C = \gamma_w (2.531 \text{ m}) = (9.81 \text{ kN/m}^3)(2.531 \text{ m}) = \textbf{24.83 kPa} \tag{6.9.11}$$

At point D:

   The pressure head is $p_{ad}/\gamma_w = -2.0$ m, thus

$$p_D = \gamma_w(-2.0 \text{ m}) = (9.81 \text{ kN/m}^3)(-2.0 \text{ m}) = -19.62 \text{ kPa} \qquad (6.9.12)$$

c) The hydraulic gradients across filters 1 and 2 are

$$i_1 = \frac{h_A - h_B}{L_1} = \frac{8.0 - 6.531}{1.0} = 1.469 \qquad (6.9.13)$$

$$i_2 = \frac{h_C - h_D}{L_1} = \frac{6.531 - 0.0}{2.0} = 3.266 \qquad (6.9.14)$$

The discharge through the two filters is thus

$$Q_1 = k_1 i_1 A_1 = (0.40 \text{ m/h})(1.469)(1.2 \text{ m}^2) = \mathbf{0.705 \text{ m}^3/h} \qquad (6.9.15)$$

$$Q_2 = k_2 i_2 A_2 = (0.18 \text{ m/h})(3.266)(1.2 \text{ m}^2) = \mathbf{0.705 \text{ m}^3/h} \qquad (6.9.16)$$

which confirms the requirement that $Q_1 = Q_2$.

d) Finally, since the seepage is directed *downward*, there is *no potential* for heave ("boiling").

EXAMPLE PROBLEM 6.10

General Remarks

This example problem illustrates the manner in which the value of the coefficient of permeability is determined from a constant-head permeability test.

Problem Statement

In a constant-head permeability test, a sample of soil 14 cm long and 6 cm in diameter discharged 1.65×10^{-3} m^3 of water in 12 min. The head difference in two piezometers located at 1 and 12 cm, respectively, from the bottom of the sample is 2.4 cm. Determine the coefficient of permeability of the soil. What is the soil type tested?

Solution

Since the total discharge volume is $Q = 1.65 \times 10^{-3}$ m^3, it follows that $Q = vAt$, where A is the cross-sectional area of the sample and t is the time over which q was measured. Finally, the velocity v is given by Darcy's law to be $v = ki$. Thus, using Eq. (6.21)

$$Q = vAt = (ki)t \Rightarrow k = \frac{Q}{Ait} \qquad (6.10.1)$$

The hydraulic gradient across the sample is computed as follows:

$$i = \frac{\Delta h}{L} = \frac{2.4 \text{ cm}}{(12 - 1) \text{ cm}} = 0.218 \tag{6.10.2}$$

Substituting all known quantities into Eq. (6.10.1) gives

$$k = \frac{Q}{Ait} = \frac{1.65 \times 10^{-3} \text{ m}^3}{\left[\frac{\pi}{4}(0.060 \text{ m})^2\right](0.218)\left(12 \text{ min}^* \frac{60 \text{ s}}{\text{min}}\right)} = \textbf{3.718} \times \textbf{10}^{-3} \textbf{ m/s} \tag{6.10.3}$$

Based on this value, it is likely that the soil is a "dirty" sand, i.e., a sand containing a relatively small percentage of fines.

EXAMPLE PROBLEM 6.11

General Remarks

This example problem illustrates the manner in which the value of the coefficient of permeability is determined from a falling-head permeability test performed on a sand.

Problem Statement

A falling-head permeability test was performed on a sample of uniform clean sand. The standpipe consisted of a graduated burette, and it was observed that 1 min was required for the water level to fall from the 0 cm^3 to the 51 cm^3 mark on the burette. The initial head was 92 cm and the final head was 43 cm. The sample was 25 cm long and had a diameter of 4.0 cm. Determine the coefficient of permeability for the sand.

Solution

As shown in Section 6.7.1, the coefficient of permeability is computed from Eq. (6.27), i.e.,

$$k = \left(\frac{aL}{A \, \Delta t}\right) \ln\left(\frac{h_1}{h_2}\right) \tag{6.11.1}$$

Since the area of the standpipe remains constant, the change in volume of the standpipe in time Δt will be

$$\Delta V = A\Delta t \Rightarrow A = \frac{\Delta V}{\Delta t} = \frac{50 \text{ cm}^2}{(92 - 43) \text{ cm}} = 1.02 \text{ cm}^2 \tag{6.11.2}$$

The cross-sectional area of the standpipe is

$$A = \frac{\pi}{4}(4.0 \text{ cm})^2 = 4\pi \text{ cm}^2 \tag{6.11.3}$$

The elapsed time between readings is 60 s. Thus, $\Delta t = 60 - 0 = 60$ s, $h_1 = 92$ cm, and $h_2 = 43$ cm. Thus,

$$k = \left(\frac{aL}{A\Delta t}\right)\ln\left(\frac{h_1}{h_2}\right) = \frac{(1.02 \text{ cm}^2)(25 \text{ cm})}{(4\pi \text{ cm}^2)(60 \text{ s})}\ln\left(\frac{92 \text{ cm}}{43 \text{ cm}}\right) = \mathbf{2.572 \times 10^{-2} \text{ cm/s}}$$
$$\tag{6.11.4}$$

EXAMPLE PROBLEM 6.12

General Remarks

This example problem illustrates the manner in which the value of the coefficient of permeability is determined from a falling-head permeability test performed on a clayey soil.

Problem Statement

A falling-head permeability test was carried out on a clay soil of diameter 10 cm and length 15 cm. In 1 h the head in the standpipe of diameter 5 mm dropped from 68.5 to 50.7 cm. Compute the coefficient of permeability for this clay.

Solution

As noted in the previous example, the coefficient of permeability is computed using Eq. (6.27), i.e.,

$$k = \left(\frac{aL}{A\Delta t}\right)\ln\left(\frac{h_1}{h_2}\right) \tag{6.12.1}$$

The cross-sectional area of the standpipe is

$$a = \frac{\pi}{4}\left(5.0 \text{ mm}^* \frac{\text{cm}}{10 \text{ mm}}\right)^2 = 0.196 \text{ cm}^2 \tag{6.12.2}$$

The cross-sectional area of the clay sample is

$$A = \frac{\pi}{4}(10.0 \text{ cm})^2 = 78.54 \text{ cm}^2 \tag{6.12.3}$$

Since the length of the sample is 15 cm, and since the elapsed time for the test is 1 h, it follows that

$$k = \left(\frac{aL}{A\Delta t}\right)\ln\left(\frac{h_1}{h_2}\right) = \frac{(0.196 \text{ cm}^2)(15 \text{ cm})}{(78.54 \text{ cm}^2)(1 \text{ h})}\ln\left(\frac{68.5 \text{ cm}}{50.7 \text{ cm}}\right)$$
$$= \mathbf{1.128 \times 10^{-2} \text{ cm/h}} \tag{6.12.4}$$

Changing to more commonly used units gives

$$k = (1.128 \times 10^{-2} \text{ cm/h}) \left(\frac{h}{60 \text{ min}} \right) \left(\frac{\text{min}}{60 \text{ s}} \right) = \mathbf{3.134 \times 10^{-6} \text{ cm/s}} \quad (6.12.5)$$

EXAMPLE PROBLEM 6.13

General Remarks

This example problem illustrates the manner in which the average or superficial velocity and seepage velocity are computed for the case of one-dimensional steady-state flow.

Problem Statement

A permeability test was performed on a compacted sample of sandy gravel. The sample was 150 mm long and the diameter of the mold was 150 mm. In 83 s the discharge under a constant head of 40 cm was 392 cm^3. The sample had a dry mass of 5300 g and its specific gravity of solids (G_s) was 2.68. Calculate a) the coefficient of permeability, b) the superficial velocity (v), and c) the seepage velocity (v_s) during the test.

Solution

a) The hydraulic gradient for the problem is given by Eq. (6.10), i.e.,

$$i = \frac{\Delta h}{L} = \frac{40 \text{ cm}}{150 \text{ mm}} \left(\frac{10 \text{ mm}}{\text{cm}} \right) = \frac{8}{3} \quad (6.13.1)$$

The discharge is related to the permeability through Darcy's law, i.e., $q = kiA$. The cross-sectional area perpendicular to the direction of flow is

$$A = \left(\frac{\pi}{4} \right) d^2 = \left(\frac{\pi}{4} \right) (150 \text{ mm})^2 \left(\frac{\text{cm}}{10 \text{ mm}} \right)^2 = 176.7 \text{ cm}^2 \quad (6.13.2)$$

In addition, from its definition, the discharge is related to the volume of flow (V) and time (t) through the relation $q = V/t$. Thus,

$$kiA = \frac{V}{t} \Rightarrow k = \frac{V}{t} \frac{1}{iA} = \frac{(392 \text{ cm}^3)}{(83 \text{ s})} \frac{1}{\left(\frac{8}{3} \right) (176.7 \text{ cm}^2)} \quad (6.13.3)$$

$$= \mathbf{0.010 \text{ cm/s}}$$

b) The superficial velocity is computed from the discharge according to

$$Q = vA \Rightarrow v = \frac{1}{A}Q = \frac{1}{176.7 \text{ cm}^2}\left(\frac{392 \text{ cm}^3}{83 \text{ s}}\right) = \textbf{0.027 cm/s} \qquad (6.13.4)$$

c) The determination of the seepage velocity requires the porosity (n). Recall the following expression for the dry unit weight (recall Case 1.4 in Chapter 1)

$$\gamma_d = \frac{W_s}{V} = \frac{G_s\gamma_w}{1+e} \Rightarrow e = \frac{G_s\gamma_w}{\gamma_d} - 1 \qquad (6.13.5)$$

The weight of the solid phase is computed from the dry mass, i.e.,

$$W_s = M_s\,g = (5300 \text{ g})\left(\frac{\text{kg}}{1000 \text{ g}}\right)(9.81 \text{ m/s}^2) = 51.99 \text{ N} \qquad (6.13.6)$$

The total volume of the saturated soil is

$$V = \frac{\pi}{4}d^2\,L = AL = (176.7 \text{ cm}^2)\left(\frac{\text{m}}{100 \text{ cm}}\right)^2(0.150 \text{ m}) = 2.651 \times 10^{-3} \text{ m}^3 \qquad (6.13.7)$$

Thus,

$$\gamma_d = \frac{W_s}{V} = \frac{51.99 \text{ N}}{2.651 \times 10^{-3} \text{ m}^3}\left(\frac{\text{kN}}{1000 \text{ N}}\right) = 19.62 \text{ kN/m}^3 \qquad (6.13.8)$$

Substituting this value for γ_d into the aforesaid expression for void ratio gives

$$e = \frac{G_s\gamma_w}{\gamma_d} - 1 = \frac{(2.68)(9.81 \text{ kN/m}^3)}{(19.62 \text{ kN/m}^3)} - 1 = 0.340 \qquad (6.13.9)$$

The porosity is thus

$$n = \frac{e}{1+e} = \frac{0.340}{1+0.340} = 0.254 \qquad (6.13.10)$$

The seepage velocity is thus

$$v_s = \frac{v}{n} = \frac{0.027 \text{ cm/s}}{0.254} = \textbf{0.106 cm/s} \qquad (6.13.11)$$

EXAMPLE PROBLEM 6.14

General Remarks

This example problem illustrates the manner in which the critical hydraulic gradient is computed.

Problem Statement

A soil has a porosity (n) of 0.40 and a specific gravity of solids (G_s) of 2.70. Determine the critical hydraulic gradient (i_c).

Solution

From Case 1.8 in Chapter 1, the saturated unit weight for a soil is given by

$$\gamma_{sat} = \frac{\gamma_w(G_s + e)}{1 + e} \tag{6.14.1}$$

Using Case 1.1 of Chapter 1, the void ratio (e) is related to the porosity (n), i.e.,

$$e = \frac{n}{1 - n} = \frac{0.40}{1 - 0.40} = 0.667 \tag{6.14.2}$$

Thus,

$$\gamma_{sat} = \frac{\gamma_w(G_s + e)}{1 + e} = \frac{(9.81\ \text{kN/m}^3)(2.70 + 0.667)}{1 + 0.667} = 19.82\ \text{kN/m}^3 \tag{6.14.3}$$

Also from Case 1.8 in Chapter 1,

$$\gamma_b = \gamma_{sat} - \gamma_w \tag{6.14.4}$$

Thus, for this soil

$$\gamma_b = \gamma_{sat} - \gamma_w = 19.82 - 9.81 = 10.01\ \text{kN/m}^3 \tag{6.14.5}$$

The critical gradient is thus

$$i_c = \frac{\gamma_b}{\gamma_w} = \frac{10.01\ \text{kN/m}^3}{9.81\ \text{kN/m}^3} = \mathbf{1.02} \tag{6.14.6}$$

Alternately, as developed in Section 6.10, the critical gradient can be determined more directly as

$$i_c = \frac{G_s - 1}{1 + e} = \frac{2.70 - 1}{1 + 0.667} = \mathbf{1.02} \tag{6.14.7}$$

EXAMPLE PROBLEM 6.15

General Remarks

This example problem investigates the variation in critical hydraulic gradient between the loose and dense configurations of a soil.

Problem Statement

The specific gravity of solids (G_s) of a sand is 2.66. The porosity (n) of the soil in its loosest and densest state is 45% and 37%, respectively. Determine the critical hydraulic gradient (i_c) for these two states.

Solution

As noted in the previous problem, the saturated unit weight for a soil is given by

$$\gamma_{sat} = \frac{\gamma_w(G_s + e)}{1 + e} \qquad (6.15.1)$$

In addition, the void ratio (e) is related to the porosity by (recall Case 1.1 of Chapter 1)

$$e = \frac{n}{1 - n} \qquad (6.15.2)$$

The critical gradient is then

$$i_c = \frac{\gamma_b}{\gamma_w} = \frac{\gamma_{sat} - \gamma_w}{\gamma_w} = \frac{\gamma_{sat}}{\gamma_w} - 1 = \frac{G_s + e}{1 + e} - 1 \qquad (6.15.3)$$

For the loosest state:

$$e_{max} = \frac{0.45}{1 - 0.45} = 0.818 \qquad (6.15.4)$$

$$i_c = \frac{2.66 + 0.818}{1 + 0.818} - 1 = \mathbf{0.91} \qquad (6.15.5)$$

For the densest state:

$$e_{min} = \frac{0.37}{1 - 0.37} = 0.587 \qquad (6.15.6)$$

$$i_c = \frac{2.66 + 0.587}{1 + 0.587} - 1 = \mathbf{1.05} \qquad (6.15.7)$$

EXAMPLE PROBLEM 6.16

General Remarks

This example problem illustrates the manner in which the critical hydraulic gradient is computed for soil stratum in which a layer in under artesian pressure.

Problem Statement

A deposit of clay lies between two layers of sand as shown in Figure Ex. 6.16. The lower sand layer is under artesian pressure. The moist unit weight of the sand above the groundwater table is 105 lb/ft^3; below the groundwater table its saturated unit weight is 120 lb/ft^3. The saturated unit weight of the clay layer is 110 lb/ft^3.

   Determine a) the vertical total stress, the pore pressure and vertical effective stress at the bottom of the clay layer and b) The height d above the ground surface of the water in the piezometer that would cause "boiling" (heave) of the upper sand layer.

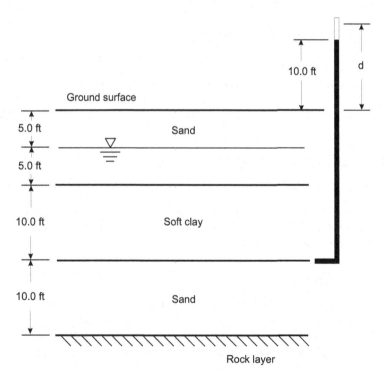

FIGURE EX. 6.16   Soil stratum with sand layer under artesian pressure.

Solution

Since the permeability of the clay will be much lower than for the sand layers, it is assumed to be impermeable in this problem.

a) At the bottom of the clay layer, the vertical total stress is

$$\sigma_v = \left(105 \text{ lb/ft}^3\right)(5 \text{ ft}) + \left(120 \text{ lb/ft}^3\right)(5 \text{ ft}) + \left(110 \text{ lb/ft}^3\right)(10 \text{ ft}) = \mathbf{2225.0 \ lb/ft^2} \tag{6.16.1}$$

The pore pressure at the bottom of the clay layer is

$$u = \gamma_w(10 + 5 + 5 + 10 \text{ ft}) = \left(62.4 \text{ lb/ft}^3\right)(30 \text{ ft}) = \mathbf{1872.0 \ lb/ft^2} \tag{6.16.2}$$

Finally, the vertical effective stress is

$$\sigma_v' = 2225.0 - 1872.0 = \mathbf{353 \ lb/ft^2} \tag{6.16.3}$$

b) To cause boiling, the effective stress at the bottom of the clay layer must go to zero. The total stress at the bottom of the clay layer is unchanged from part a. The pore pressure at the bottom of the clay layer is now

$$u = \gamma_w(10 + 5 + 5 + d \text{ ft}) = \left(62.4 \text{ lb/ft}^3\right)(20 + d \text{ ft}) \tag{6.16.4}$$

The vertical effective stress is

$$\sigma_v' = 2225.0 - \left(62.4 \text{ lb/ft}^3\right)(20 + d \text{ ft}) \tag{6.16.5}$$

Setting the vertical effective stress equal to zero gives

$$2225.0 = (62.4)(20 + d \text{ ft}) \Rightarrow d = \frac{2225.0}{62.4} - 20 = \mathbf{15.7 \ ft} \tag{6.16.6}$$

EXAMPLE PROBLEM 6.17

General Remarks

This example problem illustrates the manner in which the critical hydraulic gradient is computed so as to determine the maximum permissible upward gradient in the design of a dam.

Problem Statement

The foundation soil beneath a dam has a porosity (n) of 42% and a specific gravity of solids (G_s) equal to 2.71. To ensure safety against piping at the toe of the dam, the design specifications state that the upward gradient must not exceed 25% of the critical gradient. Determine the maximum permissible upward gradient.

Solution

Since the foundation soil is saturated, its unit weight is determined using the result of Case 1.8 in Chapter 1, i.e.,

$$\gamma_{sat} = \frac{\gamma_w(G_s + e)}{1 + e} \qquad (6.17.1)$$

The void ratio is computed form the porosity using the result of Case 1.1 in Chapter 1; i.e.,

$$e = \frac{n}{1 - n} = \frac{0.42}{1 - 0.42} = 0.724 \qquad (6.17.2)$$

The saturated unit weight of the foundation soil is thus

$$\gamma_{sat} = \frac{(9.81 \text{ kN/m}^3)(2.71 + 0.724)}{1 + 0.724} = 19.54 \text{ kN/m}^3 \qquad (6.17.3)$$

The buoyant unit weight is thus

$$\gamma_b = \gamma_{sat} - \gamma_w = (19.54 - 9.81) = 9.73 \text{ kN/m}^3 \qquad (6.17.4)$$

The critical hydraulic gradient is next computed, i.e.,

$$i_c = \frac{\gamma_b}{\gamma_w} = \frac{9.73 \text{ kN/m}^3}{9.81 \text{ kN/m}^3} = 0.99 \qquad (6.17.5)$$

The maximum permissible upward gradient is thus $0.25i_c = 0.25(0.99) =$ **0.249**.

EXAMPLE PROBLEM 6.18

General Remarks

This example problem illustrates the manner in which "quick" conditions can occur in an excavation.

Problem Statement

A large excavation was made in a stratum of stiff clay with specific gravity of solids (G_s) equal to 2.64, an average void ratio (e) of 0.745, and an average degree of saturation (S) of 86%. When the depth of the excavation (d) reached 8.0 m (Figure Ex. 6.18), it failed, as a mixture of sand and clay boiled up and rushed into the excavation. Subsequent exploratory borings indicated that a layer of sand underlay the clay, with its top surface at a depth (H) of 12.5 m below the ground surface as shown in Figure Ex. 6.18. To what height h (in meters) would water have risen above the sand layer into the piezometer before the excavation was started?

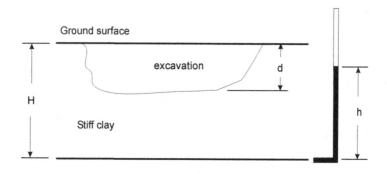

Ground surface

excavation

d

H

Stiff clay

h

Sand

Rock layer

FIGURE EX. 6.18 Site with excavation of a stiff clay layer.

Solution

The moist unit weight of the clay layer must first be determined. Using the expression developed in Case 1.7 of Chapter 1,

$$\gamma = \frac{\gamma_w(G_s + Se)}{1 + e} = \frac{(9.81 \text{ kN/m}^3)\left[2.64 + (0.86)(0.745)\right]}{1 + 0.745} = 18.44 \text{ kN/m}^3$$

$$(6.18.1)$$

The effective stress at the clay/sand interface at the time of excavation is

$$\sigma'_v = \gamma(H - d) - \gamma_w h \approx 0 \qquad (6.18.2)$$

The height to which water would rise above the sand layer into the piezometer before the excavation was started is thus

$$h = \frac{\gamma(H - d)}{\gamma_w} = \frac{(18.44 \text{ kN/m}^3)(12.5 \text{ m} - 8 \text{ m})}{9.81 \text{ kN/m}^3} = \textbf{8.46 m} \qquad (6.18.3)$$

EXAMPLE PROBLEM 6.19

General Remarks

This example problem illustrates the effect that lowering the groundwater table has on an excavation.

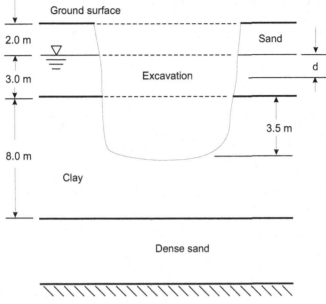

FIGURE EX. 6.19   Site with excavation of a clay layer.

Problem Statement

Figure Ex. 6.19 shows a soil profile consisting if a clay layer underlain by layer of dense sand. Above the clay layer is a layer of sand. For the portion of the upper sand layer above the groundwater table, the moisture content (w) is 18%, the degree of saturation (S) is 80%, and the specific gravity of solids (G_s) is 2.68. Below the groundwater table the moisture content is 20%. For the clay layer the moisture content is 40% and the specific gravity of solids is 2.70.

a) Determine the effective stress at the clay/dense sand interface prior to excavation.
b) Repeat part a immediately after the excavation is complete.
c) Finally, assuming that following the excavation the water remains at the elevation of the groundwater table, determine by how much the water level can be lowered without causing the effective stress at the clay/dense sand interface to go to zero.

Solution

The first step in the solution is to determine the appropriate unit weights for the soil layers. The general expression for the moist unit weight of a soil is (recall Case 1.7 in Chapter 1)

$$\gamma = \frac{\gamma_w(G_s + Se)}{1 + e} \qquad (6.19.1)$$

Since the void ratio is not given explicitly, it must be computed using the expression developed in Case 1.3 of Chapter 1, i.e., $e = G_s w / S$.

For the portion of the sand layer above the groundwater table,

$$e = \frac{G_s w}{S} = \frac{(2.68)(0.18)}{0.80} = 0.603 \qquad (6.19.2)$$

and

$$\gamma = \frac{\gamma_w (G_s + Se)}{1+e} = \frac{(9.81 \text{ kN/m}^3)\left[2.68 + (0.80)(0.603)\right]}{1+0.603} \qquad (6.19.3)$$
$$= 19.35 \text{ kN/m}^3$$

For the portion of the sand layer below the groundwater table,

$$e = \frac{G_s w}{S} = \frac{(2.68)(0.20)}{1.0} = 0.536 \qquad (6.19.4)$$

and

$$\gamma_{sat} = \frac{\gamma_w (G_s + e)}{1+e} = \frac{(9.81 \text{ kN/m}^3)[2.68 + 0.536]}{1+0.536} = 20.54 \text{ kN/m}^3 \quad (6.19.5)$$

For the clay layer,

$$e = \frac{G_s w}{S} = \frac{(2.70)(0.40)}{1.0} = 1.080 \qquad (6.19.6)$$

and

$$\gamma_{sat} = \frac{\gamma_w (G_s + e)}{1+e} = \frac{(9.81 \text{ kN/m}^3)[2.70 + 1.080]}{1+1.080} = 17.83 \text{ kN/m}^3 \quad (6.19.7)$$

a) Prior to the excavation, the vertical total stress at the clay/dense sand interface is

$$\sigma_v = \left(19.35 \text{ kN/m}^3\right)(2.0 \text{ m}) + \left(20.54 \text{ kN/m}^3\right)(3.0 \text{ m})$$
$$+ \left(17.83 \text{ kN/m}^3\right)(8.0 \text{ m}) = 243.0 \text{ kPa} \qquad (6.19.8)$$

The pore pressure at this point is

$$u = \gamma_w (11.0 \text{ m}) = \left(9.81 \text{ kN/m}^3\right)(11.0 \text{ m}) = 107.9 \text{ kPa} \qquad (6.19.9)$$

The vertical effective stress is thus

$$\sigma'_v = 243.0 - 107.9 = \mathbf{135.1 \text{ kPa}} \qquad (6.19.10)$$

b) Immediately following excavation, the water level in the excavation is assumed to remain at the elevation of the groundwater table. The vertical total stress at the clay/dense sand interface is now

$$\sigma_v = (9.81 \text{ kN/m}^3)(6.0 \text{ m}) + (17.83 \text{ kN/m}^3)(5.0 \text{ m}) = 148.0 \text{ kPa}$$

$$(6.19.11)$$

The pore pressure at this point remains

$$u = \gamma_w(11.0 \text{ m}) = (9.81 \text{ kN/m}^3)(11.0 \text{ m}) = 107.9 \text{ kPa} \qquad (6.19.12)$$

The vertical effective stress is thus

$$\sigma'_v = 148.0 - 107.9 = \mathbf{40.1 \text{ kPa}} \qquad (6.19.13)$$

c) Assuming that very soon after the excavation the water level is lowered by an amount d. The vertical total stress at the clay/dense sand interface will become

$$\sigma_v = (9.81 \text{ kN/m}^3)(6.0 \text{ m} - d) + (17.83 \text{ kN/m}^3)(5.0 \text{ m}) \qquad (6.19.14)$$

Due to the relatively low permeability of the clay, the pore pressure at this point remains

$$u = \gamma_w(11.0 \text{ m}) = (9.81 \text{ kN/m}^3)(11.0 \text{ m}) = 107.9 \text{ kPa} \qquad (6.19.15)$$

The vertical effective stress is now

$$\sigma'_v = (9.81)(6.0 - d) + 89.15 - 107.9 = (9.81)(6.0 - d) - 18.75 \text{ kPa}$$

$$(6.19.16)$$

Setting the vertical effective stress equal to zero and solving for d gives

$$d = 6.0 - \frac{18.75}{9.81} = \mathbf{4.1 \text{ m}} \qquad (6.19.17)$$

Obviously, in an actual excavation a suitable factor of safety would be applied to this value, thus reducing the amount by which the water level can be lowered.

EXAMPLE PROBLEM 6.20

General Remarks

This example problem illustrates the manner in which the superficial velocity is determined for the case of one-dimensional steady-state flow in a nonhomogeneous aquifer.

Problem Statement

Consider the 150 m length (L) of a nonhomogeneous, saturated aquifer shown in Figure Ex. 6.20. Piezometers positioned at either end of this length of the aquifer indicate that the total head loss across this portion of the aquifer is equal to 12.0 m.

The permeability of the clay layer (k_1) is 2.0×10^{-7} m/s. The permeability of the silty clay layer (k_3) is 4.0×10^{-6} m/s. The three layers have the following thicknesses: $H_1 = 3.50$ m, $H_2 = 2.25$ m, and $H_3 = 4.25$ m. Although the permeability of the silty sand layer (k_2) is unknown, the total discharge (Q) from this length of aquifer was found to be 0.083 m³/h per meter thickness (into the page).

a) Determine the permeability k_2 (in m/s) for the silty sand, and
b) Given that $G_s = 2.68$ and $\gamma_{sat} = 18.3$ kN/m³ for layer 1, compute the seepage velocity (v_s) for this layer (in m/s).

Solution

a) Let Q_1, Q_2, and Q_3 be the discharge in clay, silty, sand and silty clay layers, respectively. From the continuity of flow,

$$Q = 0.083 \text{ m}^3/\text{h} = Q_1 + Q_2 + Q_3 \qquad (6.20.1)$$

From Darcy's law and noting that the hydraulic gradient is the same across each layer (i.e., $i_1 = i_2 = i_3 = i$),

$$Q_1 = k_1 i A_1 = (2.0 \times 10^{-7} \text{ m/s})\left(\frac{12 \text{ m}}{150 \text{ m}}\right)(3.50 \text{ m})(1 \text{ m})$$
$$= 5.600 \times 10^{-8} \text{ m}^3/\text{s} \qquad (6.20.2)$$

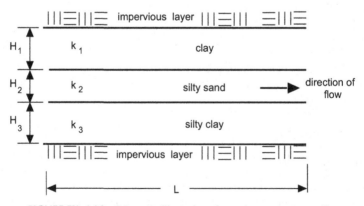

FIGURE EX. 6.20   Schematic Illustration of a nonhomogeneous aquifer.

Converting to the desired gives

$$Q_1 = (5.60 \times 10^{-8} \text{ m}^3/\text{s}) \left(\frac{60 \text{ s}}{\text{min}} \right) \left(\frac{60 \text{ min}}{\text{h}} \right) = 2.016 \times 10^{-4} \text{ m}^3/\text{h} \quad (6.20.3)$$

Similarly,

$$Q_3 = k_3 i A_3 = (4.0 \times 10^{-7} \text{ m/s}) \left(\frac{12 \text{ m}}{150 \text{ m}} \right) (4.25 \text{ m})(1 \text{ m})$$
$$= 1.360 \times 10^{-6} \text{ m}^3/\text{s} \quad (6.20.4)$$

Converting this result to the desired units gives $Q_3 = 4.896 \times 10^{-3}$ m³/s. Finally, from the continuity of flow equation,

$$Q_2 = Q - Q_1 + Q_3 = 0.083 - (2.016 \times 10^{-4} \text{ m}^3/\text{h}) - (4.896 \times 10^{-3} \text{ m}^3/\text{s})$$
$$= 7.790 \times 10^{-2} \text{m}^3/\text{h}$$

$$(6.20.5)$$

But from Darcy's law, $Q_2 = k_2 i A_2$. Solving for k_2 gives

$$k_2 \frac{Q_2}{iA_2} = \frac{(7.790 \times 10^{-2} \text{m}^3/\text{h})}{\left(\frac{12 \text{ m}}{150 \text{ m}} \right) (2.25 \text{ m})(1.0 \text{ m})} \left(\frac{\text{h}}{60 \text{ min}} \right) \left(\frac{\text{min}}{60 \text{ s}} \right) = 1.20 \times 10^{-4} \text{ m/s}$$

$$(6.20.6)$$

To check the aforesaid results, compute the equivalent horizontal permeability using the equation developed in Section 6.11.1.

$$\tilde{k}_x = \frac{\sum_{j=1}^{n} k_{xj} H_j}{\sum_{j=1}^{n} H_j} \quad (6.20.7)$$

Specializing this equation for the present case of $n = 3$ gives

$$\tilde{k}_x = \frac{\begin{array}{c} (2.0 \times 10^{-7} \text{ m/s})(3.50 \text{ m}) + (1.2 \times 10^{-4} \text{ m/s})(2.25 \text{ m}) \\ + (4.0 \times 10^{-6} \text{ m/s})(4.25 \text{ m}) \end{array}}{3.50 \text{ m} + 2.25 \text{ m} + 4.25 \text{ m}}$$

$$= 2.877 \times 10^{-5} \text{ m/s} \quad (6.20.8)$$

Thus, assuming a unit thickness into the page,

$$Q = \tilde{k}_x iA = (2.877 \times 10^{-5} \text{ m/s}) \left(\frac{12 \text{ m}}{150 \text{ m}} \right) (10.0 \text{ m})(1) \left(\frac{3600 \text{ s}}{\text{h}} \right)$$
$$= 0.083 \text{ m}^3/\text{h} \quad (6.20.9)$$

which verifies that the result for k_2 is indeed correct.

b) The seepage velocity requires knowledge of the porosity. Beginning with the expression for the saturated unit weight (recall Case 1.8 in Chapter 1)

$$\gamma = \gamma_{sat} = \frac{\gamma_w (G_s + e)}{1 + e} \quad (6.20.10)$$

and solving for the void ratio gives

$$e = \frac{G_s\, \gamma_w - \gamma_{sat}}{\gamma_{sat} - \gamma_w} \qquad (6.20.11)$$

Substituting the given values of $G_s = 2.68$, $\gamma_{sat} = 18.3$ kN/m^3 for layer 1, along with γ_w gives

$$e = \frac{(2.68)(9.81\ \text{kN/m}^3) - (18.3\ \text{kN/m}^3)}{(18.3 - 9.81)\text{kN/m}^3} = 0.941 \qquad (6.20.12)$$

The porosity is then computed from the void ratio using the relation developed in Case 1.2 of Chapter 1, i.e.,

$$n = \frac{0.941}{1 + 0.941} = 0.485 \qquad (6.20.13)$$

The superficial (Darcy) velocity for layer 1 is

$$v_1 = k_1 i = \left(2.0 \times 10^{-7}\ \text{m/s}\right)\left(\frac{12\ \text{m}}{150\ \text{m}}\right) = 1.60 \times 10^{-8}\ \text{m}^3/\text{s} \qquad (6.20.14)$$

The seepage velocity for layer 1 is thus

$$v_{s_1} = \frac{v_1}{n} = \frac{1.60 \times 10^{-8}\ \text{m/s}}{0.485} = \mathbf{3.30 \times 10^{-8}\ m/s} \qquad (6.20.15)$$

which is consistent for a clay.

EXAMPLE PROBLEM 6.21

General Remarks

This example problem illustrates the manner in which equivalent permeability values are computed for a soil deposit.

Problem Statement

Figure Ex. 6.21 shows the soil profile for a soft seabed clay located in the Canadian Beaufort Sea.

Available field data indicate that the permeability for the clay is anisotropic and varies with depth. In particular,

For layer 1: $k_{x1} = 3.2 \times 10^{-7}$ cm/s; $k_{z1} = 1.1 \times 10^{-7}$ cm/s
For layer 2: $k_{x2} = 1.7 \times 10^{-7}$ cm/s; $k_{z2} = 3.4 \times 10^{-8}$ cm/s
For layer 3: $k_{x3} = 1.4 \times 10^{-7}$ cm/s; $k_{z3} = 4.4 \times 10^{-8}$ cm/s

Compute the equivalent permeability for flow parallel and perpendicular to the clay layers.

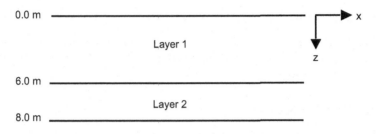

Depth below seabed

FIGURE EX. 6.21 Profile of a seabed clay stratum (not to scale).

Solution

The thicknesses of the individual layers are

$$H_1 = 6.0 \text{ m}, \quad H_2 = 2.0 \text{ m}, \quad H_3 = 10.0 \text{ m}$$

The equivalent permeability parallel to the soil layer is computed as follows (recall the discussion of Section 6.11.1):

$$\tilde{k}_x = \frac{1}{H} \sum_{j=1}^{n} k_{x_j} H_j$$

$$= \frac{1}{18.0 \text{ m}} \left[(3.2 \times 10^{-7} \text{ cm/s})(6.0 \text{ m}) + (1.7 \times 10^{-7} \text{ cm/s})(2.0 \text{ m}) \right.$$

$$\left. + (1.4 \times 10^{-7} \text{ cm/s})(10.0 \text{ m}) \right]$$

$$= \mathbf{2.03 \times 10^{-7} \text{ cm/s}}$$

$$(6.21.1)$$

The equivalent permeability perpendicular to the soil layer is computed as follows (recall the discussion of Section 6.11.2):

$$\tilde{k}_z = \frac{\sum_{j=1}^{n} H_j}{\left(\sum_{j=1}^{n} \dfrac{H_j}{k_{z_j}} \right)} = \frac{18.0 \text{ m}}{\dfrac{6.0 \text{ m}}{1.1 \times 10^{-7} \text{ cm/s}} + \dfrac{2.0 \text{ m}}{3.4 \times 10^{-8} \text{ cm/s}} + \dfrac{10.0 \text{ m}}{4.4 \times 10^{-8} \text{ cm/s}}}$$

$$= \mathbf{5.28 \times 10^{-8} \text{ cm/s}}$$

$$(6.21.2)$$

EXAMPLE PROBLEM 6.22

General Remarks

This example problem illustrates the use of Darcy's law and the continuity of flow to better understand the one-dimensional flow through two soils that are connected in series.

Problem Statement

Figure Ex. 6.22 shows two saturated soil samples that are connected in series in a tube with three vertical standpipes. Soil samples 1 and 2 can have different coefficients of permeability (k_1 and k_2), lengths (L_1 and L_2), and cross-sectional areas (A_1 and A_2). Suitable screens are provided to maintain the samples intact.

a) First develop the general equations that are associated with flow through the two soil samples.
b) Next consider the case where the *same* soil is used for both samples in the device. For $h_1 = 60.0$ cm, $h_3 = 6.0$ cm, $L_1 = 45$ cm, $L_2 = 55$ cm, $A_1 = 28$ cm², and $A_2 = 32$ cm², determine the value of h_2.
c) Finally, consider the case where two *different* soils are tested. If the discharge is equal to 0.20 cm³/s, and $h_1 = 75$ cm, $h_2 = 38$ cm, and $h_3 = 10$ cm, compute the permeability of soils 1 and 2.

Solution

a) Let Δh_1 be the head loss across soil 1. Referring to Figure Ex. 6.22,

$$\Delta h_1 = h_1 - h_2 \tag{6.22.1}$$

The hydraulic gradient for soil 1 is thus

$$i_1 = \frac{\Delta h_1}{L_1} = \frac{h_1 - h_2}{L_1} \tag{6.22.2}$$

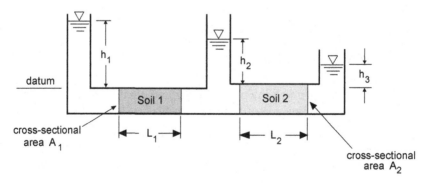

FIGURE EX. 6.22   Hypothetical device with two saturated soil samples connected in series.

The discharge through soil 1 is computed then using Darcy's law, giving

$$Q_1 = k_1 i_1 A_1 = k_1 A_1 \left(\frac{h_1 - h_2}{L_1} \right) \tag{6.22.3}$$

Next, let Δh_2 be the head loss across soil 2. Referring to Figure 6.22,

$$\Delta h_2 = h_2 - h_3 \tag{6.22.4}$$

The hydraulic gradient for soil 2 is thus

$$i_2 = \frac{\Delta h_2}{L_2} = \frac{h_2 - h_3}{L_2} \tag{6.22.5}$$

The discharge through soil 2 is computed then using Darcy's law, giving

$$Q_2 = k_2 i_2 A_2 = k_2 A_2 \left(\frac{h_2 - h_3}{L_2} \right) \tag{6.22.6}$$

Due to continuity of flow, $Q_1 = Q_2$. Equating Eqs. (6.22.3) and (6.22.6) gives

$$k_1 A_1 \left(\frac{h_1 - h_2}{L_1} \right) = k_2 A_2 \left(\frac{h_2 - h_3}{L_2} \right) \tag{6.22.7}$$

b) If the *same* soil is used for both samples in the device, $k_1 = k_2$ and Eq. (6.22.7) reduces to

$$A_1 \left(\frac{h_1 - h_2}{L_1} \right) = A_2 \left(\frac{h_2 - h_3}{L_2} \right) \tag{6.22.8}$$

Typically, h_1 and h_3 are known. Eq. (6.22.8) is then solved for h_2, giving

$$\left(\frac{A_1}{L_1} \right) h_1 + \left(\frac{A_2}{L_2} \right) h_3 = \left(\frac{A_1}{L_1} + \frac{A_2}{L_2} \right) h_2 \Rightarrow h_2 = \frac{\left(\frac{A_1}{L_1} \right) h_1 + \left(\frac{A_2}{L_2} \right) h_3}{\frac{A_1}{L_1} + \frac{A_2}{L_2}}$$

or

$$h_2 = \frac{A_1 L_2 h_1 + A_2 L_1 h_3}{A_1 L_2 + A_2 L_1} \tag{6.22.9}$$

As a check on the aforesaid result, let $L_1 = L_2 = L$ and $A_1 = A_2 = A$. Eq. (6.22.9) then reduces to

$$h_2 = \frac{AL(h_1 + h_3)}{2AL} = \frac{h_1 + h_3}{2} \tag{6.22.10}$$

which is the expected *average* of h_1 and h_3.

Substituting $h_1 = 60.0$ cm, $h_3 = 6.0$ cm, $L_1 = 45$ cm, $L_2 = 55$ cm, $A_1 = 28$ cm^2, and $A_2 = 32$ cm^2 into Eq. (6.22.9) gives

$$h_2 = \frac{(28.0 \text{ cm}^2)(55.0 \text{ cm})(60.0 \text{ cm}) + (32.0 \text{ cm}^2)(45.0 \text{ cm})(6.0 \text{ cm})}{(28.0 \text{ cm}^2)(55.0 \text{ cm}) + (32.0 \text{ cm}^2)(45.0 \text{ cm})}$$

$$= \mathbf{33.9 \text{ cm}}$$

$$(6.22.11)$$

which differs slightly from the average (33.0 cm) of h_1 and h_3.

c) Returning to Eq. (6.22.3), solving for k_1 and using the pertinent values from part b gives

$$k_1 = \frac{q_1}{A_1}\left(\frac{L_1}{h_1 - h_2}\right) = \frac{(0.20 \text{ cm}^3/\text{s})}{(28.0 \text{ cm}^2)}\left(\frac{45.0 \text{ cm}}{75.0 - 38.0 \text{ cm}}\right) \quad (6.22.12)$$

$$= \mathbf{8.69 \times 10^{-3} \text{ cm/s}}$$

In a similar fashion, from Eq. (6.22.6)

$$k_2 = \frac{q_2}{A_2}\left(\frac{L_2}{h_2 - h_3}\right) = \frac{(0.20 \text{ cm}^3/\text{s})}{(32.0 \text{ cm}^2)}\left(\frac{55.0 \text{ cm}}{38.0 - 10.0 \text{ cm}}\right) = \mathbf{1.23 \times 10^{-2} \text{ cm/s}}$$

$$(6.22.13)$$

EXAMPLE PROBLEM 6.23

General Remarks

This example problem illustrates the use of Darcy's law and the continuity of flow to better understand the one-dimensional flow through a flow channel that changes in cross-sectional area.

Problem Statement

Figure Ex. 6.23 shows a case in which water is seeping from a reservoir into an open trench a distance $(L_1 + L_2)$ away from the reservoir through a stratum of fine sand having a coefficient of permeability of k. The stratum is overlain and underlain by impervious material. The sand stratum's thickness at the reservoir (a) differs from that at the open trench (b). The total head loss from the reservoir to the open trench is denoted by Δh.

Develop the general expressions required to determine a) the head loss across each portion of the stratum and b) the discharge in each of the portions of the stratum. Assume the reservoir and trench sides to be vertical and assume the change in section of the sand stratum to be abrupt.

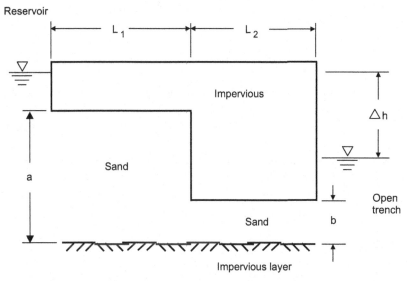

Reservoir

FIGURE EX. 6.23 Hypothetical seepage from reservoir through a confined stratum.

Solution

Denote the total head loss across the first portion of the pervious sand stratum by Δh_a; across the second portion of the stratum the total head loss is Δh_b. It follows that

$$\Delta h_a + \Delta h_b = \Delta h \qquad (6.23.1)$$

The hydraulic gradient (recall Eq. 6.10) across the first portion of the sand stratum is

$$i_a = \frac{\Delta h_a}{L_1} \qquad (6.23.2)$$

The discharge through the first portion of the stratum is then given by Darcy's law, i.e.,

$$Q_a = k i_a A_a = k\left(\frac{\Delta h_a}{L_1}\right)(a)(1) \qquad (6.23.3)$$

where a unit thickness into the page has been assumed. Similarly, the hydraulic gradient across the second portion of the sand stratum is

$$i_b = \frac{\Delta h_b}{L_2} \qquad (6.23.4)$$

The discharge through the first portion of the stratum is then given by Darcy's law, i.e.,

$$Q_b = ki_bA_b = k\left(\frac{\Delta h_b}{L_2}\right)(b)(1) \qquad (6.23.5)$$

For continuity of flow, $Q_a = Q_b$. Equating Eqs. (6.23.3) and (6.23.5) gives

$$\left(\frac{\Delta h_a}{L_1}\right)(a) = \left(\frac{\Delta h_b}{L_2}\right)(b) \qquad (6.23.6)$$

From Eq. (6.23.1), $\Delta h_b = \Delta h - \Delta h_a$. Substituting this relation into Eq. (6.23.6) and solving for Δh_a gives

$$\Delta h_a = \left(\frac{bL_1}{aL_2 + bL_1}\right)\Delta h \qquad (6.23.7)$$

Then,

$$\Delta h_b = \Delta h - \Delta h_a = \left(\frac{aL_2}{aL_2 + bL_1}\right)\Delta h \qquad (6.23.8)$$

As a check of the aforesaid expressions, substitute Eq. (6.23.7) into Eq. (6.23.3), giving

$$Q_a = k\left(\frac{\Delta h_a}{L_1}\right)(a) = \frac{ak}{L_1}\left(\frac{bL_1}{aL_2 + bL_1}\right)\Delta h = k\left(\frac{ab}{aL_2 + bL_1}\right)\Delta h \qquad (6.23.9)$$

Next, substitute Eq. (6.23.8) into Eq. (6.23.5), giving

$$Q_b = k\left(\frac{\Delta h_b}{L_2}\right)(b) = \frac{bk}{L_2}\left(\frac{aL_2}{aL_2 + bL_1}\right)\Delta h = k\left(\frac{ab}{aL_2 + bL_1}\right)\Delta h \qquad (6.23.10)$$

which verifies that indeed $Q_a = Q_b$. As an additional check, assume that $a = b$, i.e., the cross-sectional area for seepage is the same throughout the confined sand stratum. Eqs. (6.23.7) and (6.23.8) then reduce to

$$\Delta h_a = \left(\frac{L_1}{L_2 + L_1}\right)\Delta h \qquad (6.23.11)$$

and

$$\Delta h_b = \left(\frac{L_2}{L_2 + L_1}\right)\Delta h \qquad (6.23.12)$$

which are consistent with a *linear* variation of total head that would be expected for a homogeneous, prismatic confined stratum.

Finally, assume the following specific values: $\Delta h = 7.0$ m, $k = 5 \times 10^{-4}$ m/s, $L_1 = 200$ m, $L_2 = 150$ m, $a = 6.5$ m, and $b = 3.3$ m. Substituting these values into Eqs. (6.23.7) and (6.23.8) gives

$$\Delta h_a = \left[\frac{(3.3 \text{ m})(200 \text{ m})}{(6.5 \text{ m})(150 \text{ m}) + (3.3 \text{ m})(200 \text{ m})} \right] (7.0 \text{ m}) = \mathbf{2.83 \text{ m}} \quad (6.23.13)$$

Then,

$$\Delta h_b = \left[\frac{(6.5 \text{ m})(150 \text{ m})}{(6.5 \text{ m})(150 \text{ m}) + (3.3 \text{ m})(200 \text{ m})} \right] (7.0 \text{ m}) = \mathbf{4.17 \text{ m}} \quad (6.23.14)$$

Substituting the results obtained in Eq. (6.23.13) into Eq. (6.23.9) then gives

$$Q_a = (5.0 \times 10^{-4} \text{ m/s}) \left(\frac{2.83 \text{ m}}{200 \text{ m}} \right) (6.5 \text{ m}) = 4.592 \times 10^{-5} \text{ m}^3/\text{s/m} = Q_b$$

$$(6.23.15)$$

Converting this result into more commonly used units gives

$$Q_a = Q_b = (4.592 \times 10^{-5} \text{ m}^3/\text{s/m}) \left(\frac{60 \text{ s}}{\text{min}} \right) \left(\frac{60 \text{ min}}{\text{h}} \right) = \mathbf{0.165 \text{ m}^3/\text{hr}/\text{m}}$$

$$(6.23.16)$$

Chapter 7

Example Problems Involving Two-Dimensional Fluid Flow in Soils

7.0 GENERAL COMMENTS

For practical applications, the case of one-dimensional flow in soils must be extended to two- and three dimensions. This chapter considers the two-dimensional steady-state seepage[1] in saturated soils.

As noted in Chapter 6, in the case of *steady-state* flow,

- Pore fluid pressures remain constant.
- The rate of flow through the soil is a constant.
- The effective stresses remain constant.
- The soil does not deform.

The energy possessed by a particle of fluid exists in three forms, namely

1. *Pressure energy*; i.e., owing to the pressure at a point.
2. *Potential energy*; i.e., owing to the height of the point above some datum.
3. *Kinetic energy*; i.e., owing to the particle's velocity.

The aforementioned energy is typically expressed as *head* (units of *L*). Since the velocity of flow is typically small in soils and rocks (recall that the flow is laminar), the kinetic energy is thus *negligible*. The total head (*h*) is thus the sum of the elevation head (with respect to some datum) and the pressure head; i.e.,

$$h = z + \frac{p}{\gamma_w} \tag{7.1}$$

where z is the elevation head, p is the pressure at the point, and γ_w is the unit weight of water.

If, at two different points within a continuous soil mass there are *different* amounts of energy (head), there will be movement (flow) of fluid from the

---

1. Seepage is defined as the slow escape of fluid through a porous material.

Soil Mechanics. http://dx.doi.org/10.1016/B978-0-12-804491-9.00007-0

315

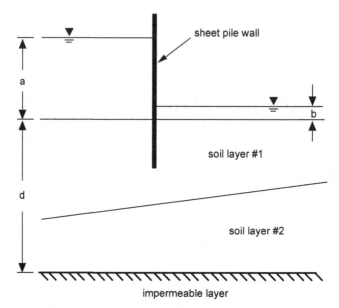

FIGURE 7.1 Schematic illustration of a sheet pile wall.

point of higher energy to the point of lower energy. When fluid flows through soils and rock, energy (head) is *lost* through friction between the fluid and the solid particles.

The topic of two- and three-dimensional flow through soils can be either *confined* or *unconfined*. In confined flow, the seepage is confined between two impermeable surfaces. Figure 7.1 shows an example of confined flow under a sheet pile wall.[2] Both the wall and the bottom boundary are impermeable. Unconfined flow problems typically have a free surface at atmospheric pressure but no impermeable layer in close proximity. Two examples of unconfined flow are earth dams and flow toward wells.

7.1 BASIC ASSUMPTIONS

The present discussion is restricted to two-dimensional steady state, confined seepage. It is based on the following assumptions:

- The soil is *saturated.*
- Darcy's law is valid.
- Each soil layer is *homogeneous* and typically *isotropic.*
- Capillary effects are negligible.

---

2. Some uses of sheet pile walls are (1) to retain a building excavation, (2) to form a wall around a marine terminal, and (3) to serve as an anchored bulkhead for a ship dock, etc.

- Both the solid phase (soil skeleton) and the pore fluid are *incompressible.*
- The volume of the pore fluid remains the same during seepage. This implies no deformation (straining) and volume changes in the soil.

7.2 GOVERNING EQUATION

The two-dimensional seepage of an incompressible fluid in a nondeforming soil is governed by the flowing equation (an elliptic partial differential equation):

$$\frac{\partial}{\partial x}\left(k_x \frac{\partial h}{\partial x}\right) + \frac{\partial}{\partial z}\left(k_z \frac{\partial h}{\partial z}\right) = 0 \tag{7.2}$$

where h is the total head, and k_x and k_z are permeability coefficients in the x- and z-directions, respectively.

Darcy's law then gives the *average* or *superficial* velocity in the x- and z-directions according to

$$v_x = -k_x \frac{\partial h}{\partial x}; \quad v_z = -k_z \frac{\partial h}{\partial z} \tag{7.3}$$

If k_x and k_z are constant, then the governing equation reduces to

$$k_x \frac{\partial^2 h}{\partial x^2} + k_z \frac{\partial^2 h}{\partial z^2} = 0 \tag{7.4}$$

Finally, if $k_x = k_z$ (i.e., permeability isotropy), then the governing equation further reduces to Laplace's equation for two-dimensional homogeneous, isotropic, steady seepage; i.e.,

$$\frac{\partial^2 h}{\partial x^2} + \frac{\partial^2 h}{\partial z^2} = 0 \tag{7.5}$$

This equation must then be solved for the total head $h = f(x,z)$.

7.3 BOUNDARY CONDITIONS

The solution of Laplace's equation requires the definition of suitable boundary conditions. Two kinds of boundary conditions apply to the problem of steady seepage, namely, the following:

- impermeable boundaries (i.e., ones where the velocity normal to the boundary is zero),
- boundaries with known total head (h).

7.4 SOLUTION OF THE GOVERNING EQUATION

The *exact* solution of Laplace's equation is typically difficult to realize. The equation must thus be solved *approximately.* Approximations are commonly obtained using either computer solutions such as the finite difference or finite element method or using hand-drawn flow nets. The topic of computer solutions is discussed elsewhere.[3] Instead, the focus herein is on approximate solution obtained using flow nets.

7.5 FLOW NETS

Flow nets consist of two sets of lines, namely, the following:

- *Flow lines*; i.e., paths taken by the moving particles of fluid.
- *Equipotential lines*; i.e., traces of *equal* total head.

Since pore fluid flows from higher energy levels to lower energy levels along paths of maximum energy gradient (for example, water flows downhill from higher elevations to lower ones, following the steepest path), the flow lines intersect the equipotential lines at right angles.

Remark: Pore fluid tends to flow along the shortest path from one point to another; when flow changes direction, it only makes *smooth* curves.

Listed here are some key points related to flow lines.

- The area between two flow lines is called a *flow channel.*
- The discharge in a flow channel is *constant.*
- Flow cannot occur across flow lines.
- Impermeable boundaries are flow lines.
- The velocity of flow is directed *normal* to the equipotential lines.

Listed here are some key points related to equipotential lines.

- Since these lines indicate points of equal total head, they represent contours of *equal energy.*
- An equipotential line cannot intersect another equipotential line.
- The groundwater table is an equipotential line.
- The difference in total head between two equipotential lines is called the *equipotential drop.*

Some important points related to the drawing of flow nets are listed.

- The upstream and downstream surfaces are equipotential lines (see Figure 7.2).
- The flow lines intersect these equipotential lines at right angles.

---

3. See, for example, Kaliakin, V.N., 2001. Approximate Solution Techniques, Numerical Modeling and Finite Element Methods. Marcel Dekker, Inc., New York, NY.

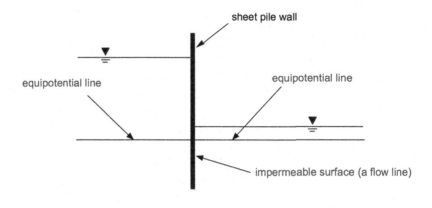

sheet pile wall

equipotential line

equipotential line

impermeable surface (a flow line)

soil layer

impermeable surface (a flow line)

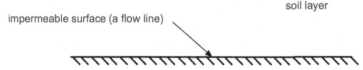

FIGURE 7.2   Schematic illustration of a sheet pile wall showing flow and equipotential lines

- The boundary of an impervious layer or surface is a flow line (see Figure 7.2).
- Equipotential lines intersect the flow lines at right angles.
- A flow line cannot intersect another flow line.
- An equipotential line cannot intersect another equipotential line.

7.6 RATE OF FLOW THROUGH FLOW NETS

To compute the discharge from a flow net, assume the simplest case of permeability isotropy; i.e., $k_x = k_z = k$.

Let N_f equal the total number of flow channels, and let N_d equal the total number of equipotential drops in a flow net.

Consider the portion of a flow net shown in Figure 7.3. The quantity of flow through a typical flow channel is computed using Darcy's law; i.e.,

$$\Delta q = k\left(\frac{h_1 - h_2}{L_1}\right) = k\left(\frac{h_2 - h_3}{L_2}\right) = L \qquad (7.6)$$

If the flow net is drawn such that $L_1 \approx L_2$, etc., then

$$h_1 - h_2 \approx h_2 - h_3 \approx L = \frac{\Delta h}{N_d} \qquad (7.7)$$

where Δh is the total head loss between the upstream and downstream sides. It follows that the head loss between any pair of equipotential lines is thus $\Delta h/N_d$.

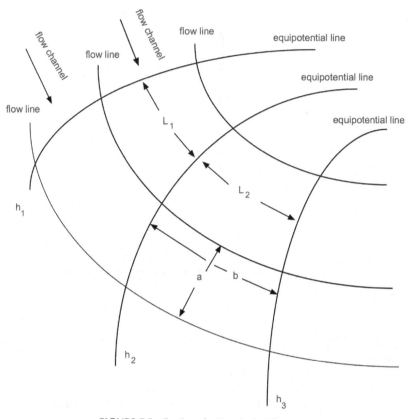

FIGURE 7.3 Portion of a hypothetical flow net.

The amount of discharge through each flow channel is $\Delta q = q/N_f$. Let a equal the width of a flow channel, and b equal the distance between equipotential lines. The hydraulic gradient between equipotential lines is thus

$$i = \frac{\Delta h/N_d}{b} \tag{7.8}$$

The discharge through each flow channel is thus

$$\Delta q = kiA = k\left(\frac{\Delta h}{bN_d}\right)a(1) = k\left(\frac{\Delta h}{N_d}\right)\left(\frac{a}{b}\right) \tag{7.9}$$

The total discharge for the flow net is thus

$$q = \Delta q N_f = k\Delta h\left(\frac{a}{b}\right)\left(\frac{N_f}{N_d}\right) \tag{7.10}$$

Remark: The ratio (a/b) is fixed by the N_f/N_d; it is thus the *same* throughout the flow net.

Remark: If N_f and N_d are selected so that $a \approx b$, the equation for total discharge (for unit dimension perpendicular to the flow net) is

$$q = k\Delta h\left(\frac{N_f}{N_d}\right)$$

EXAMPLE PROBLEM 7.1

General Remarks

This example problem investigates the topic of flow nets as applied to case of one-dimensional flow.

Problem Statement

Figure Ex. 7.1A shows a hypothetical one-dimensional flow experiment. Construct a flow net for the experiment and verify that the total discharge is identical to that predicted by Darcy's law.

Solution

Figure Ex. 7.1B (A) shows a set of equipotential lines associated with the flow experiment. Figure Ex. 7.1B (B) shows a set of flow lines for the same experiment. Finally, Figure Ex. 7.1C shows the flow net that is created by combining the aforementioned equipotential and flow lines.

The coefficient of permeability (k) for the soil sample is equal to 0.05 cm/s. The head loss (Δh) across the soil sample is equal to 0.60 m. The hydraulic gradient is thus

$$i = \frac{\Delta h}{L} = \frac{0.6 \text{ m}}{3.0 \text{ m}} = 0.20 \tag{7.1.1}$$

Using Darcy's law, the total discharge is

$$q = kiA = (0.05 \text{ cm/s})(0.20)(1.2 \text{ m})^2\left(\frac{\text{m}}{100 \text{ cm}}\right) = \mathbf{1.440 \times 10^{-4} \ m^3/s} \tag{7.1.2}$$

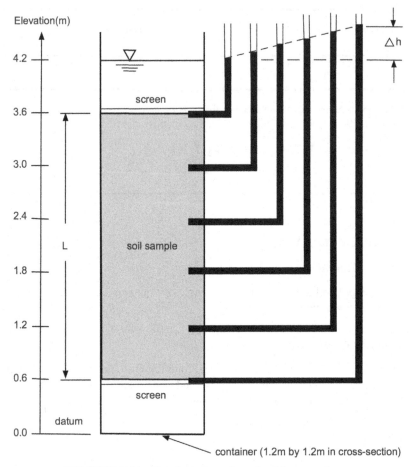

FIGURE EX. 7.1A Hypothetical one-dimensional flow experiment.

Next compute the increment in discharge flowing through a volume Δb by ΔL (Figure Ex. 7.1B) by t (directed into the plane of the paper). Again using Darcy's law, gives

$$\Delta q = k\Delta i \Delta A = k\left(\frac{1}{\Delta L}\frac{\Delta h}{N_d}\right)(\Delta bt) \tag{7.1.3}$$

where N_d is the total number of potential drops.

The total discharge is then

$$q = (\Delta q)(N_f) = k\Delta h\left(\frac{N_f}{N_d}\right)\left(\frac{\Delta b}{\Delta L}\right)t \tag{7.1.4}$$

where N_f/N_d is the "shape factor" for the flow net, and $t = 1.20$ m.

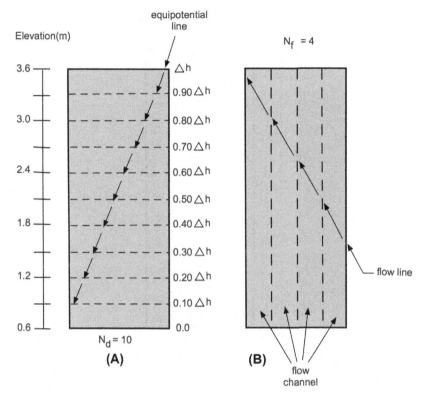

FIGURE EX. 7.1B (A) Equipotential lines and (B) Flow lines for one-dimensional flow experiment.

A square flow net is chosen (Figure Ex. 7.1C) with

$$\Delta b = \frac{1.2 \text{ m}}{4} = 0.30 \text{ m}; \quad \Delta L = \frac{(3.6 \text{ m} - 0.6 \text{ m})}{10} = 0.30 \text{ m} \quad (7.1.5)$$

Substituting for all known values gives

$$q = (0.05 \text{ cm/s})(0.60 \text{ m})\left(\frac{4}{10}\right)(1.20 \text{ m})\left(\frac{\text{m}}{100 \text{ cm}}\right) = \mathbf{1.440 \times 10^{-4} \text{ m}^3/s} \quad (7.1.6)$$

which agrees with the value computed using Darcy's law.

If the total discharge per unit length into the page is desired, then

$$\frac{q}{t} = \frac{1.440 \times 10^{-4} \text{ m}^3/s}{1.20 \text{ m}} = 1.200 \times 10^{-4} \text{ m}^3/s/m \quad (7.1.7)$$

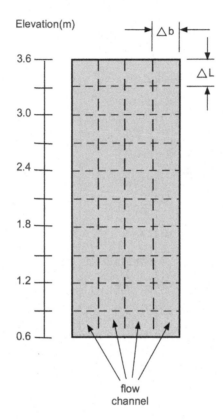

FIGURE EX. 7.1C   Flow net for hypothetical one-dimensional flow experiment.

EXAMPLE PROBLEM 7.2

General Remarks

This example illustrates the manner in which key quantities are computed from a simple flow net for two-dimensional seepage around a sheet pile.

Problem Statement

Given the flow net shown in Figure Ex. 7.2, assume the hydraulic conductivity is 10^{-4} cm/s. The sheet pile is 13 m long (into the paper). The thickness of the soil layer is 10 m and the sheet pile is driven halfway through the soil. A total head difference of 6 m of water separates both sides of the sheet pile.

Determine the following:

a) What is the total head (in meters) at Point a?

b) If the total head (in meters) at Point b is 15 m, what is the total head at Point c?

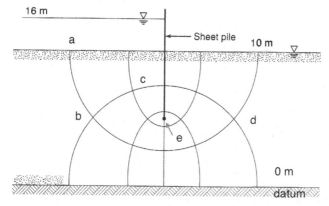

16 m

a

Sheet pile

10 m

c

b

d

e

0 m

datum

FIGURE EX. 7.2 Symmetrical flow net with sheet pile driven halfway into an aquifer.

c) What is the pressure on the sheet pile at Point e?

d) What is the total head (in meters) at Point d?

e) Compute the amount of flow under the sheet pile for its full length in units of cubic meters per day.

Solution

Note that the number of flow channels (N_f) is equal to 3, and the number of equipotential drops (N_d) is equal to 6. Assume that the decrease in total head (Δh) in between each pair of equipotential lines is approximately the same.

a) The elevation head at Point a is 10.0 m. The pressure head at this point is 6.0 m. The total head at Point a is thus $10.0 + 6.0 = \textbf{16.0 m}$.

b) Since Points b and c lie along an equipotential line, it follows that the total head at Point c is equal to that at Point b. Thus $h_c = \textbf{15 m}$.

c) The total head at Point e is

$$h_e = h_a - \left(\frac{3}{6}\right)\Delta h = 16\text{ m} - \left(\frac{3}{6}\right)(6\text{ m}) = 13\text{ m} \qquad (7.2.1)$$

The elevation head at Point e is $z_e = 5$ m. The pressure head is thus

$$\frac{p_e}{\gamma_w} = 13\text{ m} - 5\text{ m} = 8\text{ m} \qquad (7.2.2)$$

The water pressure is thus

$$p_e = \gamma_w(8\text{ m}) = \left(9.81\text{ kN/m}^3\right)(8\text{ m}) = \textbf{78.5 kPa} \qquad (7.2.3)$$

d) The total head at Point d is

$$h_d = h_a - \left(\frac{5}{6}\right)\Delta h = 16 \text{ m} - \left(\frac{5}{6}\right)(6 \text{ m}) = \mathbf{11 \text{ m}} \qquad (7.2.4)$$

e) The discharge under the sheet pile for its full length is

$$q = k\Delta h \left(\frac{N_f}{N_d}\right)(wall \ length) = (10^{-4} \text{ cm/s})(6 \text{ m})\left(\frac{3}{6}\right)\left(\frac{m}{100 \text{ cm}}\right)(13 \text{ m})$$

$$= 3.90 \times 10^{-5} \text{ m}^3/\text{s}$$

$$(7.2.5)$$

Converting to the desired units gives

$$q = (3.90 \times 10^{-5} \text{ m}^3/\text{s})\left(\frac{60 \text{ s}}{\min}\right)\left(\frac{60 \min}{\text{h}}\right)\left(\frac{24 \text{ h}}{\text{d}}\right) = \mathbf{3.37 \text{ m}^3/\text{d}} \qquad (7.2.6)$$

EXAMPLE PROBLEM 7.3

General Remarks

This example illustrates the manner in which key quantities are computed from a somewhat more complex flow net for two-dimensional seepage around a concrete structure.

Problem Statement

The concrete structure shown in Figure Ex. 7.3A is built on sandy soil and is used to retain water. A flow net has been constructed for this structure. The coefficient of permeability for the soil is 1.2×10^{-2} cm/s.

a) Compute the water uplift pressure at Points A, B, C, D, E, and F.
b) Compute the factor of safety against "boiling" of the sand.
c) Compute the total discharge per unit of length along the wall (into the paper)

Solution

The datum is located at the elevation of the base of the concrete retaining structure as shown in Figure Ex. 7.3A. Figure Ex. 7.3B shows the boundary conditions associated with the two-dimensional seepage problem. These include two flow lines (the impervious boundary and the concrete retaining structure) and two equipotential lines (the ground surface on either side of the concrete retaining structure).

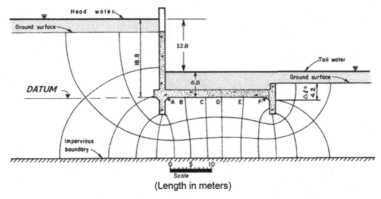

FIGURE EX. 7.3A Concrete retaining structure with flow net[4] (all lengths in meters).

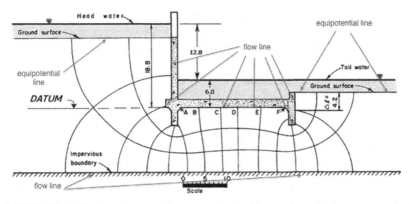

FIGURE EX. 7.3B Boundary conditions associated with two-dimensional seepage around concrete retaining structure.

a) The number of channels in the flow net (N_f) is equal to 3. The total head loss (Δh) between the head water and tail water is 12.8 m. The number of head drops (N_d) is approximately 14. The head drop between two adjacent equipotential lines is thus

$$\Delta h_L = \frac{12.8 \text{ m}}{14} = 0.914 \text{ m} \qquad (7.3.1)$$

The total head at the head water (h_{hw}) is 18.8 m.
At Point A:
The elevation head is $z_A = 0.0$.

---

4. Leshchinsky, D., 2010. Personal communication.

The number of head drops is $N \approx 5.5$. The total head is thus

$$h_A = h_{hw} - (N)(\Delta h_L) = 18.8 - (5.5)(0.914 \text{ m}) = 13.77 \text{ m} \qquad (7.3.2)$$

The pressure head is

$$\frac{p_A}{\gamma_w} = h_A - z_A = 13.77 \text{ m} - 0.0 = 13.77 \text{ m} \qquad (7.3.3)$$

The water pressure is thus

$$p_A = (9.81 \text{ kN/m}^3)(13.77 \text{ m}) = \mathbf{135.1 kPa} \qquad (7.3.4)$$

At Point B:
The elevation head is $z_B = 0.0$.
The number of head drops is $N \approx 6.0$. The total head is thus

$$h_B = h_{hw} - (N)(\Delta h_L) = 18.8 - (6.0)(0.914 \text{ m}) = 13.32 \text{ m} \qquad (7.3.5)$$

The pressure head is

$$\frac{p_B}{\gamma_w} = h_B - z_B = 13.32 \text{ m} - 0.0 = 13.32 \text{ m} \qquad (7.3.6)$$

The water pressure is thus

$$p_B = (9.81 \text{ kN/m}^3)(13.32 \text{ m}) = \mathbf{130.6 \text{ kPa}} \qquad (7.3.7)$$

At Point C:
The elevation head is $z_C = 0.0$.
The number of head drops is $N \approx 7.0$. The total head is thus

$$h_C = h_{hw} - (N)(\Delta h_L) = 18.8 - (7.0)(0.914 \text{ m}) = 12.40 \text{ m} \qquad (7.3.8)$$

The pressure head is

$$\frac{p_C}{\gamma_w} = h_C - z_C = 12.40 \text{ m} - 0.0 = 12.40 \text{ m} \qquad (7.3.9)$$

The water pressure is thus

$$p_C = (9.81 \text{ kN/m}^3)(12.40 \text{ m}) = \mathbf{121.7 \text{ kPa}} \qquad (7.3.10)$$

At Point D:
The elevation head is $z_D = 0.0$.
The number of head drops is $N \approx 8.0$. The total head is thus

$$h_D = h_{hw} - (N)(\Delta h_L) = 18.8 - (8.0)(0.914 \text{ m}) = 11.49 \text{ m} \qquad (7.3.11)$$

The pressure head is

$$\frac{p_D}{\gamma_w} = h_D - z_D = 11.49 \text{ m} - 0.0 = 11.49 \text{ m} \qquad (7.3.12)$$

The water pressure is thus

$$p_D = (9.81 \text{ kN/m}^3)(11.49 \text{ m}) = \mathbf{112.7 \text{ kPa}} \qquad (7.3.13)$$

At Point E:

The elevation head is $z_E = 0.0$.

The number of head drops is $N \approx 9.0$. The total head is thus

$$h_E = h_{hw} - (N)(\Delta h_L) = 18.8 - (9.0)(0.914 \text{ m}) = 10.57 \text{ m} \qquad (7.3.14)$$

The pressure head is

$$\frac{p_E}{\gamma_w} = h_E - z_E = 10.57 \text{ m} - 0.0 = 10.57 \text{ m} \qquad (7.3.15)$$

The water pressure is thus

$$p_E = (9.81 \text{ kN/m}^3)(10.57 \text{ m}) = \textbf{103.7 kPa} \qquad (7.3.16)$$

At Point F:

The elevation head is $z_F = 0.0$.

The number of head drops is $N \approx 9.6$. The total head is thus

$$h_F = h_{hw} - (N)(\Delta h_L) = 18.8 - (9.6)(0.914 \text{ m}) = 10.03 \text{ m} \qquad (7.3.17)$$

The pressure head is

$$\frac{p_F}{\gamma_w} = h_F - z_F = 10.03 \text{ m} - 0.0 = 10.03 \text{ m} \qquad (7.3.18)$$

The water pressure is thus

$$p_F = (9.81 \text{ kN/m}^3)(10.03 \text{ m}) = \textbf{98.4 kPa} \qquad (7.3.19)$$

b) The maximum upward hydraulic gradient will occur next to the wall between the 14th equipotential line and the ground surface. In particular, the

$$i_{exit} = \frac{\Delta h_L}{\Delta l} = \frac{0.914 \text{ m}}{4.2 \text{ m}} = 0.218 \qquad (7.3.20)$$

The factor of safety is thus

$$FS = \frac{i_c}{i_{exit}} = \frac{1.0}{0.218} = \textbf{4.6} \qquad (7.3.21)$$

c) The total discharge per unit of length along the wall is computed as follows:

$$q = k\Delta h \left(\frac{N_f}{N_d} \right) = (1.2 \times 10^{-2} \text{ cm/s}) \left(\frac{\text{m}}{100 \text{ cm}} \right) (12.8 \text{ m}) \left(\frac{3}{14} \right) \qquad (7.3.22)$$
$$= \textbf{3.291} \times \textbf{10}^{-4} \textbf{ m}^3\textbf{/s/m}$$

In more commonly used units,

$$q = (3.291 \times 10^{-4} \text{ m}^3/\text{s/m}) \left(\frac{60 \text{ s}}{\text{min}} \right) \left(\frac{60 \text{ min}}{\text{h}} \right) \left(\frac{24 \text{ h}}{\text{d}} \right) = \textbf{28.4 m}^3\textbf{/d/m}$$
$$(7.3.23)$$

Chapter 8

Example Problems Related to Compressibility and Settlement of Soils

8.0 GENERAL COMMENTS

The subsequent sections briefly review some key aspects related to compressibility and settlement of soils. When designing foundations, the geotechnical engineer seeks to answer two primary questions related to settlement; i.e., (1) how much settlement will occur and (2) how long will it take for this settlement to occur? The former question is answered in this chapter; problems related to the second question are investigated in Chapter 9.

8.1 DEFORMATION

As defined in basic Section 4.3, *deformation* is the change in any dimension of a body. When materials are loaded (stressed), they *deform* or *strain*. Deformation consists of changes in *shape* (i.e., distortion, quantified by changes in angle) and changes in *size* (e.g., changes in area or volume). Figure 8.1 shows both types of deformation.

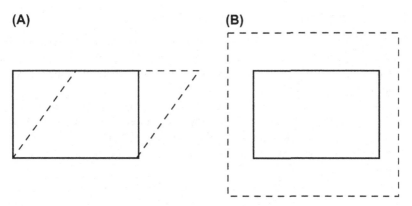

(A)

(B)

FIGURE 8.1 Deformation: (A) change in shape, (B) change in size.

Soil Mechanics. http://dx.doi.org/10.1016/B978-0-12-804491-9.00008-2

Deformations can occur immediately on loading (i.e., *instantaneously*) or over time (i.e., in a *time-dependent* manner). Due to their relatively low permeability, time-dependent deformations are particularly important for cohesive soils (i.e., clays, silty clays, etc.). The deformation of soils is typically quantified by measuring its compressibility.

8.2 COMPRESSIBILITY OF SOILS

When soil is loaded, it compresses because of *changes in volume* due to (a) deformation of the soil grains (i.e., the solid phase), (b) compression of the gas and fluid in the pores (voids), and (c) expulsion of gas and fluid from the pores.

At typical engineering load levels, the deformation of individual soil grains is *negligible*. In addition, the pore fluid is typically assumed to be *incompressible*.

In light of the aforementioned remarks, the volume of a saturated soil can thus only change as fluid is squeezed from (or drawn into) the pores (voids). As fluid flows from the innermost pores to the boundaries of a soil mass, it is assumed to be governed by Darcy's law. The flow requires a hydraulic gradient. If the gradient changes with time, the flow is *transient*; otherwise it is *steady*. In the former case, the pore fluid pressure, and thus the effective stress change with time. In summary, the variation with time of volume in a soil is governed by complex interactions between the following quantities:

- Total stress
- Pore fluid pressure
- Effective stress
- Seepage
- Compressibility of the soil

Closely related to the compressibility of soil is the subject of settlement.

8.3 SETTLEMENT

When a soil deposit is loaded by surface (foundation) stresses, it deforms. The total vertical displacement at the foundation level resulting from the load is called *settlement*. Excessive settlement may cause structural damage, especially if it occurs rapidly.

Load-induced settlement is attributed to two sources; namely the *distortion* of change in shape of the soil immediately beneath the load, and the change in void ratio (and thus volume) of the soil. The former is termed *distortion* or *contact settlement*; the latter is referred to as *compression settlement*.

In general, the total settlement (s_t) of a loaded soil has three components; i.e.,

$$s_t = s_e + s_c + s_s \tag{8.1}$$

where s_e is the *immediate settlement*, s_c is the *primary consolidation settlement*, and s_s is the *secondary consolidation settlement*. Each of these three components is now briefly described.

8.3.1 Immediate Settlement

In Section 4.5 the distribution of stresses within a soil mass resulting from loads applied to the boundary (surface) of the soil mass was briefly discussed. As the soil is loaded, it will, in general, develop distortional and volumetric strains. Associated with such strains are displacements of the soil mass; the most practically important displacement component is typically the vertical one. The vertical displacement at the level of the foundation is the *settlement*.

As noted in Section 8.3, in soils the pore fluid and solid phase are typically assumed to be incompressible. As such, volume changes in a loaded soil can only occur if fluid is able to drain from the pores. Since the permeability of cohesive soils is relatively low, it follows that in such soils the immediate settlement will take place with negligible volume change. As such, this settlement will be purely *distortional*.

In computing immediate settlements, the soil is commonly assumed to be a homogeneous, isotropic linear elastic material. The elastic strains computed using the formulas mentioned in Section 4.5 are then integrated through the depth to get displacements of the soil mass.

Although the topic of immediate elastic settlements is often covered (albeit perhaps briefly) in an introductory course on soil mechanics, it is *not* discussed further herein. Instead, it is felt that a detailed treatment of this subject belongs in a course of *foundation design*.

8.3.2 Primary Consolidation Settlement

This time-dependent component of the total settlement results from the change in volume in saturated soils as fluid is forced from the voids. To better understand the primary consolidation[1] settlement, consider a saturated soil mass that is subjected to an increase in total stress.

---

1. As mentioned in Chapter 3, soil *compaction* is an instantaneous process in which the soil is densified (i.e., its void ratio is reduced) through the expulsion of *air* from the voids. By contrast, *consolidation* is a long-term (time-dependent) process in which *fluid* is expelled from the voids of a (typically) saturated soil. The time rate of consolidation is thus greatly affected by a soil's permeability.

- Initially, since it is assumed to be incompressible, the pore fluid carries the *entire* increment in applied load (total stress). That is, the excess pore pressure is equal to the increase in total stress.
- With time, as fluid is expelled from the pores, the soil grains support more and more of the applied load. The excess pore pressure *decreases* and the effective stress *increases*. Equilibrium is always maintained, so that the sum of the excess pore pressure and the effective stress is equal to the .increment in total stress.
- The rate at which pressure dissipates depends on the rate at which fluid flows from the pores (voids). This, in turn, is a function of the permeability of the soil.
- As time approaches infinity, the excess pore pressure goes to zero, the pore pressure returns to hydrostatic conditions, and the soil skeleton fully supports the additional load.

When saturated *cohesionless* soils (e.g., gravels, sands, and silts) are loaded, the excess pore pressure dissipates *quickly* due to the *high* permeability of such soils. When saturated *clays* are loaded, the dissipation of excess pore pressure is delayed by the relatively *low* permeability of the soil. Section 8.7 provides additional details pertaining to the computation of primary consolidation settlement.

8.3.3 Secondary Consolidation Settlement

This component of the total settlement occurs at constant effective stress, with no change in pore fluid pressure; it is thus time dependent. In cohesionless soils secondary consolidation settlement is attributed to local particle crushing but is generally negligible. In cohesive soils this component of settlement is due to the presence of, and interaction between adsorbed water layers in clays. In both cases, secondary compression settlement induces changes in volume. Since the topic of secondary consolidation settlement is not extensively covered in an introductory course on soil mechanics, it is *not* discussed further herein.

Before additional details regarding settlement can be presented, it is necessary to discuss how the compressibility of soils is quantified.

8.4 QUANTIFYING SOIL COMPRESSIBILITY

As noted in Section 4.4, the solution of problems in engineering mechanics requires a mathematical description of a material's behavior. This is realized through the *constitutive relations*. In the simplest tests performed on a material (e.g., uniaxial extension of a metal, uniaxial compression of concrete, etc.), the material is typically quantified through the value of the elastic modulus (E), which represents the slope of the initial straight-line (linear elastic) portion of the stress—strain curve.

The quantification of soil compressibility is complicated by the particulate nature of the material. In particular, unlike metals, concrete, etc., a soil test specimen must somehow be confined so as to allow a test to be performed on the material.

Perhaps the simplest test apparatus that can be used to assess the compressibility of soil is the one-dimensional or oedometer consolidation test (Figure 8.2).

In the oedometer test the soil specimen is saturated, and care is taken not to let it dry out (recall Example Problem 1.16). Specimens typically have a diameter to thickness ratio of at least 3:1. No lateral displacement is permitted due to the presence of a rigid specimen ring that encircles the specimen. Drainage is provided through porous stones located on the top and bottom of the specimen. Load is applied to the specimen through a lever arm. Each loading increment is usually maintained for 24 h. Each subsequent load increment is typically doubled. A dial gauge measures compression.

Referring to Figure 8.2, and noting that only principal stresses are acting, it follows that

$$\sigma'_z \neq 0, \quad \sigma'_x = \sigma'_y = K_0 \sigma'_z \tag{8.2}$$

where K_0 is the coefficient of earth pressure at rest (recall the discussion of Section 5.5). Since no lateral displacement is permitted due to the presence of a rigid specimen ring, and since only principal strains are nonzero, it follows that

$$\varepsilon_x = \varepsilon_y = \gamma_{xy} = \gamma_{xz} = \gamma_{yz} = 0, \quad \varepsilon_z \neq 0 \tag{8.3}$$

where ε_x, ε_y, and ε_z are infinitesimal normal strains, and γ_{xy}, γ_{xz}, and γ_{yz} are engineering shear strains.

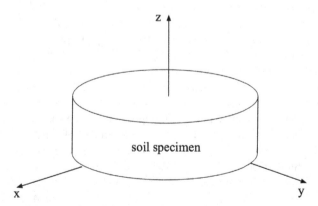

FIGURE 8.2 Schematic illustration of a soil specimen in an oedometer test.

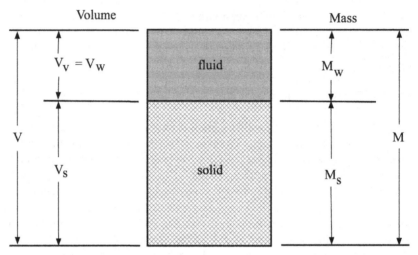

FIGURE 8.3   Phase diagram associated with a saturated soil.

The increment in volumetric strain is then

$$\Delta\varepsilon_{vol} = \frac{\Delta V}{V} = \Delta\varepsilon_x + \Delta\varepsilon_y + \Delta\varepsilon_z = \Delta\varepsilon_z \qquad (8.4)$$

Example Problem 4.6 provides additional insight into the relationship between stress and strain associated with conditions of one-dimensional compression.

To relate $\Delta\varepsilon_{vol}$ to the change in void ratio, note that for a saturated soil, $V = V_v + V_s$ (Figure 8.3).

Since the solid phase is assumed to be incompressible, the increment in volume change is equal to the increment in void ratio; i.e., $\Delta V = \Delta V_v$. Thus,

$$\Delta\varepsilon_{vol} = \frac{\Delta V}{V} = \frac{\Delta V_v}{V_s + V_v} = \frac{\Delta V_v/V_s}{1 + V_v/V_s} = \frac{\Delta e}{1 + e_0} \qquad (8.5)$$

where e_0 is the initial void ratio.

The results of oedometer tests are typically represented either as plots of void ratio versus $\sigma'_z \equiv \sigma'_v$ or as void ratio versus the logarithm of $\sigma'_z \equiv \sigma'_v$. Figure 8.4 shows the latter plot. It is important to note that the relationship between effective stress and void ratio is independent of time.

As shown in Figure 8.4, the loading portion of the response (points *a-b-c*) typically exhibits a change in slope. Removal of the load results in a *swelling* or *rebound* of the soil (points *c-d*). This portion of the response is attributed to elastic "rebound" of the particles (both bulky and platy ones), during which fluid will be drawn into the specimen. Finally, the points *d-e-f* denote the *reloading* phase of the response.

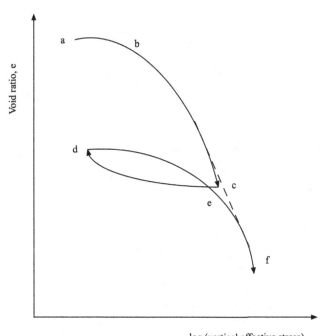

log (vertical effective stress)

FIGURE 8.4   Plot of void ratio versus logarithm of $\sigma'_z \equiv \sigma'_v$.

To facilitate the quantification of the compressibility of a soil, the straight-line assumptions shown in Figure 8.5 are made.

The compression index (C_c) is defined as the negative of the slope of the virgin consolidation curve. The swell (rebound) index (C_r) is defined as the negative of the slope of the swell/recompression curve.[2]

In light of the assumed *linear* relationship between void ratio (e) and the logarithm of vertical effective stress ($\sigma'_z \equiv \sigma'_v$), it follows that for the void ratios e_1 and e_2, and associated vertical effective stress values σ'_{v_1} and σ'_{v_2}, the slope between these two points on the virgin compression line (Figure 8.6) is

$$-C_c = \frac{e_1 - e_2}{\log \sigma'_{v_1} - \log \sigma'_{v_2}} \tag{8.6}$$

The change in void ratio is thus

$$\Delta e = e_1 - e_2 = -C_c \left(\log \sigma'_{v_1} - \log \sigma'_{v_2} \right) = C_c \log \left(\frac{\sigma'_{v_2}}{\sigma'_{v_1}} \right) \tag{8.7}$$

---

2. Both C_c and C_r are *positive* quantities, whereas the slopes of the virgin compression and swell/recompression lines are *negative*.

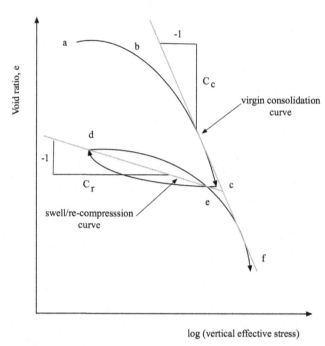

FIGURE 8.5 Definition of compression and swell/recompression indices.

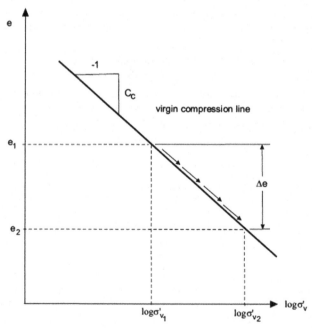

FIGURE 8.6 Calculation of slope of virgin compression line.

8.5 PRECONSOLIDATION PRESSURE

The maximum past vertical effective stress that a soil has been subjected to is called the preconsolidation pressure (σ'_p). If the current vertical effective stress (σ'_v) is equal to or greater than σ'_p, the soil is said to be *normally consolidated*. Figure 8.7 shows the location of the preconsolidation pressure in a consolidation curve.

If a soil is allowed to swell to a lower vertical effective stress (σ'_v), the soil is said to be *overconsolidated*. The degree of overconsolidation is quantified through the *overconsolidation ratio (OCR)*, which is defined as follows:

$$OCR = \frac{\sigma'_p}{\sigma'_v} \tag{8.8}$$

Thus, $OCR = 1.0$ for normally consolidated soils, $OCR > 1.0$ for overconsolidated soils, and $OCR < 1.0$ for underconsolidated soils (i.e., ones that have not yet fully consolidated).

Consider the hypothetical soil deposit shown in Figure 8.8. The moist unit weight of the upper layer is denoted by γ_1, and γ_2 denotes the saturated unit weight of the lower layer.

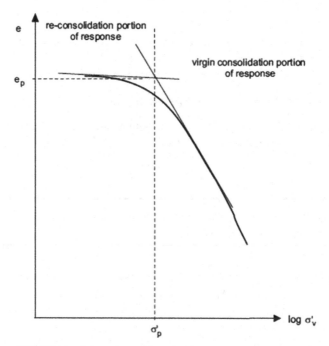

FIGURE 8.7 Schematic illustration of the preconsolidation pressure.

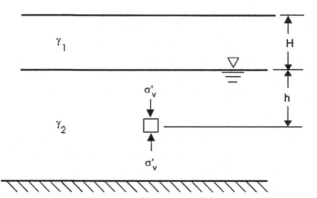

FIGURE 8.8 Hypothetical soil deposit used to explain overconsolidation.

At the point located a distance h below the top of the lower layer, the vertical total stress is

$$\sigma_v = H\gamma_1 + h\gamma_2 \qquad (8.9)$$

The pore pressure at the same point is $u = h\gamma_w$. The vertical effective stress at the same point is thus

$$\sigma'_v = \sigma_v - u = (H\gamma_1 + h\gamma_2) - h\gamma_w = H\gamma_1 + h(\gamma_2 - \gamma_w) \qquad (8.10)$$

- If $\sigma'_v = \sigma'_p$, the soil is *normally consolidated* (i.e., $OCR = 1.0$).
- If $\sigma'_v = \sigma'_p$, the soil is *overconsolidated* (i.e., $OCR > 1.0$).
- If $\sigma'_v = \sigma'_p$, the soil is *underconsolidated* (i.e., $OCR < 1.0$).

8.6 COEFFICIENT OF COMPRESSIBILITY

The coefficient of compressibility (a_v) is defined as the instantaneous slope of the void ratio (e) versus vertical effective stress (σ'_v) curve (units of $F^{-1}L^2$). That is,

$$a_v = -\frac{de}{d\sigma'_v} \qquad (8.11)$$

Remark: Since the slope of the e versus σ'_v curve will be at different points along the curve, a_v will be a function of σ'_v.

In general, for the two points (σ'_{v_1}, e_1) and (σ'_{v_2}, e_2) shown in Figure 8.9,

$$a_v = -\frac{\Delta e}{\Delta \sigma'_v} = -\frac{(e_1 - e_2)}{\left(\sigma'_{v_1} - \sigma'_{v_2}\right)} = \frac{e_2 - e_1}{\sigma'_{v_2} - \sigma'_{v_1}} \qquad (8.12)$$

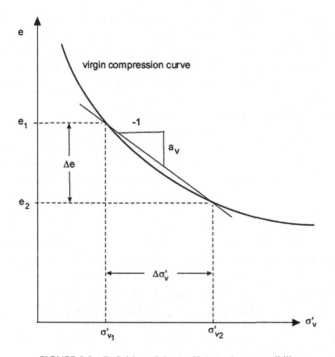

FIGURE 8.9 Definition of the coefficient of compressibility.

8.7 ULTIMATE PRIMARY CONSOLIDATION SETTLEMENT

As discussed in Section 8.3, since the solid phase is assumed to be incompressible, the increment in volume change is equal to the increment in void ratio; i.e., $\Delta V = \Delta V_v$. The volumetric strain is thus

$$\Delta \varepsilon_{vol} = \frac{\Delta V}{V} = \frac{\Delta e}{1 + e_0} \tag{8.13}$$

Let h_0 equal the initial thickness of a specimen in the oedometer and let A_0 equal its original cross-sectional area. Due to the presence of a rigid specimen ring that encircles the specimen, under one-dimensional consolidation the cross-sectional area does not change with loading. Thus,

$$\frac{\Delta V}{V} = \frac{\Delta e}{1 + e_0} = \frac{(\Delta h)(A_0)}{(h_0)(A_0)} = \frac{\Delta h}{h_0} \Rightarrow \Delta h = \left(\frac{\Delta e}{1 + e_0}\right) h_0 \tag{8.14}$$

Assume that for the same material, the volumetric strain in the laboratory is equal to that in the field. Thus, for a field deposit of saturated soil with initial thickness H_0, the average primary consolidation settlement will be

$$s_c = \Delta H = \left(\frac{\Delta e}{1 + e_0}\right) H_0 = \varepsilon_{vol} H_0 \tag{8.15}$$

where Δe and e_0 are measured in the laboratory.

The ultimate primary consolidation settlement for each soil layer is computed using the average initial vertical total stress, the average vertical total stress increment, and a vertical stress–void ratio curve for the specific soil layer as obtained from a laboratory consolidation test. The settlement for all the compressible layers is then added to obtain the total value for the point being analyzed.

The average vertical total and effective stress in each layer are the same as the initial vertical stresses at the middle of the layer because stress increases in direct proportion to the depth (recall the development presented in Chapter 5).

The average increase in vertical total and effective stress is not, however, the same as the stress at the middle of the layer because the relation between stress increase and depth is not linear. If the soil layer is rather thin and is located relatively deep in the soil deposit, it is sufficient to use the stress at the middle of the layer as the average. If the layer is thicker than a footing width and if its depth is less than twice the width, it should be divided into thinner sublayers and the average vertical stress computed for each sublayer.[3]

8.8 COEFFICIENT OF VOLUME COMPRESSIBILITY, MODIFIED COMPRESSION, AND SWELL INDICES

Since $\Delta\varepsilon_v = \Delta e/(1 + e_0)$, for uniaxial conditions, it is also possible to plot the volumetric strain ε_{vol} versus σ'_v or versus $\log(\sigma'_v)$.

- For plots of ε_{vol} versus σ'_v (see Figure 8.10),
  The slope of the resulting curve at some point is

$$m_v = -\frac{d\varepsilon_{vol}}{d\sigma'_v} \tag{8.16}$$

where m_v is the *coefficient of volume compressibility* (units of $F^{-1}L^2$). Recalling the definition of the coefficient of compressibility; i.e.,

$$a_v = -\frac{\Delta e}{\Delta\sigma'_v} \tag{8.17}$$

it follows that

$$\Delta\varepsilon_{vol} = \frac{\Delta e}{1 + e_0} = \frac{a_v(\Delta\sigma'_v)}{1 + e_0} = -m_v(\Delta\sigma'_v) \tag{8.18}$$

Thus,

$$m_v = \frac{a_v}{1 + e_0} \tag{8.19}$$

---

3. Sowers, G.B., Sowers, G.F., 1970. Introductory Soil Mechanics and Foundations. Macmillan Publishing Co., Inc., New York, NY.

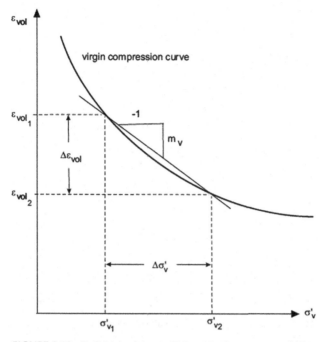

FIGURE 8.10 Definition of the coefficient of volume compressibility.

- For plots of ε_{vol} versus $\log(\sigma'_v)$ (see Figure 8.11),
  The slope of the virgin consolidation portion of the resulting curve is

$$C_{c_\varepsilon} = \frac{\Delta\varepsilon_{vol}}{\log\left(\dfrac{\sigma'_{v2}}{\sigma'_{v1}}\right)} \qquad (8.20)$$

  The quantity C_{c_ε} is called the *modified compression index*.
  Recalling the definition of the compression index; i.e.,

$$C_c = -\frac{\Delta e}{\log\left(\dfrac{\sigma'_{v2}}{\sigma'_{v1}}\right)} = -\frac{\Delta\varepsilon_{vol}(1+e_0)}{\log\left(\dfrac{\sigma'_{v2}}{\sigma'_{v1}}\right)} = (1+e_0)C_{c_\varepsilon} \qquad (8.21)$$

it follows that

$$C_{c_\varepsilon} = -\left(\frac{\Delta e}{1+e_0}\right)\frac{1}{\log\left(\dfrac{\sigma'_{v2}}{\sigma'_{v1}}\right)} = \frac{C_c}{(1+e_0)} \qquad (8.22)$$

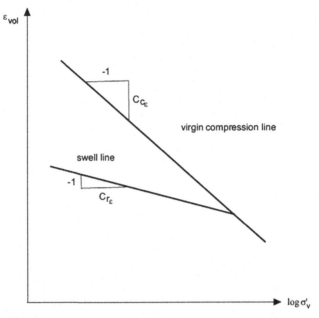

FIGURE 8.11 Definition of the modified compression and swell indices.

In a similar manner, the slope of the swell/recompression portion of the ε_{vol} versus $\log(\sigma_v')$ plot (Figure 8.11) is called the *modified swell index* C_{r_ε}. By analogy to the modified compression index, C_{r_ε} is defined as follows:

$$C_{r_\varepsilon} = \frac{C_r}{(1 + e_0)} \tag{8.23}$$

EXAMPLE PROBLEM 8.1

General Remarks

This example problem gives insight into the preconsolidation stress and the overconsolidation ratio.

Problem Statement

A sample of San Francisco Bay Mud (a marine silty clay) has an initial void ratio (e_0) of 1.42, an associated vertical effective stress (σ_{v0}') of 24.5 kPa, and an overconsolidation ratio (*OCR*) of 8.0. (a) Determine the preconsolidation stress (σ_p'), and (b) If the recompression index $C_r = 0.10$ determine the void ratio associated with σ_p'.

Solution

Figure Ex. 8.1 schematically shows the relationship between void ratio and vertical effective stress for this problem.

a) Recalling the definition of the overconsolidation ratio from Eq. (8.8); i.e.,

$$OCR = \frac{\sigma'_p}{\sigma'_{vo}} \qquad (8.1.1)$$

the preconsolidation stress is thus

$$\sigma'_p = (OCR)\left(\sigma'_{vo}\right) = (8.0)(24.5 \text{ kPa}) = \mathbf{196.0 \text{ kPa}} \qquad (8.1.2)$$

b) From the slope of the swell line in Figure Ex. 8.1,

$$e_0 - e_p = -C_r\left(\log \sigma'_{vo} - \log \sigma'_p\right) = C_r \log\left(\frac{\sigma'_p}{\sigma'_{vo}}\right) = C_r \log(OCR) \quad (8.1.3)$$

The void ratio associated with σ'_p is thus,

$$e_p = e_0 - C_r \log(OCR) = 1.42 - (0.10)\log(8.0) = \mathbf{1.33} \qquad (8.1.4)$$

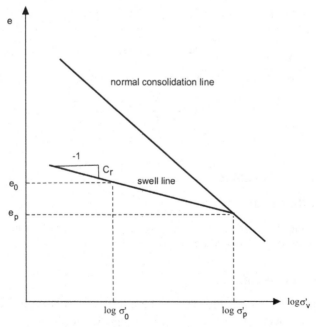

FIGURE EX. 8.1 Plot of void ratio versus logarithm of vertical effective stress.

EXAMPLE PROBLEM 8.2

General Remarks

This example problem illustrates the manner in which the change in void ratio, and thus the average ultimate primary consolidation settlement, is computed from the relative density.

Problem Statement

A well-graded sand with a minimum void ratio of 0.35 and a maximum void ratio of 0.70 has a relative density (D_r) of 39%. The specific gravity of solids (G_s) for the sand is 2.69. Determine how much a 5-m thick stratum of the sand will settle if it is densified to a relative density of 70%.

Solution

The initial void ratio (e_0) of the sand is first determined. The definition of relative density is given by Eq. (3.12);

$$D_r = \left(\frac{e_{max} - e}{e_{max} - e_{min}} \right) * 100\% \tag{8.2.1}$$

Solving for $e = e_0$ gives

$$e_0 = e_{max} - (D_r)(e_{max} - e_{min}) = 0.70 - (0.39)(0.70 - 0.35) = 0.564 \tag{8.2.2}$$

In a similar manner, the void ratio (e) at a relative density of 70% is next computed; i.e.,

$$e = e_{max} - (D_r)(e_{max} - e_{min}) = 0.70 - (0.70)(0.70 - 0.35) = 0.455 \tag{8.2.3}$$

The average ultimate settlement of the 5-m thick stratum of the sand is thus

$$\Delta H = \left(\frac{\Delta e}{1 + e_0} \right) H_0 = \left(\frac{0.564 - 0.455}{1 + 0.564} \right) (5 \text{ m}) = \mathbf{0.347 \text{ m}} = \mathbf{34.7 \text{ cm}} \tag{8.2.4}$$

EXAMPLE PROBLEM 8.3

General Remarks

This example problem presents calculations associated with volume change determination using an oedometer test.

Problem Statement

A saturated sample of a silty sand 2.54 cm thick and 10.0 cm in diameter is tested in an oedometer. After loading, the sample is compressed to a thickness of 2.05 cm. The initial void ratio (e_0) in the sample was found to be 1.420. The specific gravity of solids (G_s) for the silty sand is 2.70.

Determine (a) the initial saturated unit weight (γ_{sat}), (b) the initial moisture content (w_0), (c) the void ratio (e_1) following compression, (d) the saturated unit weight following compression, and (e) the change in moisture content due to compression.

Solution

a) From Case 1.7 of Chapter 1,

$$\gamma_{sat} = \frac{\gamma_w(G_s + e)}{1 + e} = \frac{(9.81 \text{ kN/m}^3)(2.70 + 1.420)}{1 + 1.420} = \textbf{16.70 kN/m}^3 \quad (8.3.1)$$

b) Recalling Case 1.3 of Chapter 1, it follows that for $S = 100\%$, the initial moisture content is thus

$$w = \left(\frac{Se}{G_s}\right) * 100\% \Rightarrow w_0 = \left[\frac{(1.00)(1.420)}{2.70}\right] * 100\% = \textbf{52.6\%} \quad (8.3.2)$$

c) The determination of the void ratio after compression requires knowledge of the volume change in the sample. Initially,

$$e_0 = \frac{V_v}{V_s} \Rightarrow V_v = e_0 V_s = 1.420 V_s \quad (8.3.3)$$

Due to the fact that the sample is circular in cross-section,

$$V_0 = \frac{\pi}{4}d^2 H_0 = \frac{\pi}{4}(10.0 \text{ cm})^2(2.54 \text{ cm}) = 199.5 \text{ cm}^3 \quad (8.3.4)$$

All compression in the sample is assumed to be produced by a reduction of void ratio due to the expulsion of pore fluid. In addition, since the rigid confining ring in the oedometer prevents changes in the diameter, the change in total volume is thus

$$\Delta V = \frac{\pi}{4}d^2(H_0 - H) = \frac{\pi}{4}(10.0 \text{ cm})^2(2.54 - 2.05 \text{ cm}) = 38.49 \text{ cm}^3 \quad (8.3.5)$$

where H is the thickness of the sample after compression. Recalling that

$$\Delta V = \left(\frac{\Delta e}{1 + e_0}\right) V_0 \quad (8.3.6)$$

the change in void ratio is thus

$$\Delta e = \left(\frac{\Delta V}{V_0}\right)(1 + e_0) = \left(\frac{38.49 \text{ cm}^3}{199.5 \text{ cm}^3}\right)(1 + 1.420) = 0.467 \quad (8.3.7)$$

Since, in general, $\Delta e = e_0 - e_1$, the void ratio following compression is thus

$$e_1 = e_0 - \Delta e = 1.420 - 0.467 = \mathbf{0.953} \qquad (8.3.8)$$

d) After compression,

$$\gamma_{sat} = \frac{\gamma_w(G_s + e)}{1 + e} = \frac{(9.81 \text{ kN/m}^3)(2.70 + 0.953)}{1 + 0.953} = \mathbf{18.35 \text{ kN/m}^3} \quad (8.3.9)$$

e) Finally, moisture content after compression is

$$w_1 = \left(\frac{Se}{G_s}\right) * 100\% = \left[\frac{(1.00)(0.953)}{2.70}\right] * 100\% = 35.3\% \qquad (8.3.10)$$

The change in moisture content caused by compression is thus

$$\Delta w = w_0 - w_1 = 52.6 - 35.3 = \mathbf{17.3\%} \qquad (8.3.11)$$

EXAMPLE PROBLEM 8.4

General Remarks

This example problem illustrates how compression curves are constructed and interpreted.

Problem Statement

The data given in Table Ex. 8.4 were obtained from a consolidation test on a sample of clay taken from a 14.5 m thick deposit.

The average existing overburden pressure on the clay layer is 320 kPa. It is proposed to construct a building at this site that will increase the average vertical total stress on the clay layer to 560 kPa.

a) Plot the above data on a semilogarithmic graph. Estimate the compression and swell/recompression indices.
b) Estimate the preconsolidation stress.
c) Compute the average ultimate primary consolidation settlement that the clay will undergo under the load of the proposed building.

Solution

a) Figure Ex. 8.4A shows the semilogarithmic graph of the void ratio versus the logarithm of the vertical effective stress.

TABLE EX. 8.4 Effective Stress and Void Ratio Data for a Clay

Consolidation Stress (kPa)	Void Ratio
1.960E+01	9.530E-01
3.920E+01	9.480E-01
7.840E+01	9.380E-01
1.568E+02	9.200E-01
3.136E+02	8.780E-01
6.272E+02	7.890E-01
1.254E+03	6.910E-01
3.136E+02	7.190E-01
7.840E+01	7.540E-01
1.960E+01	7.910E-01

FIGURE EX. 8.4A   Consolidation curve for data given in Table Ex. 8.4.

Figure Ex. 8.4B shows the straight-line approximations for the virgin compression and swell/recompression portions of the plot. Using data points along the former portion of the curve,

$$C_c = -\frac{(0.878 - 0.691)}{\log(313.6) - \log(1254)} = \textbf{0.311} \qquad (8.4.1)$$

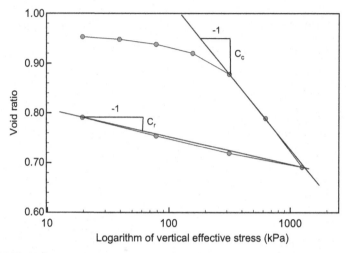

FIGURE EX. 8.4B Straight-line approximations for compression and swell/recompression portions of the consolidation curve.

In a similar manner, using data points along the swell/recompression portions of the plot gives

$$C_r = -\frac{(0.791 - 0.691)}{\log(19.6) - \log(1254)} = \textbf{0.055} \tag{8.4.2}$$

The ratio $C_r/C_c = 0.055/0.311 = 0.18$ falls within the expected range[4] of 0.10–0.20.

b) The preconsolidation stress is estimated to be approximately 206 kPa (Figure Ex. 8.4C). If necessary, this value can be refined by using Casagrande's construction.[5]

c) Due to the proposed building, the average effective stress in the clay layer will, at ultimate conditions, increase to 560 kPa. The computation of the average ultimate primary consolidation settlement requires the associated change in void ratio. This information is obtained from the consolidation curve in the manner shown in Figure Ex. 8.4D.

---

4. Holtz, R.D., Kovacs, W.D., Sheahan, T.C., 2011. An Introduction to Geotechnical Engineering. Pearson Education, Inc., Upper Saddle River, NJ.

5. Casagrande, A., 1936. Determination of the Pre-Consolidation Load and Its Practical Significance, Discussion D-34. In: Proceedings of the First International Conference on Soil Mechanics and Foundation Engineering. Cambridge, MA, III, pp. 60–64.

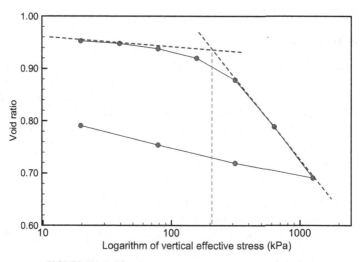

FIGURE EX. 8.4C Estimate of the preconsolidation pressure.

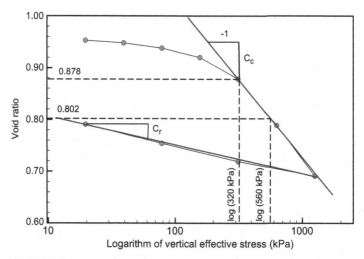

FIGURE EX. 8.4D Estimates of the void ratios from known effective stress values.

In particular, the change in void ratio is approximately equal to

$$e = 0.878 - 0.802 = 0.076 \qquad (8.4.3)$$

The average ultimate primary consolidation settlement is thus

$$s_c = \left(\frac{\Delta e}{1 + e_0}\right) H_0 = \left(\frac{0.076}{1 + 0.878}\right)(14.5 \text{ m}) = \mathbf{0.59\ m} \qquad (8.4.4)$$

EXAMPLE PROBLEM 8.5

General Remarks

This example problem illustrates how consolidation curves are constructed for use in problems involving critical state soil mechanics.[6] In critical state soil mechanics, consolidation curves are represented as plots of void ratio (e) as the ordinate versus the natural logarithm of the mean normal effective stress as the abscissa. The latter is defined as (also recall the discussion of mean stress given in Section 4.2.3)

$$p' = \frac{\sigma_1' + \sigma_2' + \sigma_3'}{3} \qquad (8.5.1)$$

where σ_1', σ_2', and σ_3' are principal effective stresses. In oedometer (K_0) tests $\sigma_2' = \sigma_3'$. In isotropic compression tests, $\sigma_1' = \sigma_2' = \sigma_3'$.

The slope of the virgin compression line in $e - \ln p'$ space is denoted by λ. The slope of the swell/recompression curve in the same space is denoted by κ. The critical state indices λ and κ are related to compression and swell indices C_c and C_r in the following manner:

$$\lambda = \frac{C_c}{2.303}; \quad \kappa = \frac{C_r}{2.303} \qquad (8.5.2)$$

Problem Statement

Using the effective stress versus void ratio data given in Tables Ex. 8.5A−C, create a separate void ratio (ordinate) versus $\ln p'$ (abscissa) plot for each of the following soils. From these plots determine values for the compression parameter λ and the swell/recompression parameter κ for each of the above soils.

Solution

Figure Ex. 8.5A shows the consolidation curve for Bay Mud. From the straight-line fits of the virgin compression and swell/recompression lines, λ and κ are found to be 0.239 and 0.049, respectively.

Figure Ex. 8.5B shows the consolidation curve for Edgar Plastic Kaolin. From the straight-line fits of the virgin compression and swell/recompression lines, λ and κ are found to be 0.098 and 0.014, respectively.

Finally, Figure Ex. 8.5C shows the consolidation curve for Kaolin. From the straight-line fits of the virgin compression and swell/recompression lines, λ and κ are found to be 0.330 and 0.061, respectively.

---

6. Schofield, A.N., Wroth, C.P., 1968. Critical State Soil Mechanics. McGraw-Hill Book Co., Inc., London, UK.

TABLE EX. 8.5A Effective Stress and Void Ratio Data for Bay Mud, as Reported by Kavazanjiana[a]

Effective Isotropic Consolidation Stress (kg/cm^2)	Void Ratio
3.034E-01	1.950E+00
5.024E-01	1.813E+00
5.022E-01	1.789E+00
9.924E-01	1.567E+00
2.262E+00	1.297E+00
2.042E+00	1.302E+00
1.001E+00	1.326E+00
5.047E-01	1.363E+00
2.493E-01	1.407E+00

[a]Kavazanjian Jr., E., 1978. A Generalized Approach to the Prediction of the Stress-Strain-Time Behavior of Soft Clay (Dissertation submitted in partial satisfaction of the requirements for the degree of Doctor of Philosophy). University of California, Berkeley, California.

TABLE EX. 8.5B Mean Normal Effective Stress and Void Ratio Data for Edgar Plastic Kaolin, as Reported by Lade[a]

p' (kPa)	Void Ratio
98.00	1.1750
147.00	1.1460
196.00	1.1140
245.00	1.0920
294.00	1.0740
343.00	1.0580
392.00	1.0400
294.00	1.0430
196.00	1.0470
98.00	1.0600
196.00	1.0510
294.00	1.0470

[a]Lade, P.V., 1990. Single-hardening model with application to NC clay. Journal of Geotechnical Engineering, ASCE 116 (3), 394—414.

TABLE EX. 8.5C Mean Normal Effective Stress and Void Ratio Data for Kaolin, as Reported by Parry and Nadarajah[a]

Consolidation Stress (kPa)	Void Ratio
245.539	1.3099
279.494	1.3046
347.803	1.2907
414.514	1.2762
486.019	1.2675
551.132	1.2634
487.617	1.3203
414.914	1.3760
314.647	1.4375
248.336	1.5484
183.222	1.6319

[a]*Parry, R.H.G., Nadarajah, V., 1973. A volumetric yield locus for lightly over-consolidated clay. Geotechnique 23 (3), 450–453.*

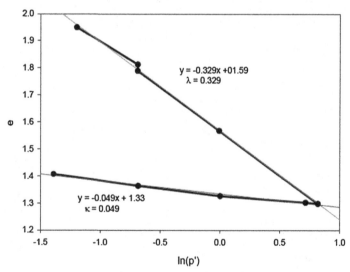

FIGURE EX. 8.5A Plot of void ratio versus natural logarithm of mean normal effective stress for Bay Mud.

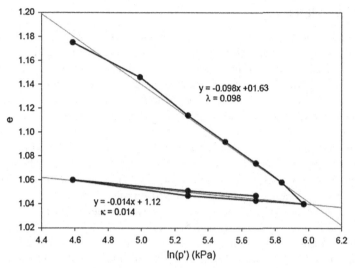

FIGURE EX. 8.5B   Plot of void ratio versus natural logarithm of mean normal effective stress for Edgar Plastic Kaolin.

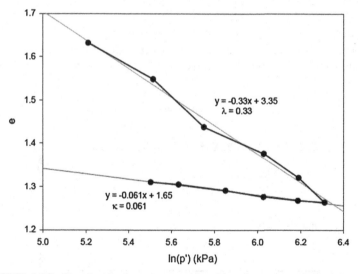

FIGURE EX. 8.5C   Plot of void ratio versus natural logarithm of mean normal effective stress for Kaolin.

EXAMPLE PROBLEM 8.6

General Remarks

This example problem illustrates how to compute the preconsolidation pressure, as well as the average ultimate primary consolidation settlement.

Problem Statement

The soil profile at a site for a proposed parking garage is shown in Figure Ex. 8.6. It consists of a 10.6 m think layer of fine sand that is underlain by a 2.40 m thick layer of overconsolidated Chicago clay. Below the clay is a deposit of coarse sand. The groundwater table was observed at a depth of 3.5 m below the ground surface. Assume the soil above the groundwater table to be *saturated*. In this problem the effect of capillarity is ignored.

The specific gravity of solids (G_s) for the fine sand is 2.70 and its average void ratio (e) is 0.780. For the Chicago clay layer, $G_s = 2.68$ and the moisture content is 45.8%. The results from a one-dimensional (oedometer) consolidation test on a sample of the same clay from the middle of the layer give a compression index (C_c) of 0.265, a recompression index (C_r) of 0.050, and an overconsolidation ratio (*OCR*) of 2.5.

The parking garage will impose a vertical stress increase of 150 kPa in the middle of the clay layer. Assume the unit weight of water to be 9.81 kN/m³.

a) First determine the average initial void ratio in the clay layer.
b) Next determine the vertical effective stress at the center of the clay layer *before* applying the garage load.
c) Then determine the preconsolidation pressure σ'_p.
d) Finally, using the information from parts (b) and (c), determine the average ultimate primary consolidation settlement (s_c) in *millimeters*.

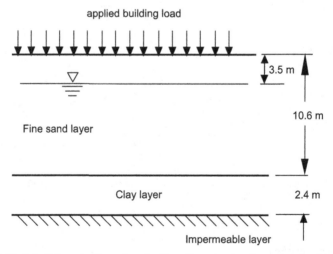

FIGURE EX. 8.6 Schematic illustration of soil profile at the site of a proposed parking garage.

Solution

a) The initial void ratio in the Chicago clay layer is obtained using the equation developed in Case 1.3 of Chapter 1; i.e.,

$$e = G_s w = (2.68)(0.458) = \mathbf{1.227} \tag{8.6.1}$$

b) Determination of the vertical effective stress at the middle of the clay layer first requires the saturated unit weights for the fine sand and Chicago clay layers. For the fine sand layer

$$\gamma_{sat} = \frac{\gamma_w(G_s + e)}{1 + e} = \frac{(9.81 \text{ kN/m}^3)(2.70 + 0.78)}{1 + 0.78} = 19.18 \text{ kN/m}^3 \tag{8.6.2}$$

Similarly, for the clay layer,

$$\gamma_{sat} = \frac{\gamma_w(G_s + e)}{1 + e} = \frac{(9.81 \text{ kN/m}^3)(2.68 + 1.227)}{1 + 1.227} = 17.21 \text{ kN/m}^3 \tag{8.6.3}$$

The vertical total stress at the center of the clay layer is thus

$$\sigma_{v0} = (19.18 \text{ kN/m}^3)(10.6 \text{ m}) + (17.21 \text{ kN/m}^3)(1.2 \text{ m}) = 223.96 \text{ kPa} \tag{8.6.4}$$

The initial pore pressure at the same point is

$$u_0 = (9.81 \text{ kN/m}^3)[(10.6 - 3.50 \text{ m}) + 1.2 \text{ m}] = 81.42 \text{ kPa} \tag{8.6.5}$$

The vertical effective stress at the center of the clay layer is thus

$$\sigma'_{v0} = \sigma_{v0} - u_0 = 223.96 - 81.42 = \mathbf{142.5 \text{ kPa}} \tag{8.6.6}$$

c) The preconsolidation pressure stress is determined from the definition of the overconsolidation ratio; i.e.,

$$OCR = \frac{\sigma'_p}{\sigma'_{v0}} \Rightarrow \sigma'_p = (OCR)\sigma'_{v0} = (2.50)(142.5 \text{ kPa}) = \mathbf{356.3 \text{ kPa}} \tag{8.6.7}$$

d) To determine the average ultimate primary consolidation settlement, note that after the load from the parking garage is applied, the effective stress at the center of the clay layer will be

$$\sigma'_v = \sigma'_{v0} + \Delta\sigma_{v0} = 142.5 + 150 = 292.5 \text{ kPa} \tag{8.6.8}$$

Since $\sigma'_v < \sigma'_p$, the clay layer is overconsolidated. The average ultimate primary consolidation settlement is thus computed as follows:

$$S_c \equiv \Delta H = \left(\frac{\Delta e}{1+e_0}\right) H_0 \qquad (8.6.9)$$

where

$$\Delta e = C_r \log\left(\frac{\sigma'_{v_0} + \Delta\sigma'_v}{\sigma'_{v_0}}\right) = (0.050)\log\left(\frac{292.5 \text{ kPa}}{142.5 \text{ kPa}}\right) = 0.016 \qquad (8.6.10)$$

Thus,

$$S_c = \left(\frac{0.016}{1+1.227}\right)(2.40 \text{ m})\left(\frac{1000 \text{ mm}}{\text{m}}\right) = \textbf{16.8 mm} \qquad (8.6.11)$$

Remark: Another way of quantifying that the clay layer is overconsolidated is through its overconsolidation ratio after the load from the parking garage is applied; i.e.,

$$OCR = \frac{\sigma'_p}{\sigma'_v} = \frac{356.3 \text{ kPa}}{292.5 \text{ kPa}} = 1.22 > 1.00$$

EXAMPLE PROBLEM 8.7

General Remarks

This example problem illustrates how to compute the preconsolidation pressure, as well as the average ultimate primary consolidation settlement.

Problem Statement

The construction plan for a particular site calls for the loading of a 15.5 m thick soft clay layer by a 4 m thick layer of fill material consisting of sand and gravel. The groundwater table is located at the surface of the clay layer.

The moist unit weight of the sand/gravel layer is 19.5 kN/m^3. At the center of the soft clay layer the moisture content (w) is 0.548 and the specific gravity of solids (G_s) is 2.69. From a consolidation test on this soil, the preconsolidation pressure was found to be 102.0 kPa. Finally, the compression index (C_c) and swell index (C_r) were found to be equal to 0.301 and 0.060, respectively.

(a) Calculate the average ultimate primary consolidation settlement that will occur after placement of the sand/gravel layer. (b) How much will the clay layer rebound (heave) when the sand/gravel layer is removed?

Assume that the sand/gravel layer is placed relatively quickly and that the deformations are one dimensional. Neglect capillary rise above the ground-water level in this layer.

Solution

Before placement of the fill, the initial void ratio or the soft clay layer is obtained using the equation developed in Case 1.3 of Chapter 1; i.e.,

$$e = G_s w = (2.70)(0.548) = 1.480 \tag{8.7.1}$$

The saturated unit weight is thus

$$\gamma_{sat} = \frac{\gamma_w (G_s + e)}{1 + e} = \frac{(9.81 \text{ kN/m}^3)(2.69 + 1.480)}{1 + 1.480} = 16.50 \text{ kN/m}^3 \tag{8.7.2}$$

At the center of the clay layer, the vertical total stress is

$$\sigma_{v_0} = \left(16.50 \text{ kN/m}^3\right)\left(\frac{15.5 \text{ m}}{2}\right) = 128.9 \text{ kPa} \tag{8.7.3}$$

The initial pore pressure at the same point is

$$u_0 = \left(9.81 \text{ kN/m}^3\right)\left(\frac{15.5 \text{ m}}{2}\right) = 76.0 \text{ kPa} \tag{8.7.4}$$

The vertical effective stress at the center of the clay layer is thus

$$\sigma'_{v_0} = \sigma_{v_0} - u_0 = 128.9 - 76.0 = 52.9 \text{ kPa} \tag{8.7.5}$$

Since σ'_{v_0} is less than the preconsolidation pressure σ'_p, it follows that the soft clay layer is overconsolidated. The associated overconsolidation ratio is

$$OCR = \frac{\sigma'_p}{\sigma'_{v_0}} = \frac{102.0 \text{ kPa}}{52.9 \text{ kPa}} = 1.93 \tag{8.7.6}$$

a) The increment in total stress attributed to the sand/gravel fill is

$$\Delta\sigma_v = \left(19.5 \text{ kN/m}^3\right)(4 \text{ m}) = 78.0 \text{ kPa} \tag{8.7.7}$$

At ultimate conditions all excess pore fluid pressure has dissipated and $\Delta\sigma'_v = \Delta\sigma_v$. As such, the new effective stress state is thus

$$\sigma'_v = \sigma'_{v_0} + \Delta\sigma'_v = 52.9 + 78.0 = 130.9 \text{ kPa} \tag{8.7.8}$$

Since $\sigma'_v > \sigma'_p$, the average primary consolidation settlement must be computed in *two* parts. In particular, the first portion of the response takes

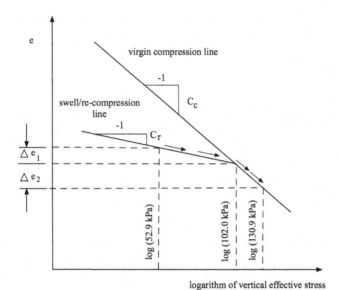

logarithm of vertical effective stress

FIGURE EX. 8.7A Consolidation curve associated with application of the fill (not to scale).

place along the swell/recompression line between $\sigma'_{v_0} = 52.9$ kPa and $\sigma'_p = 102.0$ kPa (Figure Ex. 8.7A). The associated change in void ratio is

$$\Delta e_1 = -C_r(\log 52.9 - \log 102.0) = (0.060)\log\left(\frac{102.0}{52.9}\right) = 0.017 \quad (8.7.9)$$

The second portion of the response takes place along the virgin compression line between $\sigma'_p = 102.0$ kPa and $\sigma'_v = \sigma'_{v_0} + \Delta\sigma'_v = 130.9$ kPa (Figure Ex. 8.7A). The associated change in void ratio is

$$\Delta e_2 = -C_r(\log 102.0 - \log 130.9) = (0.301)\log\left(\frac{130.9}{102.0}\right) = 0.033 \quad (8.7.10)$$

The total average ultimate primary consolidation settlement is thus

$$s_c \equiv \Delta H = \left(\frac{\Delta e_1}{1+e_0}\right)H_0 + \left(\frac{\Delta e_2}{1+e_0}\right)H_0 = \left(\frac{\Delta e_1 + \Delta e_2}{1+e_0}\right)H_0$$

$$= \left(\frac{0.017 + 0.033}{1 + 1.480}\right)(15.5 \text{ m}) = \underline{\underline{0.311 \text{ m}}} \quad (8.7.11)$$

b) The heave associated with the removal of the sand/gravel fill takes place along the swell/recompression line that begins from $\sigma'_v = 130.9$ kPa and ends at the point $\sigma'_{v_0} = 52.9$ kPa (Figure Ex. 8.7B). Thus,

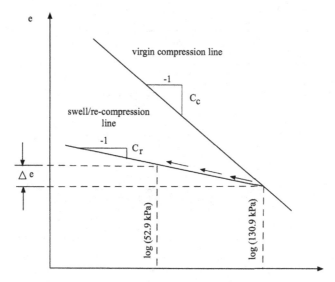

FIGURE EX. 8.7B   Consolidation curve associated with removal of the fill (not to scale).

$$s_{swell} = C_r \log\left(\frac{\sigma_v'}{\sigma_{v0}'}\right) = (0.060)\log\left(\frac{130.9}{52.9}\right) = 0.024 \text{ m} \qquad (8.7.12)$$

The residual settlement is thus

$$s_{residual} = s_c - s_{swell} = 0.311 - 0.024 = \mathbf{0.287 \text{ m}} \qquad (8.7.13)$$

EXAMPLE PROBLEM 8.8

General Remarks

This example problem illustrates the manner in which the average ultimate primary consolidation settlement is computed.

Problem Statement

Given the soil deposit shown in Figure Ex. 8.8, the groundwater table is located at a depth $d = 1$ m below the initial ground surface. A 6 m thick clay layer underlies the 4 m thick sand layer.

The moist sand above the groundwater table has a unit weight (γ) equal to 19.5 kN/m³. Below the groundwater table the sand has a saturated unit weight (γ_{sat}) of 20.0 kN/m³. The results of laboratory tests on a sample of the clay from the middle of the layer give a specific gravity of solids (G_s) equal to 2.65, a natural moisture content (w) of 60%, and a liquid limit (LL) of 75%. Estimate the average ultimate primary consolidation settlement (s_c) of the clay if the

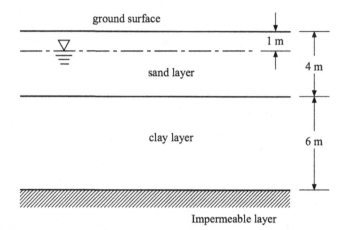

FIGURE EX. 8.8 Schematic illustration of soil deposit subjected to loading generated by raising the ground level.

ground level is raised by the addition of a 2 m thick fill of sand with a moist unit weight (γ) of 19.0 kN/m^3.

Solution

Since the clay is saturated, the initial void ratio is simply

$$e_0 = G_s w = (2.65)(0.60) = 1.590 \qquad (8.8.1)$$

The saturated unit weight of the clay layer is next computed as

$$\gamma_{sat} = \frac{\gamma_w(G_s + e)}{1 + e} = \frac{(9.81 \text{ kN/m}^3)(2.65 + 1.590)}{1 + 1.590} = 16.06 \text{ kN/m}^3 \quad (8.8.2)$$

Prior to addition of the fill, the vertical total stress at the middle of the clay layer is thus

$$\sigma_{v_0} = (19.5 \text{ kN/m}^3)(1 \text{ m}) + (20.0 \text{ kN/m}^3)(3 \text{ m}) + (16.06 \text{ kN/m}^3)(3 \text{ m})$$
$$= 127.7 \text{ kN/m}^2$$

$$(8.8.3)$$

The initial pore fluid pressure at this location is

$$u_0 = (9.81 \text{ kN/m}^3)(3 + 3 \text{ m}) = 58.9 \text{ kN/m}^2 \qquad (8.8.4)$$

The vertical effective stress at the middle of the clay layer is thus

$$\sigma'_{v_0} = 127.7 - 58.9 = 68.8 \text{ kN/m}^2 \qquad (8.8.5)$$

At "ultimate" conditions the increment in effective stress is equal to the increment in total stress attributed to placement of the fill. Thus,

$$\Delta\sigma'_v = \left(19.0 \text{ kN/m}^3\right)(2 \text{ m}) = 38.0 \text{ kN/m}^2 \qquad (8.8.6)$$

The change in void ratio associated with this stress increment will be computed as follows:

$$\Delta e = C_c \log\left(\frac{\sigma'_{v_0} + \Delta\sigma'_v}{\sigma'_{v_0}}\right) \qquad (8.8.7)$$

Lacking any additional information concerning the relationship between changes in void ratio and those in effective stress, the compression index is estimated using the empirical expression for soft clays that was proposed by Terzaghi and Peck[7]; i.e.,

$$C_c = 0.009(LL - 10) = 0.009(75 - 10) = 0.585 \qquad (8.8.8)$$

Thus,

$$\Delta e = (0.585)\log\left(\frac{68.8 + 38.0}{68.8}\right) = 0.112 \qquad (8.8.9)$$

Finally, the average ultimate primary consolidation settlement is computed as follows:

$$s_c = \left(\frac{\Delta e}{1 + e_0}\right) H_0 = \left(\frac{0.112}{1 + 1.590}\right)(6 \text{ m}) = 0.259 \text{ m} = \mathbf{259 \text{ mm}} \qquad (8.8.10)$$

EXAMPLE PROBLEM 8.9

General Remarks

This example problem illustrates the manner in which the average ultimate primary consolidation settlement is computed.

Problem Statement

A large embankment is to be built on the 5 m thick layer of normally consolidated clay that is overlain by a 2-m thick layer of sandy soil in the manner shown in Figure Ex. 8.9. The groundwater table is located at the surface of the clay layer, and capillary rise in the sandy soil is negligible.

---

7. Terzaghi, K., Peck, R.B., 1967. Soil Mechanics in Engineering Practice, second ed. John Wiley and Sons, New York, NY.

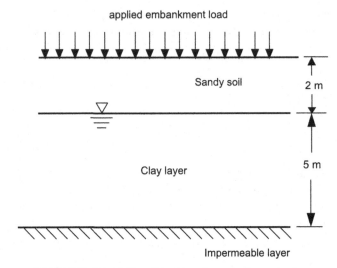

applied embankment load

Sandy soil

2 m

5 m

Clay layer

Impermeable layer

FIGURE EX. 8.9   Soil profile subjected to embankment loading.

The unit weight of the sandy soil is 18.5 kN/m^3. The clay has a specific gravity of solids (G_s) equal to 2.68 and a moisture content (w) of 50.2%. The results from a one-dimensional (oedometer) consolidation test on a sample of the same clay from the middle of the layer give an average value of 0.245 for the compression index (C_c).

If the final (ultimate) effective stress at the middle of the layer after the application of the embankment loading is 120 kN/m^2, what is the average ultimate primary consolidation settlement of the clay layer resulting from this loading?

Solution

Since the clay later is saturated, the initial void ratio is computed using the expression developed in Case 1.3 of Chapter 1; i.e.,

$$e_0 = G_s w = (2.68)(0.502) = 1.345 \tag{8.9.1}$$

The saturated unit weight of the clay layer is thus

$$\gamma_{sat} = \frac{\gamma_w(G_s + e)}{1 + e} = \frac{(9.81 \text{ kN/m}^3)(2.68 + 1.345)}{1 + 1.345} = 16.84 \text{ kN/m}^3 \tag{8.9.2}$$

Prior to construction of the large embankment, the vertical total stress at the middle of the clay layer is

$$\sigma_v = (18.5 \text{ kN/m}^3)(2 \text{ m}) + (16.84 \text{ kN/m}^3)(2.5 \text{ m}) = 79.10 \text{ kN/m}^2 \tag{8.9.3}$$

The initial pore fluid pressure at this location is

$$u = (9.81 \text{ kN/m}^3)(2.5 \text{ m}) = 24.53 \text{ kN/m}^2 \qquad (8.9.4)$$

The vertical effective stress at the middle of the clay layer is thus

$$\sigma_v' = 79.10 - 24.53 = 54.57 \text{ kN/m}^2 \qquad (8.9.5)$$

The change in void ratio at the middle of the clay layer is next computed using the given value of the compression index. In particular,

$$\Delta e = -C_c(\log \sigma_0' - \log \sigma_1') = -0.245(\log 54.57 - \log 120.0)$$
$$= 0.245 \log\left(\frac{120.0}{54.57}\right) = 0.084 \qquad (8.9.6)$$

The average ultimate primary consolidation settlement is thus

$$s_c = \left(\frac{\Delta e}{1 + e_0}\right)H_0 = \left(\frac{0.084}{1 + 1.345}\right)(5 \text{ m}) = 0.179 \text{ m} = \textbf{179 mm} \qquad (8.9.7)$$

EXAMPLE PROBLEM 8.10

General Remarks

Since the duration of consolidation tests is generally quite long, several empirical expressions have been proposed to estimate the compression index C_c from other quantities typically measured when testing soil. In this way, lengthy consolidation tests are avoided when determining C_c. This example problem investigates the effect that the choice of such an empirical expression has on the average ultimate primary consolidation settlement.

Problem Statement

The soil profile at a site for a proposed office building consists of a layer of fine sand 10.6 m thick above a layer of soft, normally consolidated 2.2 m thick clay layer in the manner shown in Figure Ex. 8.10. Underlying the clay is a layer of stiff coarse sand. The groundwater table was observed to be at a depth of 3.2 m below ground level. The soil above the ground water table is assumed to be saturated.

Laboratory tests indicate that the void ratio (e) for the sand is equal to 0.758 and that the specific gravity of solids (G_s) is 2.67. The moisture content (w) of the clay layer is 43%, its liquid limit (LL) is 45%, and G_s is 2.70.

Using three different empirical expressions for the compression index (C_c), determine the effect of these expressions on the maximum allowable vertical effective stress increase that the building can impose on the middle of the clay layer so that the primary consolidation settlement (s_c) does not exceed 90 mm.

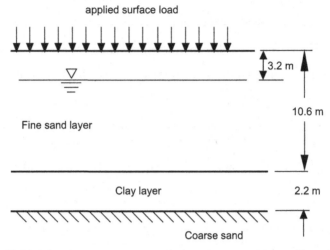

FIGURE EX. 8.10   Schematic illustration of soil profile at site of future office building.

Solution

The unit weights of the sand and clay layers are first computed. For the fine sand layer,

$$\gamma_{sat} = \frac{\gamma_w(G_s + e)}{1 + e} = (9.81 \text{ kN/m}^3)\left(\frac{2.67 + 0.758}{1 + 0.758}\right) = 19.13 \text{ kN/m}^3$$

$$(8.10.1)$$

For the saturated clay layer, the initial void ratio is

$$e_0 = G_s w = (2.70)(0.43) = 1.161 \qquad (8.10.2)$$

The saturated unit weights is thus

$$\gamma_{sat} = \frac{\gamma_w(G_s + e)}{1 + e} = (9.81 \text{ kN/m}^3)\left(\frac{2.70 + 1.161}{1 + 1.161}\right) = 17.53 \text{ kN/m}^3$$

$$(8.10.3)$$

The effective vertical stress at the mid-depth of the clay layer is next computed. The total vertical stress at this depth is

$$\sigma_v = (19.13 \text{ kN/m}^3)(10.6 \text{ m}) + (17.53 \text{ kN/m}^3)(1.1 \text{ m}) = 222.1 \text{ kN/m}^2$$

$$(8.10.4)$$

The water fluid pressure is

$$u = (9.81 \text{ kN/m}^3)(7.4 + 1.1 \text{ m}) = 83.39 \text{ kN/m}^2 \qquad (8.10.5)$$

The effective vertical stress at this depth is thus

$$\sigma'_v = 222.1 - 83.39 = 138.7 \text{ kN/m}^2 \qquad (8.10.6)$$

In general, the primary consolidation settlement is given by

$$s_c = \left(\frac{H_0}{1 + e_0}\right) C_c \log\left(\frac{\sigma'_v + \Delta\sigma'_v}{\sigma'_v}\right) \qquad (8.10.7)$$

Thus,

$$\log\left(\frac{\sigma'_v + \Delta\sigma'_v}{\sigma'_v}\right) = s_c\left(\frac{1 + e_0}{C_c H_0}\right) \qquad (8.10.8)$$

$$\frac{\sigma'_v + \Delta\sigma'_v}{\sigma'_v} = 10^{\left[s_c\left(\frac{1+e_0}{C_c H_0}\right)\right]} \qquad (8.10.9)$$

$$\Delta\sigma'_v = \sigma'_v\left\{10^{\left[s_c\left(\frac{1+e_0}{C_c H_0}\right)\right]} - 1\right\} \qquad (8.10.10)$$

Substituting all known values gives the following general expression for the allowable increment in effective vertical stress:

$$\Delta\sigma'_v = \left(138.7 \text{ kN/m}^2\right)\left\{10^{\left[(0.090 \text{ m})\left(\frac{1+1.161}{C_c(2.2 \text{ m})}\right)\right]} - 1\right\} \qquad (8.10.11)$$

- Using the empirical expression proposed by Terzaghi and Peck (1967)[8], the compression index is

$$C_c = 0.009(45 - 10) = 0.315 \qquad (8.10.12)$$

Thus,

$$\Delta\sigma'_v = \left(138.7 \text{ kN/m}^2\right)\left\{10^{\left[(0.090 \text{ m})\left(\frac{1+1.161}{(0.315)(2.2 \text{ m})}\right)\right]} - 1\right\} = \mathbf{126.0 \text{ kN/m}^2}$$

$$(8.10.13)$$

---

8. Terzaghi, K., Peck, R.B., 1967. Soil Mechanics in Engineering Practice, second ed. John Wiley and Sons, New York, NY.

- If the empirical expression proposed by Azzouz et al. (1976)[9] is used instead, the compression index is

$$C_c = 0.40(e_0 - 0.25) = 0.40(1.161 - 0.25) = 0.364 \qquad (8.10.14)$$

Thus,

$$\Delta\sigma'_v = \left(138.7 \text{ kN/m}^2\right)\left\{ 10^{\left[(0.090 \text{ m})\left(\frac{1+1.161}{(0.364)(2.2 \text{ m})}\right)\right]} - 1 \right\} = \textbf{103.9 kN/m}^2$$

$$(8.10.15)$$

- Finally, using the empirical expression proposed by Nadaraj and Srinivasa (1985)[10], the compression index is

$$C_c = 0.00234(LL)(G_s) = 0.00234(45)(2.67) = 0.281 \qquad (8.10.16)$$

Thus,

$$\Delta\sigma'_v = \left(138.7 \text{ kN/m}^2\right)\left\{ 10^{\left[(0.090 \text{ m})\left(\frac{1+1.161}{(0.281)(2.2 \text{ m})}\right)\right]} - 1 \right\} = \textbf{147.4 kN/m}^2$$

$$(8.10.17)$$

The empirical expression for C_c proposed by Azzouz et al. (1976) thus gives the most conservative results.

EXAMPLE PROBLEM 8.11

General Remarks

This problem illustrates the proper way to compute the average primary consolidation settlement for normally consolidated and overconsolidated samples.

Problem Statement

Given the soil deposit shown in Figure Ex. 8.11A. The groundwater table is located at a depth of 2 m below the ground surface. The layer of saturated sand has a thickness equal to 4 m. Finally, the thickness of the clay layer is equal to 8 m.

9. Azzouz, A.S., Krizek, R.J., Corotis, R.B., 1976. Regression analysis of soil compressibility. Soils and Foundations 16 (2), 19–29.
10. Nagaraj, T.S., Srinivasa Murthy, B.R., 1985. A critical reappraisal of compression index. Geotechnique 36 (1), 27–32.

applied surface load = 100 kPa

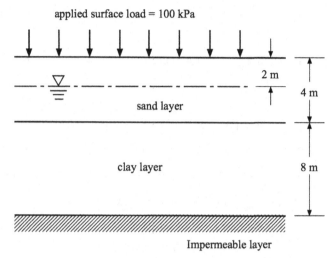

FIGURE EX. 8.11A Schematic illustration of soil profile at site of future office building (not to scale).

The moist sand above the groundwater table has a moisture content (w) of 29% and a dry unit weight (γ_d) equal to 16.0 kN/m³. Below the groundwater table the sand has a saturated unit weight of 20.0 kN/m³. Laboratory test results on a sample of the clay from the middle of the layer indicate a specific gravity of solids (G_s) equal to 2.71, a natural moisture content (w) of 29.5%, a compression index (C_c) equal to 0.270, a recompression index (C_r) equal to 0.045, and a saturated unit weight (γ_{sat}) of 19.2 kN/m³.

A uniformly distributed (over a large area) total stress increment ($\Delta\sigma$) of 100 kPa is applied at the ground surface (e.g., by a building foundation). Estimate the average ultimate primary consolidation settlement (s_c) of the clay if:

a) the clay is normally consolidated,
b) the clay is overconsolidated with the preconsolidation stress equal to 230 kPa, and,
c) the clay is overconsolidated with the preconsolidation stress equal to 190 kPa.

Solution

The moist unit weight of the sand above the groundwater table is computed from the dry unit weight and the moisture content as follows:

$$\gamma = \gamma_d(1+w) = \left(16.0 \text{ kN/m}^3\right)(1+0.29) = 20.64 \text{ kN/m}^3 \qquad (8.11.1)$$

Prior to addition of the applied total stress, the vertical total stress at the middle of the clay layer is thus

$$\sigma_{v_0} = (20.64 \text{ kN/m}^3)(2 \text{ m}) + (20.0 \text{ kN/m}^3)(4 \text{ m}) + (19.2 \text{ kN/m}^3)(4 \text{ m})$$
$$= 198.1 \text{ kN/m}^2$$

(8.11.2)

The initial pore fluid pressure at this location is

$$u_0 = (9.81 \text{ kN/m}^3)(4 + 4 \text{ m}) = 78.5 \text{ kN/m}^2 \qquad (8.11.3)$$

The vertical effective stress at the middle of the clay layer is thus

$$\sigma'_{v_0} = 198.1 - 78.5 = 119.6 \text{ kN/m}^2 \qquad (8.11.4)$$

Since the clay is saturated, the initial void ratio is computed as follows:

$$e_0 = G_s w = (2.71)(0.295) = 0.800 \qquad (8.11.5)$$

Before proceeding with the solution, it is important to note that initially the excess pore fluid pressure is equal to the applied total stress increment ($\Delta\sigma$). However, at the *ultimate* condition where s_c is computed, all excess pore pressure has dissipated, so the increment in effective stress is equal to $\Delta\sigma$.

a) When the clay is normally consolidated (i.e., $OCR = 1.0$), the effective stress state remains on the virgin compression line, and the current vertical effective stress is the preconsolidation stress (Figure Ex. 8.11B). The average ultimate primary consolidation settlement is computed as follows:

$$s_c = \left(\frac{\Delta e}{1 + e_0}\right) H_0 \qquad (8.11.6)$$

where

$$\Delta e = -C_c \left[\log\left(\sigma'_{v_0}\right) - \log\left(\sigma'_{v_0} + \Delta\sigma'_v\right)\right] = C_c \log\left(\frac{\sigma'_{v_0} + \Delta\sigma'_v}{\sigma'_{v_0}}\right)$$

$$= (0.270)\log\left(\frac{119.6 + 100}{119.6}\right) = 0.071$$

(8.11.7)

Thus,

$$s_c = \left(\frac{\Delta e}{1 + e_0}\right) H_0 = \left(\frac{0.071}{1 + 0.800}\right)(8 \text{ m}) = \mathbf{0.317 \text{ m}} \qquad (8.11.8)$$

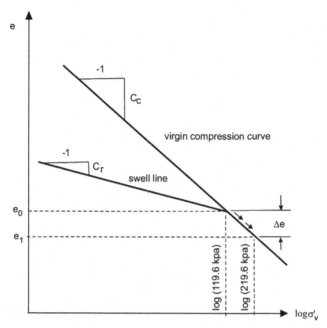

FIGURE EX. 8.11B   Consolidation curve for Case (a) (normally consolidated soil) (not to scale).

b) Next consider the case when the clay is overconsolidated with the pre-consolidation stress equal to 230 kN/m^2. Since

$$\sigma'_{v_0} + \Delta\sigma'_v = 119.6 + 100.0 = 219.6 < 230 \text{ kN/m}^2 \qquad (8.11.9)$$

the effective stress state remains on the swell/recompression curve (Figure Ex. 8.11C). It follows that

$$\Delta e = C_r \log\left(\frac{\sigma'_{v_0} + \Delta\sigma'_v}{\sigma'_{v_0}}\right) = (0.045)\log\left(\frac{119.6 + 100}{119.6}\right) = 0.012 \quad (8.11.10)$$

Thus,

$$s_c = \left(\frac{\Delta e}{1 + e_0}\right)H_0 = \left(\frac{0.012}{1 + 0.800}\right)(8 \text{ m}) = \mathbf{0.053 \text{ m}} \qquad (8.11.11)$$

c) Finally, consider the case when the clay is overconsolidated with the preconsolidation stress equal to 190 kN/m^2. Since

$$\sigma'_{v_0} + \Delta\sigma'_v = 119.6 + 100.0 = 219.6 > 190 \text{ kN/m}^2 \qquad (8.11.12)$$

the first portion of the vertical effective stress (119.6 $\leq \sigma'_v \leq$ 190 kN/m^2) state will lie on the swell/recompression curve, and the second portion (190 $\leq \sigma'_v \leq$ 219.6 kN/m^2) will lie on the virgin compression curve. From Figure Ex. 8.11D,

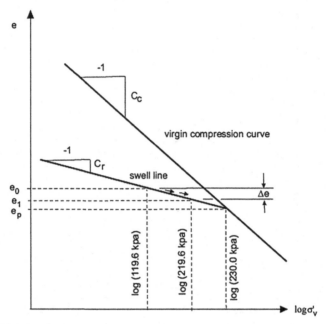

FIGURE EX. 8.11C   Consolidation curve for Case (b) (overconsolidated soil) (not to scale).

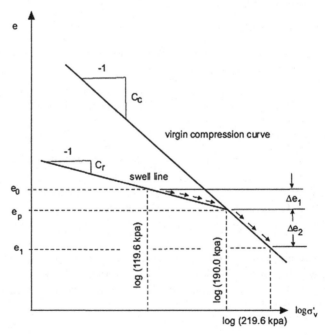

FIGURE EX. 8.11D   Consolidation curve for Case (c) (overconsolidated soil) (not to scale).

$$\Delta e_1 = C_r \log\left(\frac{190.0}{119.6}\right) = (0.045)\log\left(\frac{190.0}{119.6}\right) = 0.009 \qquad (8.11.13)$$

and

$$\Delta e_2 = C_c \log\left(\frac{219.6}{190.0}\right) = (0.270)\log\left(\frac{219.6}{190.0}\right) = 0.017 \qquad (8.11.14)$$

Thus,

$$s_c = \left(\frac{\Delta e_1 + \Delta e_2}{1 + e_0}\right) H_0 = \left(\frac{0.009 + 0.017}{1 + 0.800}\right)(8.0 \text{ m}) = \textbf{0.116 m} \qquad (8.11.15)$$

EXAMPLE PROBLEM 8.12

General Remarks

This example problem provides insight into the coefficient of volume compressibility for isotropic, linear elastic material idealizations.

Problem Statement

Consider an isotropic, linear elastic material idealization. Derive expressions for the coefficient of volume compressibility (m_v) associated with (a) one-, (b) two-, and (c) three-dimensional states of deformation.

Solution

For purposes of this example problem, the definition of m_v given in Eq. (8.16) is rewritten in incremental form as

$$m_v = \frac{\Delta \varepsilon_{vol}}{\Delta \sigma'_v} \qquad (8.12.1)$$

a) One-dimensional deformation:

Taking the z-direction as the direction of loading and deformation, it follows that $\varepsilon_z \neq 0$ and $\varepsilon_x = \varepsilon_y = 0$. Specializing the constitutive relations (recall Eqs. 4.59 and 4.63), written in direct form, for the case of one-dimensional compression gives

$$\varepsilon_x = \frac{1}{E}[\sigma_x - \nu(\sigma_y + \sigma_z)] = 0 \qquad (8.12.2)$$

$$\varepsilon_y = \frac{1}{E}[\sigma_y - \nu(\sigma_x + \sigma_z)] = 0 \qquad (8.12.3)$$

$$\varepsilon_z = \frac{1}{E}[\sigma_z - \nu(\sigma_x + \sigma_y)] \tag{8.12.4}$$

From Eq. (8.12.2),

$$\sigma_x = \nu(\sigma_y + \sigma_z) \tag{8.12.5}$$

Substituting this result into Eq. (8.12.3) gives

$$\sigma_y = \nu(\sigma_x + \sigma_z) = \nu[\nu\sigma_y + (1+\nu)\sigma_z] \tag{8.12.6}$$

Solving for σ_y gives

$$\sigma_y = \left(\frac{\nu}{1-\nu}\right)\sigma_z \tag{8.12.7}$$

Substituting this result into Eq. (8.12.2) gives

$$\sigma_x = \nu(\sigma_y + \sigma_z) = \left(\frac{\nu^2}{1-\nu}\right)\sigma_z + \nu\sigma_z = \left(\frac{\nu}{1-\nu}\right)\sigma_z = \sigma_y \tag{8.12.8}$$

The volumetric strain associated with one-dimensional compression is

$$\varepsilon_{vol} = \varepsilon_x + \varepsilon_y + \varepsilon_z = \varepsilon_z = \frac{1}{E}[\sigma_z - \nu(\sigma_x + \sigma_y)] \tag{8.12.9}$$

Substituting Eq. (8.12.8) for σ_x and σ_y into Eq. (8.12.9) gives

$$\varepsilon_{vol} = \frac{1}{E}[\sigma_z - \nu(\sigma_x + \sigma_y)] = \frac{1}{E}\left[\sigma_z - 2\nu\left(\frac{\nu}{1-\nu}\right)\sigma_z\right] \tag{8.12.10}$$

Writing Eq. (8.12.10) in incremental form, and noting that $\Delta\sigma'_v \equiv \Delta\sigma_z$, Eq. (8.12.1) becomes

$$m_v = \frac{\Delta\varepsilon_{vol}}{\Delta\sigma'_v} = \frac{\frac{1}{E}\left[1 - \left(\frac{2\nu^2}{1-\nu}\right)\right]\sigma_z}{\sigma_z} = \frac{1}{E}\left[\frac{(1+\nu)(1-2\nu)}{(1-\nu)}\right] \tag{8.12.11}$$

which is the desired expression.

b) Two-dimensional deformation:
Assuming conditions of plane strain and taking the z-direction as the thickness direction, it follows that $\varepsilon_z = 0$ and $\varepsilon_x \neq 0$, $\varepsilon_y \neq 0$. From Eq. (8.12.4)

$$\varepsilon_z = \frac{1}{E}[\sigma_z - \nu(\sigma_x + \sigma_y)] = 0 \Rightarrow \sigma_z = \nu(\sigma_x + \sigma_y) \tag{8.12.12}$$

The volumetric strain associated with two-dimensional compression is

$$\varepsilon_{vol} = \varepsilon_x + \varepsilon_y + \varepsilon_z = \varepsilon_x + \varepsilon_y = \frac{1}{E}[\sigma_x - \nu(\sigma_y + \sigma_z)] + \frac{1}{E}[\sigma_y - \nu(\sigma_x + \sigma_z)]$$

$$= \frac{(1-\nu)}{E}(\sigma_x + \sigma_y) - \frac{2\nu}{E}\sigma_z \tag{8.12.13}$$

Substituting Eq. (8.12.12) into Eq. (8.12.13) and combining like terms gives

$$\varepsilon_{vol} = \frac{1}{E}[\sigma_x - v(\sigma_y + \sigma_z)] + \frac{1}{E}[\sigma_y - v(\sigma_x + \sigma_z)]$$
$$= \frac{1}{E}(1 + v)(1 - 2v)(\sigma_x + \sigma_y) \tag{8.12.14}$$

For two-dimensional deformation $\Delta\sigma'_v$ is replaced by

$$\frac{1}{2}(\Delta\sigma_x + \Delta\sigma_y)$$

Writing Eq. (8.12.14) in incremental form, Eq. (8.12.1) becomes

$$m_v = \frac{\dfrac{1}{E}(1 + v)(1 - 2v)(\Delta\sigma_x + \Delta\sigma_y)}{\dfrac{1}{2}(\Delta\sigma_x + \Delta\sigma_y)} = \frac{2(1 + v)(1 - 2v)}{E} \tag{8.12.15}$$

which is the desired expression.

c) Three-dimensional deformation:

The volumetric strain associated with three-dimensional compression is

$$\varepsilon_{vol} = \varepsilon_x + \varepsilon_y + \varepsilon_z = \frac{1}{E}[\sigma_x - v(\sigma_y + \sigma_z)] + \frac{1}{E}[\sigma_y - v(\sigma_x + \sigma_z)]$$
$$+ \frac{1}{E}[\sigma_z - v(\sigma_x + \sigma_y)] = \frac{(1 - 2v)}{E}(\sigma_x + \sigma_y + \sigma_z) \tag{8.12.16}$$

For three-dimensional deformation $\Delta\sigma'_v$ is replaced by the mean stress (recall Eq. 4.5); i.e.,

$$\frac{1}{3}(\Delta\sigma_x + \Delta\sigma_y + \Delta\sigma_z)$$

Writing Eq. (8.12.16) in incremental form, Eq. (8.12.1) becomes

$$m_v = \frac{\dfrac{(1 - 2v)}{E}(\Delta\sigma_x + \Delta\sigma_y + \Delta\sigma_z)}{\dfrac{1}{3}(\Delta\sigma_x + \Delta\sigma_y + \Delta\sigma_z)} = \frac{3(1 - 2v)}{E} \tag{8.12.17}$$

which is the desired expression.

EXAMPLE PROBLEM 8.13

General Remarks

This example problem illustrates the fact the ultimate primary consolidation settlement can also be computed using the coefficient of volume compressibility (m_v). It is important to note, however, that unlike C_c, which is a constant, m_v varies with stress level.

Problem Statement

A proposed multistory department store is expected to increase the vertical stress at the middle of a 2.5-m-thick clay layer by 120 kPa. If $m_v = 3.8 \times 10^{-4}$ m^2/kN, compute the ultimate primary consolidation settlement of the clay layer.

Solution

Eq. (8.13) gives the following relation between the increment in volumetric strain and the increment in void ratio:

$$\Delta \varepsilon_{vol} = \frac{\Delta V}{V} = \frac{\Delta e}{1 + e_0} \qquad (8.13.1)$$

As in the previous example problem, the definition of m_v given in Eq. (8.16) is rewritten in incremental form as

$$m_v = \frac{\Delta \varepsilon_{vol}}{\Delta \sigma'_v} \qquad (8.13.2)$$

Combining Eqs. (8.13.2) and (8.13.1) gives

$$\Delta \varepsilon_{vol} = m_v \Delta \sigma'_v = \frac{\Delta e}{1 + e_0} \qquad (8.13.3)$$

For a field deposit of saturated soil with initial thickness H_0, the average primary consolidation settlement is given by Eq. (8.15); i.e.,

$$s_c = \left(\frac{\Delta e}{1 + e_0} \right) H_0 \qquad (8.13.4)$$

Substituting Eq. (8.13.3) into Eq. (8.13.4) gives

$$s_c = \left(m_v \Delta \sigma'_v \right) H_0 \qquad (8.13.5)$$

Substituting the given values into Eq. (8.13.5) gives

$$s_c = \left(3.8 \times 10^{-4} \ \text{m}^2/\text{kN} \right) \left(120 \ \text{kN/m}^2 \right) (2.5 \ \text{m}) = \textbf{0.114 m} = \textbf{114 mm}$$

$$(8.13.6)$$

Chapter 9

Example Problems Related to Time Rate of Consolidation

9.0 GENERAL COMMENTS

When designing foundations, the geotechnical engineer seeks to answer two primary questions: How much settlement will occur, and how long will it take for this settlement to occur? The former question is investigated in Chapter 8, while this chapter investigates problems related to the second question.

9.1 FUNDAMENTAL DEFINITIONS

As mentioned in Chapter 3, soil *compaction* is an instantaneous process in which the soil is densified (i.e., its void ratio is reduced) through the expulsion of *air* from the voids. By contrast, *consolidation* is a long-term (time-dependent) process in which *fluid* is expelled from the voids of a (typically) saturated soil.

Remark: The time rate of consolidation is greatly affected by a soil's permeability.

9.2 TERZAGHI'S ONE-DIMENSIONAL CONSOLIDATION THEORY

The birth of soil mechanics as a modern engineering discipline occurred in the 1920s. One of the cornerstones of this discipline was the theory proposed by the Austrian engineer Karl Terzaghi for the consolidation of saturated fine-grained soils under applied loads[1]. Terzaghi had been studying the phenomenon of reduction in void space of soils underlying foundations. He correctly perceived that the time-dependent settlement from consolidation of these soils was due to the time-dependent expulsion of fluid from the soil skeleton

---

1. Terzaghi, K., 1925. Erdbaumechanik auf Bodenphysilascher Grundlage. Franz Deuticke, Vienna.

Soil Mechanics. http://dx.doi.org/10.1016/B978-0-12-804491-9.00009-4
377

(a porous medium) as the voids decreased in size. The rate at which these movements took place was dictated by the permeability of the soil. Thus in response to an applied load, the soil skeleton was idealized to act like a large sponge.

Terzaghi's consolidation theory was based on the following assumptions:

- The consolidating layer is horizontal, of infinite extent laterally and of constant thickness.
- Throughout the consolidating layer the soil is *homogeneous* and is *completely saturated* (i.e., it is a two-phase material—no air is present in the pores).
- Due to the above homogeneity, the constitutive relations and permeability do not vary spatially or with time.
- Compared to the soil mass as a whole, the compressibility of the soil grains and of the pore fluid is *negligible* (i.e., several orders of magnitude lower). This implies that the deformation of the soil mass is due *entirely* to changes in volume that result due to the forcing out of free fluid from the pores.
- Deformation occurs only in the direction of load application; i.e., the soil is restrained against lateral deformation.
- The load is applied in only one direction and remains constant for all time. Thus, since the total stress is given, if the pore pressure is computed, the effective stress will be determined from the difference between the total stress and pore pressure.
- The time rate of consolidation depends only on the low permeability of the soil; viscoelastic properties of the soil skeleton are not considered.
- During consolidation the pore fluid flows only in the direction of load application; the free surface boundary offers no resistance to the flow of pore fluid from the soil.
- Darcy's law (recall the discussion of Section 6.6) describes the flow of pore fluid; i.e., the flow is proportional to the gradient of pore pressure. The coefficient of permeability is assumed to remain constant (recall the third assumption above).
- The strains in the soil skeleton are controlled by the effective stresses, with a linear constitutive relation.
- Strain, displacement, velocity, and stress increments are assumed to be small. As a result, the change in thickness of the consolidating layer is negligible.
- Inertia terms are neglected; i.e., a static analysis is assumed.

According to Terzaghi's consolidation theory, the response to an increment in load is realized in the following manner:

- Initially, since it is assumed to be incompressible, the pore fluid carries the *entire* increment in applied load (total stress). That is, the excess pore pressure is equal to the increase in total stress.

- With time, as fluid is expelled from the pores, the soil grains support more and more of the applied load. The excess pore pressure *decreases* and the effective stress *increases*. Equilibrium is always maintained, so that the sum of the excess pore pressure and the effective stress is equal to the increment in total stress.
- The rate at which pressure dissipates depends on the rate at which fluid flows from the pores (voids). This, in turn, is a function of the permeability of the soil.
- As time approaches infinity, the excess pore pressure goes to zero, the pore pressure returns to hydrostatic conditions, and the soil skeleton fully supports the additional load.

Remark: When saturated *cohesionless* soils (e.g., gravels, sands, and silts) are loaded, the excess pore pressure dissipates *quickly* due to the *high* permeability of the soil.

Remark: When saturated *clays* are loaded, the dissipation of excess pore pressure is delayed by the relatively *low* permeability of the soil.

9.2.1 Governing Differential Equation

Terzaghi's one-dimensional consolidation theory is governed by the following *parabolic* partial differential equation (a one-dimensional transient diffusion equation):

$$\frac{\partial u}{\partial t} = c_v \frac{\partial^2 u}{\partial z^2} \tag{9.1}$$

where $u =$ the excess pore pressure, $t =$ time, and $c_v =$ the *coefficient of consolidation*, which is defined as follows:

$$c_v = \frac{k(1 + e_0)}{\gamma_w a_v} = \frac{k}{\gamma_w} \frac{1}{m_v} \tag{9.2}$$

where k is the *Darcy permeability* (units of Lt^{-1}), e_0 is the initial void ratio, γ_w is the unit weight of water (units of FL^{-3}) and, as defined as in Chapter 8, a_v is the *coefficient of compressibility* (units of $F^{-1}L^2$) and m_v is the *coefficient of volume compressibility* (units of L^2F^{-1}); i.e.,

$$a_v = -\frac{de}{d\sigma_v'}; \quad m_v = -\frac{d\varepsilon_{vol}}{d\sigma_v'} \tag{9.3}$$

The solution of Terzaghi's equation requires the specification of boundary conditions and an initial condition. To facilitate the problem statement, let H_{dr} equal to the length of the maximum drainage path.

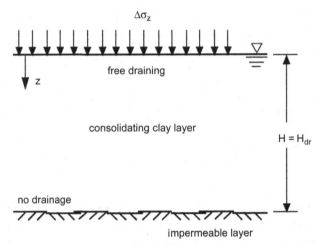

FIGURE 9.1 Schematic illustration of a single consolidating layer with single drainage.

For the case of *uniform* initial excess pore pressure with a *single drainage* boundary, it follows that $H_{dr} = H$ (Figure 9.1).

The *boundary conditions* associated with the problem, which are valid for all t, are as follows:

- Zero excess pore pressure: $u(0,t) = 0$
- Zero flow across the boundary: $\frac{\partial u}{\partial t}(H, t) = 0$

The *initial condition* is $u(z,0) = \Delta\sigma_z \equiv \sigma_v$, for all z ($0 \leq z \leq H$) at $t = 0$.

For the case of *uniform* initial excess pore pressure with a *double drainage*, it follows that $H_{dr} = H/2$ (Figure 9.2).

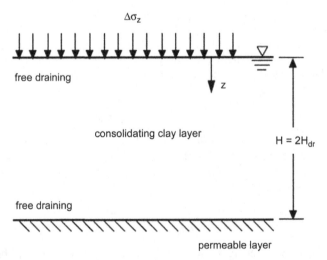

FIGURE 9.2 Schematic illustration of a single consolidating layer with double drainage.

The *boundary conditions* associated with the problem, which are valid for all t, are as follows:

- Zero excess pore pressure: $u(0,t) = 0$
- Zero excess pore pressure: $u(H,t) = 0$

The *initial condition* is again $u(z,0) = \Delta\sigma_z \equiv \sigma_v$, for all z $(0 \leq z \leq H)$ at $t = 0$.

9.2.2 Separation of Variables Solution

The exact solution of the equation governing Terzaghi's one-dimensional theory can rather easily be obtained by using the method of separation of variables[2]. The result is

$$u(z,t) = \sum_{n=0}^{\infty} \frac{2\Delta\sigma_z}{M_n} \exp\left\{ - (M_n)^2 T_v \right\} \sin\left(\frac{M_n z}{H_{dr}}\right) \tag{9.4}$$

where

$$M_n = \left(n + \frac{1}{2}\right)\pi; \quad n = 0, 1, 2, \ldots \tag{9.5}$$

and H_{dr} is as previously defined. Finally, the nondimensional "time factor" is given by

$$T_v = \frac{c_v t}{(H_{dr})^2} \tag{9.6}$$

9.2.3 Local Degree of Consolidation

The *local degree of consolidation* (or "consolidation ratio") is defined as follows:

$$U_z(z,t) = \left[\frac{u(z,0) - u(z,t)}{u(z,0)}\right] \times 100\% = \left[1 - \frac{u(z,t)}{u(z,0)}\right] \times 100\% \tag{9.7}$$

Substituting for $u(z,t)$ and noting that $u(z,0) = \Delta\sigma_z$, the local degree of consolidation can also be written as

$$U_z(z,t) = 1 - \sum_{n=0}^{\infty} \frac{2}{M_n} \exp\left\{ - (M_n)^2 T_v \right\} \sin\left(\frac{M_n z}{H_{dr}}\right) \tag{9.8}$$

---

2. Approximate solutions of the governing equation can be obtained using the finite difference and finite element methods.

From the definition of the local degree of consolidation, it is evident that

- at $t = 0$, $u(z,0) = u(z,t) \Rightarrow U_z(z,0) = 0\%$
- as $t \rightarrow \infty$, $u(z,t) \rightarrow 0 \Rightarrow U_z(z,\infty) \rightarrow 100\%$

The variation of $U_z(z,t)$ with depth (z) and time is often represented graphically as plots of $U_z(z,t)$ versus normalized depth (z/H_{dr}) for different values of T_v.

9.2.4 Average Degree of Consolidation

The *average degree of consolidation* is defined by the following integral expression:

$$\overline{U}_z = \frac{1}{H_{dr}} \int_0^{H_{dr}} U_z(z,t)dz \tag{9.9}$$

or

$$\overline{U}_z = 1 - \sum_{n=0}^{\infty} \frac{2}{(M_n)^2} \exp\left\{ - (M_n)^2 T_v \right\} \tag{9.10}$$

where M_n and T_v are as previously defined. The average degree of consolidation is commonly reported as a *percent*. Figure 9.3 graphically shows the relation between T_v and $\overline{U}_z$.

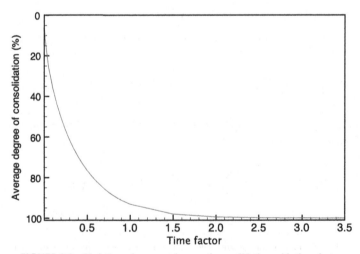

FIGURE 9.3   Variation of average degree of consolidation with time factor.

Before leaving the discussion of $\overline{U}_z$, it is timely to note the following approximations for the time factor[3]:

For $\overline{U}_z < 60\%$:

$$T_v = \frac{\pi}{4}\left(\frac{\overline{U}_z\%}{100}\right)^2 \tag{9.11}$$

For $\overline{U}_z \geq 60\%$:

$$T_v = 1.781 - 0.933 \log_{10}\left(100 - \overline{U}_z\%\right) \tag{9.12}$$

Table 9.1 quantifies the accuracy of using Eq. (9.11) to approximate T_v. The first column of this table contains a given value of $\overline{U}_z$. The second column gives the approximate value of T_v obtained using Eq. (9.11). The third column lists the value of $\overline{U}_z$ obtained using Eq. (9.10), with five terms, for the value of T_v given in second column. The final column in Table 9.1 lists the percentage relative error between the results obtained using Eqs. (9.10) and (9.11). It is

TABLE 9.1 Comparison of Results Obtained Using Eqs. (9.11) and (9.10)

$\overline{U}_z$	T_v Using Eq. (9.11)	$\overline{U}_z$ From Eq. (9.10) (Five-Term-Solution)	Relative Error (%)
5.0	0.002	5.89	15.11
10.0	0.008	10.18	1.77
15.0	0.018	15.14	0.92
20.0	0.031	19.87	0.65
25.0	0.049	24.98	0.08
30.0	0.071	30.07	0.23
35.0	0.096	34.96	0.11
40.0	0.126	40.05	0.12
45.0	0.159	44.98	0.04
50.0	0.196	49.91	0.18
55.0	0.238	54.90	0.18
56.0	0.246	55.79	0.38
57.0	0.255	56.76	0.42
58.0	0.264	57.72	0.49
59.0	0.273	58.65	0.60

---

3. Taylor, D.W., 1948. Fundamentals of Soil Mechanics. Wiley, New York, NY.

TABLE 9.2 Comparison of Results Obtained Using Eqs. (9.12) and (9.10)

$\overline{U}_z$	T_v Using Eq. (9.12)	$\overline{U}_z$ From Eq. (9.10) (Five-Term-Solution)	Relative Error (%)
60.0	0.286	59.96	0.07
65.0	0.340	64.96	0.06
70.0	0.403	70.01	0.01
75.0	0.477	75.02	0.03
80.0	0.567	79.99	0.01
85.0	0.684	85.01	0.01
90.0	0.848	90.00	0.00
95.0	1.129	95.00	0.00

evident that, except for low values of $\overline{U}_z$, approximation obtained using Eq. (9.11) is quite accurate.

Table 9.2 quantifies the accuracy of using Eq. (9.12) to approximate T_v. It is evident that approximation obtained using Eq. (9.12) is quite accurate.

EXAMPLE PROBLEM 9.1

General Remarks

This example problem illustrates the manner in which laboratory results are related to field consolidation.

Problem Statement

The time to reach 40% consolidation of a two-way drained laboratory oedometer specimen 25.4 mm thick is 220 s. Determine the time required for 65% average degree of consolidation of a 10-m thick layer of the same soil. This layer is subjected to the *same* loading conditions as the oedometer sample and is underlain by an impermeable rocky surface.

Solution

Since $\overline{U}_z = 40\% < 60\%$, for 40% consolidation the dimensionless "time factor" is computed as follows:

$$T_{v_{40}} = \frac{\pi}{4}\left(\frac{\overline{U}_z\%}{100}\right)^2 = \frac{\pi}{4}\left(\frac{40}{100}\right)^2 = 0.126 \tag{9.1.1}$$

The coefficient of consolidation is next computed from the definition of the time factor. Noting that in the laboratory oedometer $H_{dr} = (25.4/2) = 12.7$ mm, it follows that

$$T_{v_{40}} = \frac{c_v t_{40}}{(H_{dr})^2} \Rightarrow c_v = \frac{T_{v_{40}}(H_{dr})^2}{t_{40}} = \frac{(0.126)\left[\left(\dfrac{25.4 \text{ mm}}{2}\right)\left(\dfrac{m}{1000 \text{ mm}}\right)\right]^2}{220 \text{ s}}$$

$$= 9.213 \times 10^{-8} \text{ m}^2/\text{s}$$

(9.1.2)

Converting this to the more commonly used units of meters per year gives

$$c_v = \left(9.213 \times 10^{-8} \frac{\text{m}^2}{\text{s}}\right)\left(\frac{60 \text{ s}}{\text{min}}\right)\left(\frac{60 \text{ min}}{\text{h}}\right)\left(\frac{24 \text{ h}}{\text{d}}\right)\left(\frac{365 \text{ d}}{\text{yr}}\right) = 2.905 \text{ m}^2/\text{yr}$$

Since the *same* soil is found in the 10-m thick layer, for an average consolidation of $\overline{U}_z = 65\% (> 60\%)$, the time factor is approximated by

$$\begin{aligned} T_v &= 1.781 - 0.933 \log_{10}(100 - \overline{U}_z\%) \\ &= 1.781 - 0.933 \log_{10}(100 - 65\%) \\ &= 0.340 \end{aligned}$$

(9.1.3)

The time required for 65% average degree of consolidation is next computed from the definition of the time factor (recall Eq. 9.1.1). Noting that for the actual soil deposit $H_{dr} = 10.0$ m, it follows that

$$T_{v_{65}} = \frac{c_v t_{65}}{(H_{dr})^2} \Rightarrow t_{65} = \frac{T_{v_{65}}(H_{dr})^2}{c_v} = \frac{(0.340)(10 \text{ m})^2}{2.905 \text{ m}^2/\text{yr}} = \textbf{11.7 yr}$$

Remark: Since it is not feasible to perform a consolidation test on the actual soil deposit, the material from the field is tested in an oedometer. The link to the behavior of the soil in the field is then realized through the material parameter c_v, which is assumed to hold for both the oedometer test and the soil deposit.

EXAMPLE PROBLEM 9.2

General Remarks

This example problem also illustrates the manner in which laboratory results are related to field consolidation.

Problem Statement

The time required for a clay layer to achieve 98% consolidation is 10 years. What time would be required to achieve the same percent consolidation if the layer were *twice* as thick, *four* times more permeable, and *three* times more compressible?

Solution

From the definition of the time factor,

$$T_{v98} = \frac{c_v \, t_{98}}{(H_{dr})^2} \Rightarrow t_{98} = \frac{T_{v98}(H_{dr})^2}{c_v} = T_{v98}(H_{dr})^2 \left(\frac{\gamma_w m_v}{k}\right) = 10 \text{ years} \quad (9.2.1)$$

The time required to achieve 98% consolidation for the *modified* soil layer is

$$t_{98}^* = T_{v98}(2H_{dr})^2 \left(\frac{3m_v}{4k}\right)\gamma_w = 3T_{v98}(H_{dr})^2 \left(\frac{\gamma_w m_v}{k}\right) = 3t_{98} \quad (9.2.2)$$

Thus,

$$t_{98}^* = 3(10 \text{ years}) = \textbf{30 years}$$

EXAMPLE PROBLEM 9.3

General Remarks

This example problem illustrates the manner in which ultimate primary consolidation settlements are computed when laboratory results are related to field consolidation.

Problem Statement

A proposed building will increase the average total stress on a 20 ft thick clay stratum from 1.50 to 1.70 tons[4] per square foot. Field investigations indicate that an impermeable rock layer underlies this stratum.

An undisturbed sample of the clay was tested in an oedometer. When the total stress applied to the sample was increased from 1.50 to 1.70 tons per square foot, the following readings were taken:

- At time $t = 0$, the height of the sample was equal to 1 in.
- At $t = 20$ min, the height of the sample decreased to 0.935 in.

---

4. Recall that 1 ton = 2000 lbs.

The final height of the sample, at equilibrium under the 1.70 tons per square foot loading, was equal to 0.870 in. Determine (a) the ultimate settlement of the proposed building, and (b) the time required for one-half of this settlement to occur.

Solution

a) Recalling the general remarks made in Chapter 8, the ultimate settlement of the proposed building is computed based on the following equations:

$$\left(\frac{\Delta H}{H}\right)_{lab} = \left(\frac{\Delta e}{1+e_0}\right)_{lab}; \quad \left(\frac{\Delta H}{H}\right)_{field} = \left(\frac{\Delta e}{1+e_0}\right)_{field} \quad (9.3.1)$$

Since the clay sample is assumed to be undisturbed, Δe and e_0 are the same for both the laboratory sample and the field deposit. Thus,

$$\left(\frac{\Delta H}{H}\right)_{lab} = \left(\frac{\Delta H}{H}\right)_{field} \Rightarrow \Delta H_{field} = \left(\frac{\Delta H}{H}\right)_{lab} H_{field} \quad (9.3.2)$$

Substituting the laboratory data and recalling that the clay stratum is only drained at its upper boundary gives

$$\Delta H_{field} = \left(\frac{\Delta H}{H}\right)_{lab} H_{field} = \left(\frac{1.000 \text{ in.} - 0.870 \text{ in.}}{1.000 \text{ in.}}\right)(20 \text{ ft}) = \mathbf{2.60 \text{ ft}} \quad (9.3.3)$$

b) From the definition of the time factor,

$$T_{v_{50}} = \left[\frac{c_v t_{50}}{(H_{dr})^2}\right]_{lab} = \left[\frac{c_v t_{50}}{(H_{dr})^2}\right]_{field} \quad (9.3.4)$$

Since the material tested in the oedometer is identical to that found in the field, it follows that c_v is the *same* for both the laboratory sample and the field deposit. Thus,

$$\left[\frac{t_{50}}{(H_{dr})^2}\right]_{lab} = \left[\frac{t_{50}}{(H_{dr})^2}\right]_{field} \quad (9.3.5)$$

Conveniently, the height of the sample at $t = 20$ min (0.935 in.) is equal to *one-half* of the final height. Thus, noting that an oedometer permits drainage at both boundaries (implying that $H_{dr} = 1.000/2 = 0.500$ in.), the time required for one-half of the ultimate settlement to occur in the field is thus

$$t_{50_{field}} = \left[\frac{t_{50}}{(H_{dr})^2}\right]_{lab}(H_{dr})^2_{field} = \left[\frac{20.0 \text{ min}}{(0.500 \text{ in.})^2}\right](20.0 \text{ ft})^2\left(\frac{12 \text{ in.}}{\text{ft}}\right)^2 \quad (9.3.6)$$

$$= \mathbf{4.608 \times 10^6 \text{ min}}$$

Converting to more useful units of days gives the final result:

$$\left(4.608 \times 10^6 \text{ min}\right) \left(\frac{1 \text{ h}}{60 \text{ min}}\right) \left(\frac{1 \text{ d}}{24 \text{ h}}\right) \left(\frac{1 \text{ yr}}{365 \text{ d}}\right) = \textbf{8.8 years}$$

EXAMPLE PROBLEM 9.4

General Remarks

This example problem illustrates the manner in which the coefficient of consolidation is computed from the results of a conventional oedometer test.

Problem Statement

In performing a conventional oedometer test on a 2.54-cm thick sample of San Francisco Bay mud, the time for 80% average consolidation was found to be 48 min. Compute the coefficient of consolidation (c_v).

Solution

Once again recalling the definition of the time factor, the coefficient of consolidation is computed as follows:

$$T_{v80} = \frac{c_v t_{80}}{(H_{dr})^2} \Rightarrow c_v = \frac{T_{v80}(H_{dr})^2}{t_{80}} \tag{9.4.1}$$

Since $\bar{U}_z = 80\% > 60\%$,

$$T_{v80} = 1.781 - 0.933 \log(100 - 80\%) = 0.567 \tag{9.4.2}$$

Thus, since $t_{80} = 48$ min, and since the oedometer is doubly drained,

$$c_v = \frac{(0.567)\left(\dfrac{2.54 \text{ cm}}{2}\right)^2}{48 \text{ min}} = \textbf{1.846} \times \textbf{10}^{-2} \textbf{ cm}^2\textbf{/min} \tag{9.4.3}$$

Since it is more common to represent c_v in units of cm²/s and m²/yr, the above value is converted accordingly; i.e.,

$$c_v = \left(1.846 \times 10^{-2} \text{ cm}^2/\text{min}\right)\left(\frac{\text{min}}{60 \text{ s}}\right) = \textbf{3.076} \times \textbf{10}^{-4} \textbf{ cm}^2\textbf{/s}$$

and

$$c_v = \left(1.846 \times 10^{-2} \text{ cm}^2/\text{min}\right)\left(\frac{\text{m}}{100 \text{ cm}}\right)^2\left(\frac{60 \text{ min}}{\text{h}}\right)\left(\frac{24 \text{ h}}{\text{d}}\right)\left(\frac{365 \text{ d}}{\text{yr}}\right)$$

$$= \textbf{9.701} \times \textbf{10}^{-1} \textbf{ m}^2\textbf{/yr}$$

EXAMPLE PROBLEM 9.5

General Remarks

This example problem presents calculations that are made for a known ulti-mate primary consolidation settlement.

Problem Statement

A normally consolidated clay layer is 6.0 m thick and is drained along one boundary. Following the application of an applied load, the ultimate primary consolidation settlement was found to be 182 mm.

a) What is the average degree of consolidation for the clay layer when the settlement is 55 mm?

b) If the average coefficient of consolidation (c_v) for the clay is 20×10^{-4} cm^2/s, how many days will it take for 50% of the settlement to occur?

c) How long will it take for 50% of the settlement to occur if the clay layer drains at both the top and bottom boundaries?

Solution

a) When the settlement is equal to 55 mm, the average degree of consolida-tion is simply

$$\overline{U}_z = \left(\frac{55 \text{ mm}}{182 \text{ mm}}\right) \times 100\% = \textbf{30.2\%}$$

b) Since $\overline{U}_z = 50\% < 60\%$,

$$T_{v50} = \frac{\pi}{4}\left(\frac{50}{100}\right)^2 = \frac{\pi}{16} = 0.196 \qquad (9.5.1)$$

From the definition of the time factor, the desired time is computed as follows:

$$T_{v50} = \frac{c_v t_{50}}{(H_{dr})^2} \Rightarrow t_{50} = \frac{T_{v50}(H_{dr})^2}{c_v} \qquad (9.5.2)$$

Substituting all known values into Eq. (9.5.2) gives

$$t_{50} = \frac{T_{v50}(H_{dr})^2}{c_v} = \frac{\left(\frac{\pi}{16}\right)(6.0 \text{ m})^2}{(20.0 \times 10^{-4} \text{ cm}^2/\text{s})\left(\frac{\text{m}}{100 \text{ cm}}\right)^2} = 3.534 \times 10^7 \text{ s} \quad (9.5.3)$$

Converting to the desired units gives

$$t_{50} = (3.534 \times 10^7 \text{ s}) \left(\frac{\text{min}}{60 \text{ s}}\right) \left(\frac{\text{h}}{60 \text{ min}}\right) \left(\frac{\text{d}}{24 \text{ h}}\right) = \textbf{409 d}$$

c) For the case of double drainage, $H_{dr} = 6.0/2 = 3.0$ m. Since the only term in Eq. (9.5.2) that changes is the value of H_{dr}, it follows that the time to reach 50% settlement will be reduced by a factor of *four*. In particular,

$$t_{50} = \frac{T_{v_{50}}(H_{dr})^2}{c_v} = \frac{\left(\frac{\pi}{16}\right)\left(\frac{6.0 \text{ m}}{2}\right)^2}{(20.0 \times 10^{-4} \text{ cm}^2/\text{s})\left(\frac{\text{m}}{100 \text{ cm}}\right)^2} = 8.836 \times 10^6 \text{ s} = \textbf{102 d}$$

$$(9.5.4)$$

EXAMPLE PROBLEM 9.6

General Remarks

This example problem illustrates the manner in which both ultimate primary consolidation settlement and excess pore pressure dissipation are computed. As such, it ties together the discussion of settlement presented in Chapter 8 with the topic of primary consolidation.

Problem Statement

Consider the soil profile shown in Figure Ex. 9.6. The uniform stress increase in the clay layer, due to a surcharge of large areal extent, is 28 kPa.

The moist unit weight of the sand layer is 21.8 kN/m³. For the clay layer, the saturated unit weight is 18.5 kN/m³.

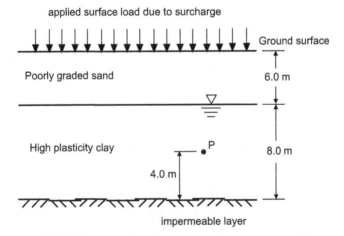

FIGURE EX. 9.6   Soil profile under consideration (not to scale).

Consolidation tests performed on samples extracted from the center of the clay layer indicate that the effective preconsolidation stress is 180 kPa. In addition, the initial void ratio is $e_0 = 1.21$, $C_c = 0.420$, $C_r = 0.085$, and $c_v = 0.0015$ m²/d.

a) Determine the vertical total stress, pore pressure, and vertical effective stress at point P for the following stages of consolidation: (1) before application of the surcharge load, (2) immediately after application of the surcharge load, and (3) after completion of primary consolidation.

b) Compute the ultimate primary consolidation settlement.

c) Compute the primary consolidation settlement 360 days after application of the surcharge load.

Solution

a) The vertical total stress, pore pressure, and vertical effective stress at point P are computed below.

1) Before application of the surcharge load:
   The vertical total stress at point P is

$$\sigma_v = \left(21.8 \text{ kN/m}^3\right)(6.0 \text{ m}) + \left(18.5 \text{ kN/m}^3\right)(4.0 \text{ m}) = \mathbf{204.8 \text{ kN/m}^2}$$

$$(9.6.1)$$

   The pore fluid pressure is

$$u = \left(9.81 \text{ kN/m}^3\right)(4.0 \text{ m}) = \mathbf{39.24 \text{ kN/m}^2} \tag{9.6.2}$$

   The vertical effective stress is thus

$$\sigma_v' = \sigma_v - u = 204.8 - 39.24 = \mathbf{165.6 \text{ kN/m}^2} \tag{9.6.3}$$

2) Immediately after application of the surcharge load:
   Due to the relatively low permeability of the clay, the entire surcharge load (i.e., increase in vertical total stress) is taken by the pore pressure. The vertical total stress at point P is this

$$\sigma_v = 204.8 + 28.0 = \mathbf{232.8 \text{ kN/m}^2} \tag{9.6.4}$$

   The pore fluid pressure is

$$u = 39.24 + 28.0 = \mathbf{67.24 \text{ kN/m}^2} \tag{9.6.5}$$

The vertical effective stress is thus *unchanged* from that computed for the above case. Thus,

$$\sigma'_v = \sigma_v - u = 232.8 - 67.24 = \textbf{165.6 kN/m}^2 \qquad (9.6.6)$$

3) After completion of primary consolidation:
   The vertical total stress at point P is unchanged from part (2) above; i.e.,

$$\sigma_v = \textbf{232.8 kN/m}^2 \qquad (9.6.7)$$

The pore fluid pressure returns to the hydrostatic value; i.e.,

$$u = \textbf{39.24 kN/m}^2 \qquad (9.6.8)$$

The vertical effective stress is thus

$$\sigma'_v = \sigma_v - u = 232.8 - 39.24 = \textbf{193.6 kN/m}^2 \qquad (9.6.9)$$

b) Since the initial vertical effective stress of 165.6 kN/m² is less than the preconsolidation stress of $\sigma'_p = 180$ kPa, at point P the clay is initially overconsolidated; the overconsolidation ratio is equal to 180/165.6 = 1.09. Recalling the discussion of ultimate primary consolidation settlement given in Chapter 8, the first part of the settlement response involves the swell/recompression line. The increment in vertical effective stress for this part of the response is

$$\Delta\sigma'_{v_1} = 180.0 - 165.5 = 14.4 \text{ kPa} \qquad (9.6.10)$$

The associated reduction in void ratio is thus

$$\Delta e_1 = C_r\left(\log \sigma'_p - \log 165.6\right) = C_r \log\left(\frac{180.0}{165.6}\right)$$
$$= (0.085)\log\left(\frac{180.0}{165.6}\right) = 0.003 \qquad (9.6.11)$$

The second part of the settlement response involves the virgin compression line. The increment in vertical effective stress for this part of the response is

$$\Delta\sigma'_{v_2} = 28.0 - 14.4 = 13.6 \text{ kPa} \qquad (9.6.12)$$

The associated reduction in void ratio is thus

$$\Delta e_2 = C_c \log\left(\frac{\sigma'_v + \Delta\sigma'_v}{\sigma'_v}\right) = (0.420)\log\left(\frac{180.0 + 13.6}{180.0}\right) = 0.013 \quad (9.6.13)$$

The total reduction in void ratio is thus

$$\Delta e = \Delta e_1 + \Delta e_2 = 0.003 + 0.013 = 0.016 \qquad (9.6.14)$$

The ultimate primary consolidation settlement is thus

$$\Delta H = \left(\frac{\Delta e}{1+e_0}\right)H_0 = \left(\frac{0.016}{1+1.21}\right)(8.0 \text{ m}) = \mathbf{0.058\ m} \qquad (9.6.15)$$

c) Computation of the primary consolidation settlement 360 days after application of the surcharge load requires knowledge of the local degree of consolidation at point P. The first step in this process is the computation of the time factor; i.e.,

$$T_v = \frac{c_v t}{(H_{dr})^2} = \frac{(0.0015 \text{ m}^2/\text{d})(365 \text{ d})}{(4.0 \text{ m})^2} = 0.034 \qquad (9.6.16)$$

At a depth of 4.0 m in the clay, $z/H_{dr} = 4.0/8.0 = 0.50$. From either a figure of z/H_{dr} versus U_z, or from a series solution, for $T_v = 0.034$, $U_z = 0.208 = 20.8\%$. The associated settlement is thus

$$s_c = (0.208)(0.058 \text{ m}) = \mathbf{0.012\ m} \qquad (9.6.17)$$

EXAMPLE PROBLEM 9.7

General Remarks

This example problem illustrates the manner in which laboratory results are related to field consolidation.

Problem Statement

A laboratory test was performed on a 25.4 mm thick, doubly drained oedometer sample of Boston blue clay. Based on the readings taken in the test, the time (t_{50}) for 50% consolidation was 8.75 min. The laboratory sample was taken from a saturated layer of Boston blue clay that was 18 m thick. The layer was subjected to a loading that was very similar to that imposed in the laboratory test.

a) Determine the coefficient of consolidation (c_v).

b) Assuming *single* drainage, how long will it take until the 18 m thick clay layer consolidates 60%?

c) Compute the ratio of the excess pore pressure 15 years after application of the load (i.e., $u(z,15)$) to the initial excess pore pressure (i.e., $u(z,0)$) at a depth (z) of 9 m.

d) If the following void ratio–effective stress data was obtained from the aforementioned oedometer test,

$$e_1 = 1.24, \quad \sigma'_v = 200 \text{ kPa}$$
$$e_2 = 1.09, \quad \sigma'_v = 400 \text{ kPa}$$

determine the average coefficient of permeability (k) for the Boston blue clay.

e) Subsequent field explorations appear to indicate that the 18 m thick clay layer may actually be *doubly* drained. How long (in years) will it take until this layer consolidates 60%?

Solution

a) For the laboratory sample the average degree of consolidation $(\overline{U}_z)$ is 50%. The associated dimensionless time factor (T_v) is thus computed from the following approximation:

$$T_{v_{50}} = \frac{\pi}{4}\left(\frac{\overline{U}_z}{100}\right)^2 = \frac{\pi}{4}\left(\frac{50}{100}\right)^2 = 0.196 \tag{9.7.1}$$

The coefficient of consolidation is then computed from the definition of the time factor; i.e.,

$$T_v = \frac{c_v t}{(H_{dr})^2} \Rightarrow c_v = \frac{T_{v_{50}}(H_{dr})^2}{t_{50}} = \frac{(0.196)(12.7 \text{ mm})^2}{8.75 \text{ min}} = 3.613 \text{ mm}^2/\text{min} \tag{9.7.2}$$

Converting this to the more commonly used units of meters per year gives

$$c_v = \left(3.613\frac{\text{mm}^2}{\text{min}}\right)\left(\frac{\text{m}}{1000 \text{ mm}}\right)^2\left(\frac{60 \text{ min}}{\text{h}}\right)\left(\frac{24 \text{ h}}{\text{d}}\right)\left(\frac{365 \text{ d}}{\text{yr}}\right) = \mathbf{1.90 \text{ m}^2/\text{yr}}$$

b) For the 18 m thick soil deposit, $\overline{U}_z = 60\%$. The associated time factor is thus computed from the approximation

$$T_{v_{60}} = 1.781 - 0.933 \ \log(100 - 60\%) = 0.286 \tag{9.7.3}$$

From the definition of the dimensionless time factor (recall Eq. 9.7.2) and noting that the clay layer is drained only along one boundary (i.e., $H_{dr} = H = 18.0$ m), it follows that

$$t_{60} = \frac{T_{v_{60}}(H_{dr})^2}{c_v} = \frac{(0.286)(18.0 \text{ m})^2}{1.90 \text{ m}/\text{yr}} = \mathbf{48.8 \text{ yr}} \tag{9.7.4}$$

c) After 15 years,

$$T_{v_{15}} = \frac{c_v t_{15}}{(H_{dr})^2} = \frac{(1.90 \text{ m}^2/\text{yr})(15 \text{ yr})}{(18.0 \text{ m})^2} = 0.088 \tag{9.7.5}$$

From a plot of the local degree of consolidation (U_z) versus normalized depth (z/H_{dr}), for $T_v = 0.088$ and $z/H_{dr} = 0.50$, interpolation gives an U_z value approximately equal to 0.22. Since

$$U_z(9 \text{ m}, 15 \text{ yr}) = 1 - \frac{u(9 \text{ m}, 15 \text{ yr})}{u(9 \text{ m}, 0)} \tag{9.7.6}$$

it follows that

$$\frac{u(9 \text{ m}, 15 \text{ yr})}{u(9 \text{ m}, 0)} = 1 - 0.22 = \mathbf{0.78} \tag{9.7.7}$$

d) The average coefficient of permeability is determined by recalling the definition of the coefficient of consolidation; i.e.,

$$c_v = \frac{k(1 + e_0)}{a_v \gamma_w} \Rightarrow k = \frac{c_v a_v \gamma_w}{1 + e_0} \tag{9.7.8}$$

The coefficient of compressibility value to be used in Eq. (9.7.8) is estimated from the given void ratio—effective stress data; i.e.,

$$a_v = -\frac{\Delta e}{\Delta \sigma_v'} = -\frac{(1.24 - 1.09)}{(200 - 400)\text{kPa}} = 7.50 \times 10^{-4} \text{ m}^2/\text{kN} \tag{9.7.9}$$

Thus,

$$k = \frac{c_v a_v \gamma_w}{1 + e_0} = \frac{(1.90 \text{ m}^2/\text{yr})(7.50 \times 10^{-4} \text{ m}^2/\text{kN})(9.81 \text{ kN/m}^2)}{1 + 1.24} \tag{9.7.10}$$
$$= 6.241 \times 10^{-3} \text{ m/yr}$$

Converting this to the more commonly used units of centimeters per second gives

$$k = \left(6.241 \times 10^{-3} \frac{\text{m}}{\text{yr}}\right) \left(\frac{100 \text{ cm}}{\text{m}}\right) \left(\frac{\text{yr}}{365 \text{ d}}\right) \left(\frac{1 \text{ d}}{24 \text{ h}}\right) \left(\frac{1 \text{ h}}{60 \text{ min}}\right) \left(\frac{1 \text{ min}}{60 \text{ s}}\right)$$
$$= \mathbf{1.979 \times 10^{-8} \text{ cm/s}}$$

e) For a *doubly* drained layer, $H_{dr} = H/2 = 9$ m. In addition, for $\overline{U}_z = 60\%$, $T_{v_{60}}$ is again equal to 0.286 (it is *independent* of the drainage path). Thus,

$$t_{60} = \frac{T_{v_{60}}(H_{dr})^2}{c_v} = \frac{(0.286)(9.0 \text{ m})^2}{1.90 \text{ m/yr}} = \mathbf{12.2 \text{ yr}} \tag{9.7.11}$$

EXAMPLE PROBLEM 9.8

General Remarks

This example problem illustrates the manner in which both ultimate primary consolidation settlement and excess pore pressure dissipation are computed. As such, it ties together the discussion of settlement presented in Chapter 8 with the topic of primary consolidation.

Problem Statement

Consider the soil profile shown in Figure Ex. 9.8. The uniform stress increase in the clay layer, due to a surcharge of large areal extent, is 25 kPa.

The moist unit weight of the sand layer above and below the groundwater table is 18.0 kN/m^3. For the clay layer, the saturated unit weight is 20.5 kN/m^3.

Consolidation tests performed on samples extracted from the center of the clay layer (point P) indicate that the initial void ratio is $e_0 = 1.40$, $C_c = 0.38$, and $C_r = 0.08$. A piezometer reading, taken 380 days after load application, was a height of $d = 4.7$ m.

Compute the following quantities:

a) The ultimate primary consolidation settlement assuming that the preconsolidation pressure is equal to 60 kPa.
b) The ultimate primary consolidation settlement assuming that the preconsolidation pressure is equal to 90 kPa.
c) The coefficient of consolidation (c_v).
d) The primary consolidation settlement 380 days after load application assuming that the preconsolidation pressure is equal to 60 kPa.
e) The height of the water (d) in the piezometer 1000 days after load application.

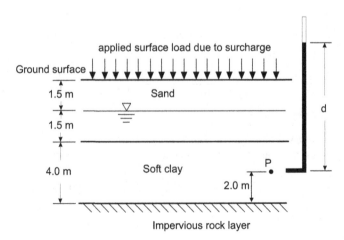

FIGURE EX. 9.8   Soil profile under consideration (not to scale).

Solution

To answer parts (a) and (b) of this problem, the effective stress at point P must be computed prior to load application. The vertical total stress at point P is

$$\sigma_v = (18.0 \text{ kN/m}^3)(3.0 \text{ m}) + (20.5 \text{ kN/m}^3)(2.0 \text{ m}) = 95.0 \text{ kPa} \quad (9.8.1)$$

The pore fluid pressure corresponds to hydrostatic conditions at point P. Thus,

$$u = (9.81 \text{ kN/m}^3)(1.5 + 2.0 \text{ m}) = 34.4 \text{ kPa} \quad (9.8.2)$$

The vertical effective stress is thus

$$\sigma'_v = \sigma_v - u = 95.0 - 34.3 = 60.7 \text{ kPa} \quad (9.8.3)$$

a) Since the initial vertical effective stress at point P is greater than the preconsolidation pressure of 60 kPa, the clay is normally consolidated. Using the approach presented in Chapter 8, the reduction in void ratio associated with the applied surcharge loading of 25 kPa is

$$\Delta e = C_c \log\left(\frac{\sigma'_v + \Delta\sigma'_v}{\sigma'_v}\right) = (0.38)\log\left(\frac{60.7 + 25.0}{60.7}\right) = 0.057 \quad (9.8.4)$$

The ultimate primary consolidation settlement is thus

$$\Delta H = \left(\frac{\Delta e}{1 + e_0}\right) H_0 = \left(\frac{0.057}{1 + 1.40}\right)(4.0 \text{ m}) = \mathbf{0.095 \text{ m}} \quad (9.8.5)$$

b) When the preconsolidation pressure is equal to 90 kPa, prior to application of the surcharge load, the vertical effective stress will be less than this value. Consequently, clay will be overconsolidated.

The sum of the initial effective stress at point P and the increment of effective stress is

$$\sigma'_v + \Delta\sigma'_v = 60.7 + 25.0 = 85.7 \text{ kPa} \quad (9.8.6)$$

Since this value is also less than 90 kPa, it follows that the clay remains overconsolidated throughout the primary consolidation settlement.

The reduction in void ratio associated with the applied surcharge loading of 25 kPa is thus

$$\Delta e = C_s \log\left(\frac{\sigma_v' + \Delta\sigma_v'}{\sigma_v'}\right) = (0.08)\log\left(\frac{60.7 + 25.0}{60.7}\right) = 0.012 \qquad (9.8.7)$$

The ultimate primary consolidation settlement is thus

$$\Delta H = \left(\frac{\Delta e}{1 + e_0}\right) H_0 = \left(\frac{0.012}{1 + 1.40}\right)(4.0 \text{ m}) = \mathbf{0.020 \text{ m}} \qquad (9.8.8)$$

c) As noted earlier, the hydrostatic pore pressure at point P is 34.4 kPa. This corresponds to a height of water (d_{hydro}) in the piezometer of 3.5 m. Since d is equal to 4.7 m 380 days after application of the surcharge load, it follows that the height associated with the excess pore pressure is

$$d_{excess} = d_{380} - d_{hydro} = 4.7 - 3.5 = 1.2 \text{ m} \qquad (9.8.9)$$

The excess pore pressure 380 days after load application is thus

$$u_{excess} = (1.2 \text{ m})(9.81 \text{ kN/m}^3) = 11.8 \text{ kPa} \qquad (9.8.10)$$

Recalling the expression for the local degree of consolidation; i.e.,

$$U_z(z, t) = \frac{u_0 - u_{excess}}{u_0} = \frac{25.0 - 11.8}{25.0} = 0.529 \qquad (9.8.11)$$

For single drainage, $H_{dr} = 4.0$ m; thus $z/H_{dr} = 2.0$ m/4.0 m $= 0.50$. From a plot of U_z versus normalized depth (z/H_{dr}), for $U_z = 0.529$ and $z/H_{dr} = 0.50$, interpolation gives a time factor T_v value approximately equal to 0.263. This result can likewise be obtained from the series solution for U_z (recall Eq. 9.8). The coefficient of consolidation is then computed from the definition of the time factor; i.e.,

$$T_v = \frac{c_v t}{(H_{dr})^2} \Rightarrow c_v = \frac{T_v (H_{dr})^2}{t} = \frac{(0.263)(4.0 \text{ m})^2}{380 \text{ d}} = \mathbf{0.011 \text{ m}^2/\text{d}} \qquad (9.8.12)$$

d) Since the preconsolidation pressure is equal to 60 kPa, the clay will be normally consolidated. Recalling that 380 days after load application the local degree of consolidation is 0.529, the primary consolidation settlement at this time will be

$$s_c(380 \text{ d}) = (0.529)(0.095 \text{ m}) = \mathbf{0.050 \text{ m}} \qquad (9.8.13)$$

where the result of Eq. (9.8.5) has been used.

e) To determine the height of the water in the piezometer 1000 days after load application, compute the associated time factor; i.e.,

$$T_v = \frac{c_v t}{(H_{dr})^2} = \frac{(0.011 \text{ m}^2/\text{d})(1000 \text{ d})}{(4.0 \text{ m})^2} = 0.688 \qquad (9.8.14)$$

From a plot of U_z versus normalized depth (z/H_{dr}), or from the series solution for U_z (recall Eq. 9.8), a value of U_z equal to 0.835 is obtained for $T_v = 0.688$ and $z/H_{dr} = 0.50$. Recalling Eq. (9.8.11),

$$U_z(z,t) = \frac{u_0 - u_{excess}}{u_0} \Rightarrow u_{excess} = u_0(1 - U_z)$$

$$= (25.0 \text{ kPa})(1 - 0.835) = 4.13 \text{ kPa}$$

(9.8.15)

Recalling Eq. (9.8.10), the height of water in the piezometer associated with the excess pore pressure is thus

$$d_{excess} = \frac{u_{excess}}{\gamma_w} = \frac{4.13 \text{ kPa}}{9.81 \text{ kN/m}^3} = 0.42 \text{ m}$$

(9.8.16)

Since the height of water in the piezometer associated with hydrostatic is 3.5 m, it follows that the total height 1000 days after load application will be

$$d_{1000 \text{ d}} = d_{hydro} + d_{excess} = 3.50 + 0.42 = \mathbf{3.92 \text{ m}}$$

(9.8.17)

EXAMPLE PROBLEM 9.9

General Remarks

This example problem illustrates the manner in which field data are used to solve consolidation problems.

Problem Statement

A 32 ft layer of clay at a site is underlain by an impermeable layer and is overlain by a 20 ft thick layer of pervious sand, gravel, and nonplastic silt. The groundwater table is located at the surface of the clay layer.

At some time in the past a 12-ft thick fill was placed above the sand–gravel–silt layer. The moist unit weight of the fill is found to be 133.5 lb/ft³.

The owner of the site desires to construct some light industrial buildings on the fill. It is thus necessary to ascertain if the clay is fully consolidated under the weight of the fill. To help in answering this question, a piezometer was installed 9 ft below the top of the clay layer as shown in Figure Ex. 9.9. The water in the piezometer tube rose to an elevation 6.7 ft above the groundwater table. The fact that excess pore fluid pressure is measured in the clay layer indicates that it has *not* fully consolidated.

Based on the results of several consolidation tests on samples from the same clay layer, average coefficient of consolidation (c_v) was found to be 4×10^{-3} cm²/s.

a) Determine the percentage degree of consolidation associated with the current conditions within the clay layer.

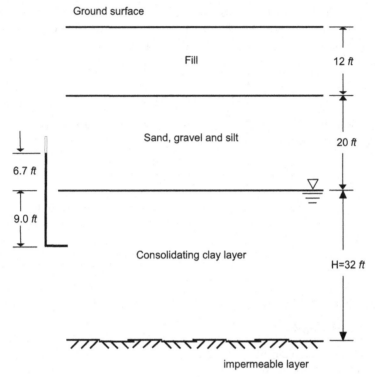

Ground surface

Fill

12 ft

Sand, gravel and silt

20 ft

6.7 ft

9.0 ft

Consolidating clay layer

H=32 ft

impermeable layer

FIGURE EX. 9.9 Schematic illustration of the consolidating clay layer, the sand, gravel and silt soil, and the fill (not to scale).

b) If the time factor T_v is equal to 0.30 for the calculated degree of consolidation, determine the age of the fill; i.e., how long it has been present at the site?

Solution

a) Expressed as a percentage, the average degree of consolidation ($\overline{U}_z$) is

$$\overline{U}_z = \left(1 - \frac{\Delta u}{\Delta \sigma}\right) \times 100\% \qquad (9.9.1)$$

where $\Delta \sigma$ is the increase in total stress, and u is the pore fluid pressure. For the given conditions,

$$\overline{U}_z = \frac{\Delta \sigma - u}{\Delta \sigma} = \left[1 - \frac{(6.7 \text{ ft})(62.4 \text{ lb/ft}^3)}{(133.7 \text{ lb/ft}^3)(12.0 \text{ ft})}\right] \times 100\% = \mathbf{73.9\%} \qquad (9.9.2)$$

b) If the time factor T_v is equal to 0.30 for the calculated degree of consolidation, the age of the fill is determined from the definition of the time factor; i.e.,

$$T_{v30} = \frac{c_v \, t_{30}}{(H_{dr})^2} \Rightarrow t_{30} = \frac{T_{v30}(H_{dr})^2}{c_v} = \frac{(0.30)\left[(32 \text{ ft})(30.5 \text{ cm/ft})\right]^2}{(4.0 \times 10^{-3} \text{ cm/s})} = \textbf{7.144} \times \textbf{10}^7 \text{ s}$$

(9.9.3)

Converting to the more convenient units of days gives

$$t_{30} = \left(7.144 \times 10^7 \text{ s}\right)\left(\frac{\min}{60 \text{ s}}\right)\left(\frac{1 \text{ h}}{60 \min}\right)\left(\frac{d}{24 \text{ h}}\right) = \textbf{827 d}$$

EXAMPLE PROBLEM 9.10

General Remarks

This example problem investigates the spatial and temporal variation in excess pore fluid pressure that is predicted by Terzaghi's consolidation theory. In particular, the problem illustrates the manner in which, at a given time, excess pore pressures are computed at different depths.

Problem Statement

Consider the soil deposit shown in Figure Ex. 9.10A. The construction of a large embankment results in a uniform vertical total stress increase of 100 kPa.

The 3-m thick layer of coarse sand has a saturated unit weight of 20.0 kN/m³. The 8-m think layer of soft silty clay has a saturated unit weight of 14.0 kN/m³ and a coefficient of consolidation equal to 20 m²/year.

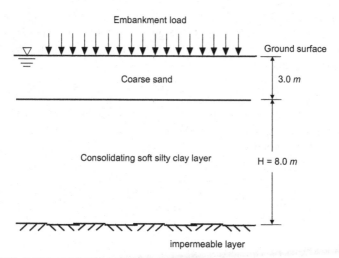

FIGURE EX. 9.10A Soft clay layer with single drainage subjected to embankment load (not to scale).

Compute the following quantities:

a) The excess pore pressure that would be measured at depths of 2, 4, 6, and 8 m in the soft silty clay layer 3 months, 6 months, and 1 year after the embankment load was applied; and,
b) The average degree of consolidation of the clay layer 3, 6, and 9 months following loading. Assume that the embankment was constructed relatively quickly.

Solution
a) After 3 months, the time factor will be

$$T_v = \frac{c_v t}{(H_{dr})^2} = \frac{\left(20\frac{m^2}{yr}\right)(3\text{ mos})\left(\frac{yr}{12\text{ mos}}\right)}{(8.0\text{ m})^2} = 0.078 \qquad (9.10.1)$$

Although a plot of the local degree of consolidation (U_z) versus normalized depth (z/H_{dr}) in the clay layer can be used to estimate suitable values of U_z for the given $T_v = 0.078$, the relatively low magnitude of the latter may lead to somewhat inaccurate results. Consequently, it is advantageous to use the series solution for excess pore pressure $\Delta u(z,t)$ and for the local degree of consolidation U_z; i.e., recalling Eqs. (9.4) and (9.8),

$$\Delta u(z, t) = \sum_{n=0}^{\infty} \frac{2\Delta\sigma_z}{M_n} \exp\left[-(M_n)^2 T_v\right] \sin\left(\frac{M_n z}{H_{dr}}\right) \qquad (9.10.2)$$

$$U_z(z, t) = 1 - \sum_{n=0}^{\infty} \frac{2}{M_n} \exp\left[-(M_n)^2 T_v\right] \sin\left(\frac{M_n z}{H_{dr}}\right) \qquad (9.10.3)$$

where $M_n = \pi\left(n + \frac{1}{2}\right)$.

For example, in the case of $z = 2.0$ m, $z/H_{dr} = 0.25$. Using three terms in the series solution for $T_v = 0.078$ gives $\Delta u = 47.32$ kPa and $U_z = 0.527$. The change in vertical effective stress at a depth of 2.0 m and time of 3 months is thus

$$\Delta\sigma'_z = \Delta\sigma_z - \Delta u(2.0\text{ m}, 3\text{ mos}) = 100.0 - 47.32 = 52.68\text{ kPa} \qquad (9.10.4)$$

Table Ex. 9.10A summarizes the results obtained for the other three depths under consideration in this problem.
For $t = 6$ months.

$$T_v = \frac{c_v t}{(H_{dr})^2} = \frac{\left(20\frac{m^2}{yr}\right)(6\text{ mos})\left(\frac{yr}{12\text{ mos}}\right)}{(8.0\text{ m})^2} = 0.156 \qquad (9.10.5)$$

TABLE EX. 9.10A Change in Vertical Effective Stress and Excess Pore Pressure at Various Depths 3 Months After Load Application

z (m)	z/H_{dr}	U_z	$\Delta\sigma'_v$ (kPa)	Δu (kPa)
2.0	0.25	0.527	52.68	47.32
4.0	0.50	0.206	20.57	79.43
6.0	0.75	0.059	5.91	94.09
8.0	1.00	0.023	2.27	97.73

Again using three terms in the series solution for this value of T_v gives the results shown in Table Ex. 9.10B.

Finally, for $t = 1$ year,

$$T_v = \frac{c_v t}{(H_{dr})^2} = \frac{\left(20\frac{m^2}{yr}\right)(1\ yr)}{(8.0\ m)^2} = 0.313 \qquad (9.10.6)$$

Again using three terms in the series solution for this value of T_v gives the results shown in Table Ex. 9.10C.

b) The series solution for the average degree of consolidation is given by Eq. (9.10); i.e.,

$$\overline{U}_z = 1 - \sum_{n=0}^{\infty} \frac{2}{M_n} \exp\left[-(M_n)^2 T_v\right] \qquad (9.10.7)$$

where, as before, $M_n = \pi\left(n + \frac{1}{2}\right)$.

TABLE EX. 9.10B Change in Vertical Effective Stress and Excess Pore Pressure at Various Depths 6 Months After Load Application

z (m)	z/H_{dr}	U_z	$\Delta\sigma'_v$ (kPa)	Δu (kPa)
2.0	0.25	0.656	65.61	34.39
4.0	0.50	0.378	37.79	62.21
6.0	0.75	0.205	20.46	79.54
8.0	1.00	0.147	14.68	85.32

TABLE EX. 9.10C Change in Vertical Effective Stress and Excess Pore Pressure at Various Depths 1 Year After Load Application

z (m)	z/H_dr	U_z	$\Delta\sigma_v'$ (kPa)	Δu (kPa)
2.0	0.25	0.775	77.45	22.55
4.0	0.50	0.584	58.38	41.62
6.0	0.75	0.457	45.68	54.32
8.0	1.00	0.412	41.22	58.78

For $t = 3$ months, as computed in part a), $T_v = 0.078$. Using three terms in the series solution for this value of T_v gives $\overline{U}_z = \mathbf{31.5\%}$.

For $t = 6$ months, $T_v = 0.156$. Using three terms in the series solution for this value of T_v gives $\overline{U}_z = \mathbf{44.6\%}$.

Finally, for $t = 1$ year, $T_v = 0.323$. Using three terms in the series solution for this value of T_v gives $\overline{U}_z = \mathbf{62.6\%}$.

To better understand the relation between the increments in excess pore pressure and vertical total and effective stresses computed in part a), using the approach described in Chapter 5, compute the effective stress variation with depth below the ground surface *before* construction of the embankment. At a depth of 3.0 m:

The vertical total stress is

$$\sigma_v = \left(20.0 \text{ kN/m}^3\right)(3.0 \text{ m}) = \mathbf{60.0 \text{ kPa}} \tag{9.10.8}$$

The pore fluid pressure is

$$u = \left(9.81 \text{ kN/m}^3\right)(3.0 \text{ m}) = \mathbf{29.5 \text{ kPa}} \tag{9.10.9}$$

The vertical effective stress is thus

$$\sigma_v' = \sigma_v - u = 60.0 - 29.4 = \mathbf{30.6 \text{ kPa}} \tag{9.10.10}$$

At a depth of 5.0 m (i.e., 2.0 m below the surface of the clay layer): The vertical total stress is

$$\sigma_v = 60.0 \text{ kPa} + \left(14.0 \text{ kN/m}^3\right)(2.0 \text{ m}) = \mathbf{88.0 \text{ kPa}} \tag{9.10.11}$$

The pore fluid pressure is

$$u = \left(9.81 \text{ kN/m}^3\right)(5.0 \text{ m}) = \mathbf{49.1 \text{ kPa}} \tag{9.10.12}$$

The vertical effective stress is thus

$$\sigma_v' = \sigma_v - u = 88.0 - 49.1 = \mathbf{38.9 \text{ kPa}} \tag{9.10.13}$$

At a depth of 7.0 m (i.e., 4 m below the surface of the clay layer):
The vertical total stress is

$$\sigma_v = 88.0 \text{ kPa} + (14.0 \text{ kN/m}^3)(2.0 \text{ m}) = \textbf{116.0 kPa} \qquad (9.10.14)$$

The pore fluid pressure is

$$u = (9.81 \text{ kN/m}^3)(7.0 \text{ m}) = \textbf{68.7 kPa} \qquad (9.10.15)$$

The vertical effective stress is thus

$$\sigma_v' = \sigma_v - u = 116.0 - 68.7 = \textbf{47.3 kPa} \qquad (9.10.16)$$

At a depth of 9 m (i.e., 6 m below the surface of the clay layer):
The vertical total stress is

$$\sigma_v = 116.0 \text{ kPa} + (14.0 \text{ kN/m}^3)(2.0 \text{ m}) = \textbf{144.0 kPa} \qquad (9.10.17)$$

The pore fluid pressure is

$$u = (9.81 \text{ kN/m}^3)(9.0 \text{ m}) = \textbf{88.3 kPa} \qquad (9.10.18)$$

The vertical effective stress is thus

$$\sigma_v' = \sigma_v - u = 144.0 - 88.3 = \textbf{55.7 kPa} \qquad (9.10.19)$$

At a depth of 11 m (i.e., 8 m below the surface of the clay layer):
The vertical total stress is

$$\sigma_v = 144.0 \text{ kPa} + (14.0 \text{ kN/m}^3)(2.0 \text{ m}) = \textbf{172.0 kPa} \qquad (9.10.20)$$

The pore fluid pressure is

$$u = (9.81 \text{ kN/m}^3)(11.0 \text{ m}) = \textbf{107.9 kPa} \qquad (9.10.21)$$

The vertical effective stress is thus

$$\sigma_v' = \sigma_v - u = 172.0 - 107.9 = \textbf{64.1 kPa} \qquad (9.10.22)$$

Figure Ex. 9.10B shows the variation with depth of the vertical total stress, pore pressure, and vertical effective stress before construction of embankment.

Assuming that the embankment is constructed relatively quickly, at $t = 0$, the vertical total stress throughout the soil deposit increases by 100 kPa (Figure Ex. 9.10C). Due to the relatively low permeability of the clay layer, the pore pressure also uniformly increases by the increment $\Delta u = 100$ kPa.

At $t = 3$ months, 6 months, and 1 year, the vertical total stress throughout the soil deposit remains unchanged. The excess pore pressure (Δu) decreases to the values listed in Tables Ex. 9.10A–C, respectively. Figure Ex. 9.10D shows the change in excess pore pressure with depth and time. As $t \rightarrow \infty$,

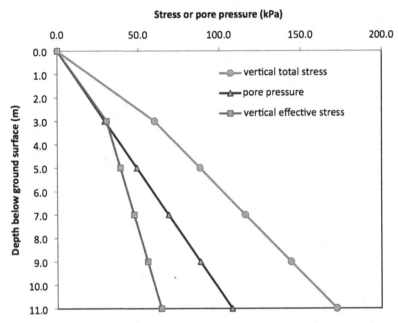

FIGURE EX. 9.10B Variation with depth of vertical total stress, pore fluid pressure, and vertical effective stress before construction of embankment.

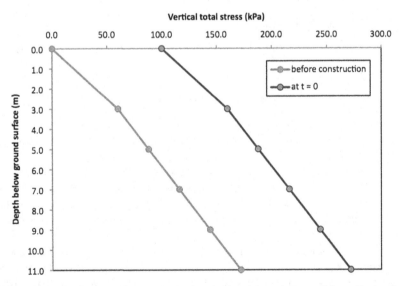

FIGURE EX. 9.10C Variation with depth of vertical total stress before and immediately after construction of embankment.

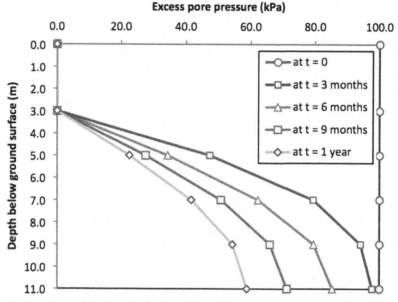

FIGURE EX. 9.10D   Variation with depth and time of increment in excess pore pressure.

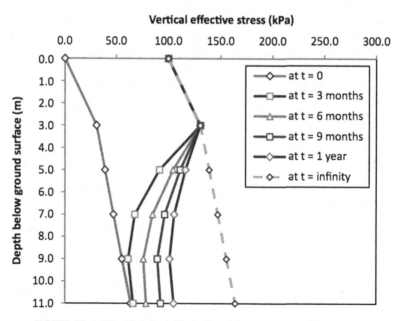

FIGURE EX. 9.10E   Variation with depth and time of vertical effective stress.

the excess pore pressure in the clay layer goes to zero. Since the permeability of the coarse sand layer would be much higher than for the clay, the excess pore pressure in the former goes to zero very quickly.

As the excess pore pressure decreases, the vertical effective stress increases by the increments $\Delta\sigma'_v$ listed in Tables Ex. 9.10A–C. Figure Ex. 9.10E shows the increase in vertical effective stress with depth and time. As $t \to \infty$, $\Delta u \to 0$ and the vertical effective stress distribution approaches the vertical total stress distribution (Figure Ex. 9.10E).

EXAMPLE PROBLEM 9.11

General Remarks

This example problem investigates the amount and rate of primary consolidation settlement that occurs when the groundwater table is *lowered* by a certain amount.

Problem Statement

A 60 ft thick clay layer has double drainage and is normally consolidated under its present overburden. From test results performed on samples from the same depth, the void ratio (e_0) was found to be 1.120, the coefficient of compressibility (a_v) and consolidation (c_v) were estimated to be 3.25×10^{-5} ft²/lb and 7.50×10^{-6} ft²/s, respectively.

If the groundwater table in the soil layer overlying the clay is to be *lowered* by 5 ft,

a) determine the average primary consolidation settlement of the clay layer; and

b) the time that will elapse for 50% and 90% of this settlement to occur.

Solution

A 5 ft drop in the groundwater table will result in a change in the pore pressure at the mid-depth of the clay layer of

$$\Delta u = -(62.4 \text{ lb/ft}^3)(5 \text{ ft}) = -312.0 \text{ lb/ft}^2 \qquad (9.11.1)$$

Assuming that the lowering of the groundwater table does not change the moisture content of the soil above the groundwater table, it follows that the total stress will remain *unchanged*. Consequently, the associated change in the vertical effective stress will thus be

$$\Delta\sigma'_v = \Delta\sigma_v - \Delta u = 0 - (-312.0 \text{ lb/ft}^2) = 312.0 \text{ lb/ft}^2 \qquad (9.11.2)$$

a) From the definition of the coefficient of compressibility,

$$a_v = \frac{\Delta e}{\Delta \sigma_v'} \Rightarrow \Delta e = a_v \Delta \sigma_v' \qquad (9.11.3)$$

where it is understood that the void ratio is decreasing. The average ultimate primary consolidation settlement of the ground surface is thus

$$
\begin{aligned}
s_c &= \left(\frac{\Delta e}{1+e_0}\right) H_0 = \left(\frac{a_v \Delta \sigma_v'}{1+e_0}\right) H_0 \\
&= \frac{(3.25 \times 10^{-5}\ \text{ft}^2/\text{lb})(312.0\ \text{lb}/\text{ft}^2)}{1+1.120}(60\ \text{ft}) \\
&= \mathbf{0.287\ ft = 3.44\ in.}
\end{aligned}
\qquad (9.11.4)
$$

b) From the general definition of the time factor,

$$T_v = \frac{c_v t}{(H_{dr})^2} \Rightarrow t = \frac{T_v (H_{dr})^2}{c_v} \qquad (9.11.5)$$

For 50% consolidation,

$$T_{v50} = \frac{\pi}{4}\left(\frac{50}{100}\right)^2 = 0.196 \qquad (9.11.6)$$

Thus, since the clay later has double drainage,

$$t_{50} = \frac{T_{v50}(H_{dr})^2}{c_v} = \frac{(0.196)\left(\dfrac{60\ \text{ft}}{2}\right)^2}{7.50 \times 10^{-6}\ \text{ft}^2/\text{s}} = \mathbf{2.352 \times 10^7\ s} \qquad (9.11.7)$$

Converting to the more convenient units of *years* gives

$$t_{50} = (2.352 \times 10^7\ \text{s})\left(\frac{\text{min}}{60\ \text{s}}\right)\left(\frac{1\ \text{h}}{60\ \text{min}}\right)\left(\frac{\text{d}}{24\ \text{h}}\right)\left(\frac{\text{yr}}{365\ \text{d}}\right) = \mathbf{0.75\ yr}$$

The associated settlement will thus be

$$s_{50} = (0.50)s_c = (0.50)(0.287\ \text{ft}) = 0.144\ \text{ft} = 1.72\ \text{in.} \qquad (9.11.8)$$

For 90% consolidation,

$$T_{v90} = 1.781 - 0.933\ \log_{10}(100 - 90) = 0.848 \qquad (9.11.9)$$

Thus,

$$t_{90} = \frac{T_{v90}(H_{dr})^2}{c_v} = \frac{(0.848)\left(\dfrac{60 \text{ ft}}{2}\right)^2}{7.50 \times 10^{-6} \text{ ft}^2/\text{s}} = 1.018 \times 10^8 \text{ s} = 3.23 \text{ yr} \quad (9.11.10)$$

The associated settlement will thus be

$$s_{90} = (0.90)s_c = (0.90)(0.287 \text{ ft}) = 0.258 \text{ ft} = 3.10 \text{ in.} \quad (9.11.11)$$

EXAMPLE PROBLEM 9.12

General Remarks

This problem represents a rather "standard" example related to the time rate of consolidation of fine-grained soils. It includes the determination of the ultimate primary consolidation settlement (see also example problems presented in Chapter 8), the excess pore pressure after a given period of time, and the time required to reach a specific fraction of the ultimate settlement.

Problem Statement

A load increment of 1200 lb/ft² is applied to a 10 ft thick layer of clay that is underlain by an impermeable layer of rock in the manner shown in Figure Ex. 9.12.

The following properties were determined from laboratory tests on the clay: $G_s = 2.69$, $w = 35\%$, $c_v = 30$ ft²/yr, and $k = 0.12$ ft/yr.

a) Calculate the ultimate primary consolidation settlement of the clay layer.
b) What is the approximate excess pore pressure at a depth of 5 ft 2 years after the start of consolidation?
c) How long will it take to reach 90% of the ultimate settlement?

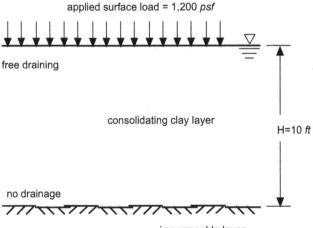

FIGURE EX. 9.12 Profile consisting of a single consolidating soil layer.

Solution

a) The determination of the ultimate primary consolidation settlement requires knowledge of the initial void ratio and the change in void ratio caused by the load increment. The void initial ratio is computed from the given information. In particular, since the clay layer is saturated (a basic assumption of Terzaghi's consolidation theory)

$$e_0 = G_s w = (2.69)(0.35) = 0.942 \tag{9.12.1}$$

The change in void ratio is also computed from the given information. In particular, since c_v and k were provided, and since the initial void ratio was computed earlier, the coefficient of compressibility is computed as follows:

$$c_v = \frac{k(1 + e_0)}{a_v \gamma_w} \Rightarrow a_v = \frac{k(1 + e_0)}{c_v \gamma_w} = \frac{(0.12 \text{ ft/yr})(1 + 0.942)}{(30 \text{ ft}^2/\text{yr})(62.4 \text{ lb/ft}^3)} \tag{9.12.2}$$
$$= 1.245 \times 10^{-4} \text{ ft}^2/\text{lb}$$

Since $a_v = -\Delta e/\Delta \sigma'$, it follows that

$$\Delta e = -a_v \, \Delta \sigma' = -(1.245 \times 10^{-4} \text{ ft}^2/\text{lb})(1200 \text{ lb/ft}^2) = -0.149 \tag{9.12.3}$$

The ultimate primary consolidation settlement is thus

$$\Delta H = \left(\frac{\Delta e}{1 + e_0}\right) H = \left(\frac{0.149}{1 + 0.942}\right)(10 \text{ ft}) = \textbf{0.767 ft} \tag{9.12.4}$$

b) From Figure Ex. 8.7 it is evident that $H = H_{dr} = 10$ ft. The dimensionless time factor associated with 2 years is thus

$$T_v = \frac{c_v t}{(H_{dr})^2} = \frac{(0.12 \text{ ft/yr})(2 \text{ yr})}{(10 \text{ ft})^2} = 0.60 \tag{9.12.5}$$

From a plot of consolidation ratio (U_z) versus normalized depth (z/H_{dr}), for $T_v = 0.60$ and $z/H_{dr} = 0.50$, $U_z = 0.80$. Since

$$U_z(z, t) = 1 - \frac{u(z, t)}{u(z, 0)} \tag{9.12.6}$$

it follows that

$$u(5 \text{ ft}, 2 \text{ yr}) = (1 - U_z)u(5 \text{ ft}, 0) = (1 - 0.80)u(5 \text{ ft}, 0) = 0.20u(5 \text{ ft}, 0) \tag{9.12.7}$$

The initial excess pore pressure at a depth of 5 ft (i.e., at the mid-depth of the clay layer) is

$$u(5 \text{ ft}, 0) = \gamma_w(5 \text{ ft}) = (62.4 \text{ lb/ft}^3)(5 \text{ ft}) = 312.0 \text{ lb/ft}^2 \qquad (9.12.8)$$

Thus, the excess pore pressure at a depth of 5 ft after 2 years is

$$u(5 \text{ ft}, 2 \text{ yr}) = (0.20)(312.0 \text{ lb/ft}^2) = \textbf{62.4 lb/ft}^2 \qquad (9.12.9)$$

c) The time factor associated with an average degree of consolidation $(\overline{U}_z)$ of 90% is computed from the approximation for $\overline{U}_z > 60\%$; i.e.,

$$T_{v90} = 1.781 - 0.933 \log(100 - 90\%) = 0.848 \qquad (9.12.10)$$

From the definition of the time factor,

$$T_{v90} = \frac{c_v t_{90}}{(H_{dr})^2} \Rightarrow t_{90} = \frac{T_{v90}(H_{dr})^2}{c_v} = \frac{(0.848)(10 \text{ ft})^2}{(0.12 \text{ ft/yr})} = \textbf{2.8 yr} \qquad (9.12.11)$$

EXAMPLE PROBLEM 9.13

General Remarks

This example problem illustrates the manner in which excess pore pressure and effective stresses are computed during primary consolidation.

Problem Statement

A 22 m thick normally consolidated clay layer is located below a granular fill of large areal extent that is 3.5 m thick. The moist unit weight of the fill is 17.8 kN/m^3. The fill was placed on the clay layer many years ago. Consequently the clay is assumed to have finished settling in response to the fill. A dense, though permeable, sandy gravel is found below the clay layer. The groundwater table is located at the top of the clay layer, and the submerged (buoyant) unit weight of this layer is 9.31 kN/m^3. The capillary rise in the granular fill is thought to be negligible.

Recently, due to construction activities, the fill had a load of 150 kPa applied to it over a large areal extent.

a) Determine the vertical total stress, pore pressure, and vertical effective stress at a depth of 16 m below the ground surface *before* application of the load.

b) Consolidation tests performed on 2.54 cm thick doubly drained samples of the clay indicate that $t_{50} = 10.6$ min for a load increment close to that of the actual-loaded clay layer. Determine the coefficient of consolidation (c_v) in units of square meters per year.

c) Compute the excess pore pressure, total pore pressure, and effective stress at a depth of 16 m below the ground surface $t = 10$ years after application of the load.

d) If the clay layer were *singly* drained, from the top only, at what time (in years) would a value of $T_v = 0.20$ be reached at a depth of 16 m below the ground surface? What is the effective stress in the clay layer at this depth at this time?

Solution

a) The saturated unit weight of the clay layer is computed using the results of Case 1.8 of Chapter 1; i.e.,

$$\gamma_b = \gamma_{sat} - \gamma_w \Rightarrow \gamma_{sat} = \gamma_b + \gamma_w = 9.31 + 9.81 = 19.12 \text{ kN/m}^3 \quad (9.13.1)$$

Before the load is applied, the vertical total stress at a depth of 16 m below the ground surface is thus

$$\sigma_v = \left(17.8 \text{ kN/m}^3\right)(3.5 \text{ m}) + \left(19.12 \text{ kN/m}^3\right)(16.0 - 3.5 \text{ m}) = \textbf{301.3 kPa}$$
$$(9.13.2)$$

The excess pore pressure at the same point is

$$u = \left(9.81 \text{ kN/m}^3\right)(16.0 - 3.5 \text{ m}) = \textbf{122.6 kPa} \quad (9.13.3)$$

Finally, before the load is applied, the vertical effective stress at a depth of 16 m below the ground surface is thus

$$\sigma_v' = 301.3 - 112.6 = \textbf{178.7 kPa} \quad (9.13.4)$$

b) From the laboratory results,

$$T_{50} = \frac{c_v \, t_{50}}{\left(H_{dr}\right)^2} \Rightarrow c_v = \frac{T_{50}\left(H_{dr}\right)^2}{t_{50}} \quad (9.13.5)$$

Since $U < 60\%$,

$$T_v = \frac{\pi}{4}\left(\frac{\overline{U_z}\%}{100}\right)^2 = \frac{\pi}{4}\left(\frac{50}{100}\right)^2 = \frac{\pi}{16} = 0.196 \quad (9.13.6)$$

Substituting Eq. (9.13.6) into Eq. (9.13.5) gives

$$c_v = \frac{\left(\frac{\pi}{16}\right)\left(\frac{2.54 \text{ cm}}{2} * \frac{m}{100 \text{ cm}}\right)^2}{10.5 \text{ min}} = 3.016 \times 10^{-6} \text{ m/min} \qquad (9.13.7)$$

Converting to the desired units gives

$$c_v = \left(3.016 \times 10^{-6} \text{ m/min}\right)\left(\frac{60 \text{ min}}{h}\right)\left(\frac{24 \text{ h}}{d}\right)\left(\frac{365 \text{ d}}{yr}\right) = \mathbf{1.59 \text{ m}^2/yr}$$

c) For the actual site,

$$T_v = \frac{c_v t}{(H_{dr})^2} = \frac{(1.59 \text{ m}^2/yr)(10 \text{ yr})}{(11.0 \text{ m})^2} = 0.131 \qquad (9.13.8)$$

At a depth of $16-3.5 = 12.5$ m in the clay, $z/H_{dr} = 12.5/11.0 = 1.136$. From either a figure of z/H_{dr} versus U_z, or from a series solution, $U_z = 0.118$. Noting that the initial excess pore pressure is equal to the applied fill load; i.e., $u_0 = 150.0$ kPa, the *excess* pore pressure after 10 years is thus

$$U_z(12.5 \text{ m}, 10 \text{ yr}) = 1 - \frac{u(12.5 \text{ m}, 10 \text{ yr})}{u_0}$$

$$\Rightarrow u(12.5 \text{ m}, 10 \text{ yr}) = u_0(1 - U_z) = (150.0 \text{ kPa})(1 - 0.118) = \mathbf{132.3 \text{ kPa}}$$
$$(9.13.9)$$

Recalling Eq. (9.13.3), the total pore pressure is thus

$$u = 122.6 + 132.3 = 254.9 \text{ kPa} \qquad (9.13.10)$$

Recalling Eq. (9.13.2) and including the applied fill load, the vertical total stress at a depth of 16 m below the ground surface is

$$\sigma_v = 301.3 + 150.0 = 451.3 \text{ kPa} \qquad (9.13.11)$$

The vertical effective stress at a depth of 16 m below the ground surface is thus

$$\sigma'_v = 451.3 - 254.9 = \mathbf{196.4 \text{ kPa}} \qquad (9.13.12)$$

d) For single drainage, $H_{dr} = 22.0$ m. Thus, for $T_v = 0.20$,

$$T_v = \frac{c_v t}{(H_{dr})^2} \Rightarrow t = \frac{T_v(H_{dr})^2}{c_v} = \frac{(0.20)(22.0 \text{ m})^2}{1.59 \text{ m}^2/yr} = \mathbf{60.9 \text{ yr}} \qquad (9.13.13)$$

At a depth of $16-3.5 = 12.5$ m in the clay, $z/H_{dr} = 12.5 \text{ m}/22.0 \text{ m} = 0.568$. From either a figure of z/H_{dr} versus U_z, or from a series solution, $U_z = 0.393$. Noting that the initial excess pore pressure is equal to the applied fill load; i.e., $u_0 = 150.0$ kPa, the *excess* pore pressure after 10 years is thus

$$u(z, t) = u_0(1 - U_z) = (150.0 \text{ kPa})(1 - 0.393) = 91.1 \text{ kPa} \qquad (9.13.14)$$

The total pore pressure is again the sum of the hydrostatic and excess pore pressures; i.e.,

$$u = 122.6 + 91.1 = 213.7 \text{ kPa} \qquad (9.13.15)$$

The vertical effective stress at a depth of 16 m below the ground surface is thus

$$\sigma_v' = 451.3 - 213.7 = \mathbf{237.6 \text{ kPa}} \qquad (9.13.16)$$

EXAMPLE PROBLEM 9.14

General Remarks

This example problem illustrates the manner in which excess pore pressure and primary consolidation settlements are computed in the presence of a sand seam that forms a free-draining boundary in a clay layer.

Problem Statement

Consider the soil profile shown in Figure Ex. 9.14A. The clay layer has an average initial void ratio (e_0) of 1.20 and a coefficient of consolidation (c_v) of 7.0×10^{-4} cm²/s. A surcharge load of large areal extent uniformly increases the stress within the clay layer by 75 kPa. As a result of the surcharge loading, the void ratio decreases by 0.126 at the end of primary consolidation.

a) Compute the ultimate primary consolidation settlement of the clay layer.
b) Compute the primary consolidation settlement 3 years after application of the surcharge load.
c) Compute the excess pore fluid pressure at the middle and bottom of the clay layer 4 years after application of the surcharge load.
d) Subsequent field investigations indicate the presence of a thin sand seam located 2 m above the bottom of the clay layer (Figure Ex. 9.14B). This seam serves as a free-draining boundary. Compute the total settlement of the two clay sublayers 3 years after application of the surcharge load.

Solution

a) Based on the given information, the ultimate primary consolidation settlement of the clay layer is computed as follows:

$$\Delta H = \left(\frac{\Delta e}{1 + e_0}\right) H_0 = \left(\frac{0.126}{1 + 1.20}\right)(8.0 \text{ m}) = \mathbf{0.458 \text{ m}} \qquad (9.14.1)$$

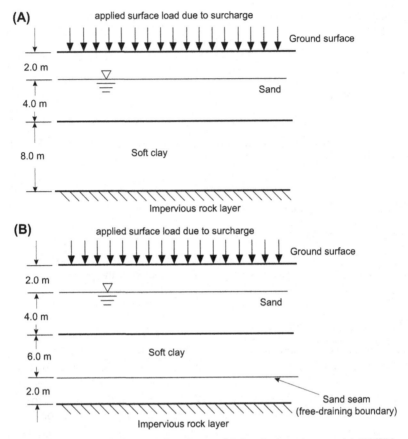

FIGURE EX. 9.14   Soil profile consisting of a consolidating clay layer (not to scale). (A) Without a sand seam and (B) with a sand seam.

b) The calculation of the primary consolidation settlement 3 years after application of the surcharge load requires the value of the time factor; i.e.,

$$T_v = \frac{c_v t}{(H_{dr})^2} \tag{9.14.2}$$

The coefficient of consolidation is first converted to units of meters per year

$$c_v = \left(7.0 \times 10^{-4} \text{ cm}^2/\text{s}\right) \left(\frac{\text{m}}{100 \text{ cm}}\right)^2 \left(\frac{60 \text{ s}}{\text{min}}\right) \left(\frac{60 \text{ min}}{\text{h}}\right) \left(\frac{24 \text{ h}}{\text{d}}\right) \left(\frac{365 \text{ d}}{\text{yr}}\right)$$

$$= 2.208 \text{ m}^2/\text{yr}$$

Substituting this value, along with the other known quantities, into Eq. (9.14.2) gives

$$T_v = \frac{c_v t}{(H_{dr})^2} = \frac{(2.208 \text{ m}^2/\text{yr})(3 \text{ yr})}{(8.0 \text{ m})^2} = 0.103 \qquad (9.14.3)$$

Assuming an average degree of consolidation ($\overline{U}_z$) less than 60%

$$T_v = \frac{\pi}{4}\left(\frac{\overline{U}_z\%}{100}\right)^2 \Rightarrow \overline{U}_z = \sqrt{\frac{4T_v}{\pi}} = \sqrt{\frac{4(0.103)}{\pi}} = 0.362 \qquad (9.14.4)$$

The value of $\overline{U}_z = 36.2\% < 60\%$ verifies the above assumption[5].

The primary consolidation settlement 3 years after application of the surcharge load is thus

$$s_c(3 \text{ yr}) = (0.362)(0.458 \text{ m}) = \mathbf{0.166 \text{ m}} \qquad (9.14.5)$$

c) To compute the excess pore fluid pressure at the middle and bottom of the clay layer 4 years after application of the surcharge load, again begin with the time factor; i.e.,

$$T_v = \frac{c_v t}{(H_{dr})^2} = \frac{(2.208 \text{ m}^2/\text{yr})(4 \text{ yr})}{(8.0 \text{ m})^2} = 0.138 \qquad (9.14.6)$$

At the middle of the clay layer, $z/H_{dr} = 4.0/8.0 = 0.50$. From either a figure of z/H_{dr} versus U_z, or from a series solution, $U_z = 0.346$. Noting that the initial excess pore pressure is equal to the applied surcharge load; i.e., $u_0 = 75.0$ kPa, the *excess* pore pressure after 4 years is thus

$$u(4.0 \text{ m}, 4 \text{ yr}) = u_0(1 - U_z) = (75.0 \text{ kPa})(1 - 0.346) = \mathbf{49.1 \text{ kPa}} \quad (9.14.7)$$

At the bottom of the clay layer, $z/H_{dr} = 8.0/8.0 = 1.00$. From either a figure of z/H_{dr} versus U_z, or from a series solution, $U_z = 0.114$. The *excess* pore pressure after 4 years is thus

$$u(4.0 \text{ m}, 4 \text{ yr}) = u_0(1 - U_z) = (75.0 \text{ kPa})(1 - 0.114) = \mathbf{66.5 \text{ kPa}} \quad (9.14.8)$$

d) The presence of a thin sand seam is present in the clay layer (Figure Ex. 9.14B) divides the layer into two sublayers. The upper sublayer is drained along both of its boundaries, thus, $H_{dr} = 6.0/2 = 3.0$ m. The lower sublayer is drained only along its top boundary, thus $H_{dr} = 2.0$ m.

Assuming the same initial void ratio and same decrease in void ratio at the end of primary consolidation, the ultimate primary consolidation settlement in the two sublayers is

---

5. This result could have likewise been obtained using a series solution (recall Eq. 9.10).

$$\Delta H_{upper} = \left(\frac{\Delta e}{1 + e_0}\right) H_{upper} = \left(\frac{0.126}{1 + 1.20}\right)(6.0 \text{ m}) = 0.344 \text{ m} \qquad (9.14.9)$$

$$\Delta H_{lower} = \left(\frac{\Delta e}{1 + e_0}\right) H_{lower} = \left(\frac{0.126}{1 + 1.20}\right)(2.0 \text{ m}) = 0.115 \text{ m} \qquad (9.14.10)$$

The coefficient of consolidation is assumed to be the *same* in the two sublayers. Three years after application of the surcharge load, the time factor for the upper sublayer is thus

$$T_{v_{upper}} = \frac{c_v t}{(H_{dr})^2} = \frac{(2.208 \text{ m}^2/\text{yr})(3 \text{ yr})}{(3.0 \text{ m})^2} = 0.736 \qquad (9.14.11)$$

Assuming an average degree of consolidation $(\overline{U}_z)$ greater than 60%,

$$T_v = 1.781 - 0.933 \log_{10}\left(100 - \overline{U}_z\%\right) \Rightarrow \overline{U}_z = 100 - 10^{\left(\frac{1.781 - 0.736}{0.933}\right)} = 86.8\%$$
$$(9.14.12)$$

The settlement of the upper sublayer after 3 years is thus

$$s_{C_{upper}} = (0.868)(0.344 \text{ m}) = 0.299 \text{ m} \qquad (9.14.13)$$

Three years after application of the surcharge load, the time factor for the lower sublayer is thus

$$T_{v_{upper}} = \frac{c_v t}{(H_{dr})^2} = \frac{(2.208 \text{ m}^2/\text{yr})(3 \text{ yr})}{(2.0 \text{ m})^2} = 1.656 \qquad (9.14.14)$$

Again assuming an average degree of consolidation $(\overline{U}_z)$ greater than 60%,

$$\overline{U}_z = 100 - 10^{\left(\frac{1.781 - 1.656}{0.933}\right)} = 98.6\% \qquad (9.14.15)$$

The settlement of the lower sublayer after 3 years is thus

$$s_{C_{lower}} = (0.986)(0.115 \text{ m}) = 0.113 \text{ m} \qquad (9.14.16)$$

The total settlement 3 years after application of the surcharge load is thus

$$s_{C_{total}} = s_{C_{upper}} + s_{C_{lower}} = 0.299 + 0.113 = \mathbf{0.412 \ m} \qquad (9.14.17)$$

To put this result into context, the total settlement associated with the sand seam is

$$\left(\frac{0.412}{0.458}\right) \times 100\% = 90\%$$

of the total settlement that would occur without the sand seam (recall the result obtained in Eq. 9.14.1).

Chapter 10

Example Problems Related to Shear Strength of Soils

10.0 GENERAL COMMENTS

The subject of shear strength of soils typically proves rather challenging for undergraduates. Indeed, most students only truly understand this subject in a second, often graduate level, course that focuses on shear strength. For this reason, the material presented in this chapter is limited to basic aspects related to shear strength of soils.

10.1 SHEAR STRENGTH OF SOILS

"Strength" typically refers to some limiting stress state that, if exceeded, will result in some type of *failure*. At failure there is typically large deformation (e.g., plastic flow, rupture, etc.).

The shear strength of a soil represents its resistance to shear stresses. It is a measure of the material's resistance to deformation by continuous displacement of its individual particles. Shear strength in soils thus depends primarily on interparticle interactions and is associated with *effective stress* (recall the discussion of Chapter 5). Shear failures occur when the stresses between the particles are such that the particles slide or roll past each other.

10.2 FACTORS CONTROLLING SHEAR STRENGTH OF SOILS

The stress—strain relationship of soils, and therefore their shearing strength, is affected by the following factors:

1. Soil composition (basic soil material)
   - mineralogy
   - grain size and grain size distribution
   - particle shape
   - pore fluid type and ion content (for cohesive soils).
2. Microstructure of the soil
   - undisturbed, disturbed, remolded

Soil Mechanics. http://dx.doi.org/10.1016/B978-0-12-804491-9.00010-0

- microfabric, i.e., the geometric arrangement of particles and voids (e.g., flocculated, dispersed, cemented, etc.).
3. Initial state of the soil defined by
   - initial void ratio
   - initial effective stress and shear stress state
   - overconsolidation ratio.
4. Loading history
   - stress path
   - type of loading (e.g., static, dynamic, monotonic, cyclic)
   - history of load application.

10.3 VOLUME CHANGE CHARACTERISTICS

For engineering materials such as steel, wood, polymers, etc., volume changes typically do *not* influence yielding and strength (i.e., failure).

In geomaterials (i.e., soil and rock), applied stresses induce changes in volume. For example, loose sands and normally consolidated clays will *contract* (i.e., undergo a reduction in void ratio) when sheared. Dense sands and overconsolidated clays will primarily dilate (i.e., undergo a increase in void ratio) when sheared. Strength and volume change are thus *interrelated* for geomaterials.

10.4 IMPORTANCE OF SHEAR STRENGTH OF GEOMATERIALS

The safety of any geostructure is dependent on the strength of the geomaterial. If the material fails, the structure founded on it can become overstressed and collapse. An understanding of shear strength is thus fundamental to analyzing different classes of soil stability problems such as (1) the lateral earth pressure acting on retaining walls, (2) the stability of slopes, (3) the bearing capacity of footings, etc.

10.5 MOHR'S FAILURE CRITERION

The principal stresses associated with failure in a soil, as well as the angle of the plane in which the failure will occur, are commonly determined using Mohr's circles such as those shown in Figure 10.1 (Section 4.2.5.4 gives additional details pertaining to Mohr's circles of stress[1]). Mohr used his well-known graphical representation of stress at a point to devise a strength theory that could be adapted to various stress conditions and thus be brought into better agreement with experimental observations. Mohr assumed that of all the

---

1. Named in honor of Otto Mohr (1835−1918).

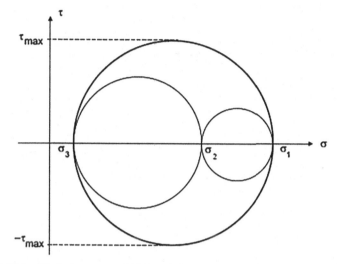

FIGURE 10.1   Principal stresses and Mohr's circles associated with a three-dimensional stress state.

planes having the same magnitude of normal stress the weakest one, on which failure is most likely to occur, is that with the *maximum* shear stress.

The failure is thus by shear, but the critical shear stress is governed by the normal effective stress acting on the potential failure plane. It is thus necessary to consider only the Mohr's circle that corresponds to the maximum difference in principal stress (i.e., the difference between the major and minor principal stress). If there are a sufficient number of such circles, an envelope of failure points can be drawn (Figure 10.2).

Stated in another way, Mohr's criterion states that failure will occur along that plane for which the ratio of shear stress to normal stress reaches a *critical limiting value*, i.e.,

$$\tau_{ff} = f(\sigma'_{ff}) \tag{10.1}$$

where τ_{ff} is the shear strength of the material and σ'_{ff} is the normal effective stress on the failure plane (Figure 10.3).

For any stress states for which there is no experimental data, it may be assumed with sufficient accuracy that the limiting circle will also touch this failure envelope. Although the failure envelopes for soils and rock are generally curved, for simplicity they are commonly replaced by a *straight* line. The most common straight line assumption is based on Coulomb's friction hypothesis. When combined with Mohr's failure criterion, it gives the so-called *Mohr–Coulomb* failure criterion.

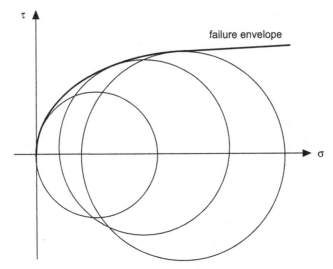

FIGURE 10.2 Schematic illustration of a failure envelope constructed using Mohr's circles.

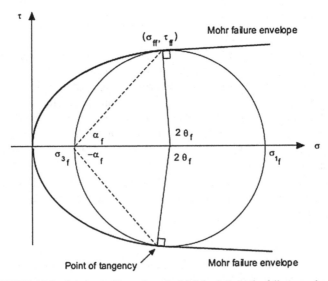

FIGURE 10.3 Schematic illustration of a Mohr's circle at the failure envelope.

10.6 MOHR–COULOMB FAILURE CRITERION

Coulomb's friction hypothesis[2], which states that a *linear* relationship exists between the shear stress at failure and the associated with normal stress,

---

2. C.A. Coulomb (1736–1806).

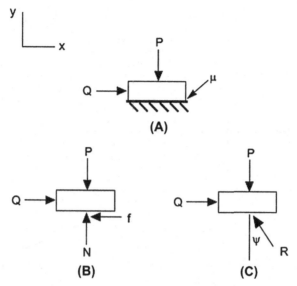

FIGURE 10.4   Schematic illustration of Coulomb's friction law.

represents the simplest approximation of a failure envelope. To better understand this approximation, briefly recall the Coulomb's friction law.

Figure 10.4A shows a block resting on a rough surface. This roughness is characterized by a coefficient of static friction μ.

Using the free-body diagram shown in Figure 10.4B, the following equations of equilibrium are written:

$$\sum F_x = Q - f = 0 \Rightarrow f = Q \tag{10.2}$$

$$\sum F_y = N - P = 0 \Rightarrow N = P \tag{10.3}$$

Then, at the point of sliding (i.e., failure) along the interface, $Q = f = \mu N = \mu P$.

The forces N and f can also be replaced by a resultant force R and an angle of "obliquity" ψ. From the free-body diagram shown in Figure 10.4C, the following equations of equilibrium are written:

$$\sum F_x = Q - R \sin \psi = 0 \Rightarrow Q = R \sin \psi \tag{10.4}$$

$$\sum F_y = R \cos \psi - P = 0 \Rightarrow R = \frac{P}{\cos \psi} \tag{10.5}$$

Thus,

$$Q = R \sin \psi = \left(\frac{P}{\cos \psi}\right) \sin \psi = P \tan \psi \tag{10.6}$$

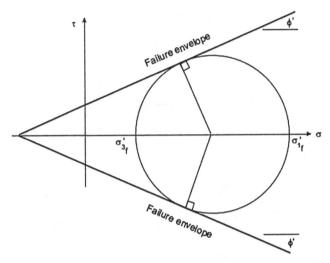

FIGURE 10.5 Schematic illustration of the Mohr–Coulomb failure criterion.

Then, at the point of sliding (i.e., failure) along the interface, $\psi = \phi$, where ϕ is angle of internal friction. Thus, $Q = P \tan \phi$, indicating that when the friction is mobilized along the interface, $\mu = \tan \phi$.

Due to its simplicity, the Coulomb's hypothesis is commonly used in conjunction with Mohr's graphical representation to determine the combination of shear and normal stress that will cause a failure of the material. Figure 10.5 shows the so-called Mohr–Coulomb failure criterion.

Stated mathematically, the Mohr–Coulomb failure criterion is written as

$$\tau_{ff} = \sigma'_{ff} \tan \phi' + c' \tag{10.7}$$

where τ_{ff} and σ'_{ff} are as defined earlier, c' is the "strength parameter" or "cohesion intercept" (units of FL^{-2}), and ϕ' represents the effective angle of internal friction.

10.6.1 Obliquity Relations

It is typically advantageous to write the Mohr–Coulomb failure criterion in terms of principal stresses. Consider the Mohr's circle at failure shown in Figure 10.6.

For the triangle having OC as its hypotenuse in Figure 10.6,

$$\sin \phi' = \frac{R}{c' \cot \phi' + \frac{1}{2}\left(\sigma'_{1_f} + \sigma'_{3_f}\right)} = \frac{\frac{1}{2}\left(\sigma'_{1_f} - \sigma'_{3_f}\right)}{c' \cot \phi' + \frac{1}{2}\left(\sigma'_{1_f} + \sigma'_{3_f}\right)} \tag{10.8}$$

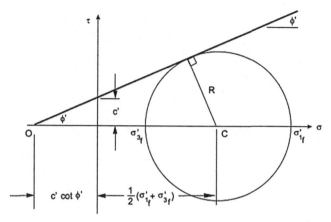

FIGURE 10.6 Mohr–Coulomb failure criterion.

or

$$\frac{1}{2}\left(\sigma'_{1_f} - \sigma'_{3_f}\right) = c' \cos \phi' + \frac{1}{2}\left(\sigma'_{1_f} + \sigma'_{3_f}\right)\sin \phi' \qquad (10.9)$$

Eq. (10.9) is commonly rewritten as

$$\sigma'_{1_f}(1 - \sin \phi') = \sigma'_{3_f}(1 + \sin \phi') + 2c' \cos \phi' \qquad (10.10)$$

or

$$\sigma'_{1_f} = \sigma'_{3_f}\left(\frac{1 + \sin \phi'}{1 - \sin \phi'}\right) + 2c'\left(\frac{\cos \phi'}{1 - \sin \phi'}\right) \qquad (10.11)$$

If $c' = 0$, then Eq. (10.11) is typically rewritten as

$$\frac{\sigma'_{1_f}}{\sigma'_{3_f}} = \frac{1 + \sin \phi'}{1 - \sin \phi'} \quad \text{or} \quad \frac{\sigma'_{3_f}}{\sigma'_{1_f}} = \frac{1 - \sin \phi'}{1 + \sin \phi'} \qquad (10.12)$$

Eq. (10.12) is commonly rewritten in terms of the angle of internal friction as

$$\sin \phi' = \frac{\sigma'_{1_f} - \sigma'_{3_f}}{\sigma'_{1_f} + \sigma'_{3_f}} \qquad (10.13)$$

Using the trigonometric identities

$$\frac{1 + \sin \alpha}{1 - \sin \alpha} = \tan^2\left(\frac{\pi}{4} + \frac{\alpha}{2}\right); \quad \frac{\cos \alpha}{1 - \sin \alpha} = \tan\left(\frac{\pi}{4} + \frac{\alpha}{2}\right) \qquad (10.14)$$

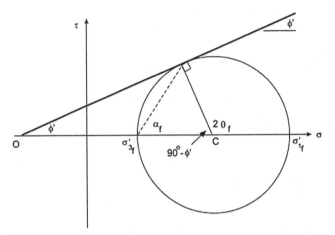

FIGURE 10.7 Mohr–Coulomb failure criterion: relation among various angles.

Eq. (10.11) is also written as

$$\sigma'_{1_f} = \sigma'_{3_f} \tan^2\left(\frac{\pi}{4} + \frac{\phi'}{2}\right) + 2c' \tan\left(\frac{\pi}{4} + \frac{\phi'}{2}\right) \tag{10.15}$$

where ϕ' must be in radians. If $c' = 0$, then Eq. (10.15) reduces to

$$\frac{\sigma'_{1_f}}{\sigma'_{3_f}} = \tan^2\left(\frac{\pi}{4} + \frac{\phi'}{2}\right) \quad \text{or} \quad \frac{\sigma'_{3_f}}{\sigma'_{1_f}} = \tan^2\left(\frac{\pi}{4} - \frac{\phi'}{2}\right) \tag{10.16}$$

To relate the orientation of the failure plane to the effective angle of internal friction ϕ', consider Figure 10.7.

From the horizontal line OC,

$$2\theta_f + (90 - \phi') = 180° \Rightarrow \theta_f = 45° + \frac{\phi'}{2} \tag{10.17}$$

where ϕ' is measured in degrees. If ϕ' is measured in radians, then

$$\theta_f = \frac{\pi}{4} + \frac{\phi'}{2} \tag{10.18}$$

Next consider the angles shown in Figure 10.8.

The triangle from Figure 10.8 is redrawn in Figure 10.9.

The center and radius of the Mohr's circle are given by Eqs. (4.32) and (4.38), respectively, i.e.,

$$C = \frac{1}{2}(\sigma_x + \sigma_y); \quad R = \sqrt{\left[\frac{1}{2}(\sigma_x - \sigma_y)\right]^2 + (\tau_{xy})^2} \tag{10.19}$$

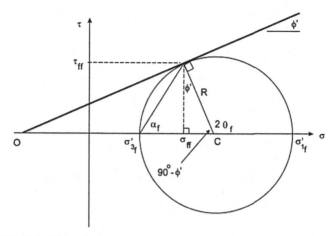

FIGURE 10.8   Mohr–Coulomb failure criterion: additional insight into relevant angles.

FIGURE 10.9   Expanded view of triangle from Figure 10.8 in terms of angles.

From Figure 10.8, it is evident that $d = C - \sigma'_{3_f} = R$. Using the *law of sines* for the triangle shown in Figure 10.9 gives

$$\frac{\sin \alpha_f}{R} = \frac{\sin \gamma}{d} = \frac{\sin \gamma}{R} \Rightarrow \gamma = \alpha_f \qquad (10.20)$$

Noting that, for the triangle shown in Figure 10.9, $2\alpha_f + (90° - \phi') = 180°$, it follows that

$$\alpha_f = 45° + \frac{\phi'}{2} = \theta_f \qquad (10.21)$$

Finally, consider the triangle shown in Figure 10.10.

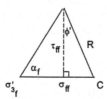

FIGURE 10.10   Expanded view of triangle from Figure 10.8 in terms of stresses.

For the triangle having R as its hypotenuse,

$$\sin \phi' = \frac{C - \sigma'_{ff}}{R} \Rightarrow \sigma'_{ff} = C - R \sin \phi' \qquad (10.22)$$

Substituting for C and R in terms of principal stress at failure gives

$$\sigma'_{ff} = \frac{1}{2}\left(\sigma'_{1_f} + \sigma'_{3_f}\right) - \frac{1}{2}\left(\sigma'_{1_f} - \sigma'_{3_f}\right)\sin \phi' \qquad (10.23)$$

For the same triangle,

$$\tan \phi' = \frac{C - \sigma'_{ff}}{\tau_{ff}} = \frac{R \sin \phi'}{\tau_{ff}} \Rightarrow \tau_{ff} = \frac{R \sin \phi'}{\tan \phi'} = R \cos \phi' \qquad (10.24)$$

Next consider the triangle containing the angle α_f. Here

$$\tan \alpha_f = \frac{\tau_{ff}}{\sigma'_{ff} - \sigma'_{3_f}} = \frac{R \cos \phi'}{(C - R \sin \phi') - \sigma'_{3_f}} \qquad (10.25)$$

where Eq. (10.22) has been used.

Writing Eq. (10.19) for C in terms of principal stresses at failure and noting that on the σ-axis

$$C = \frac{1}{2}\left(\sigma'_{1_f} - \sigma'_{3_f}\right) \qquad (10.26)$$

Eq. (10.25) becomes

$$\tan \alpha_f = \frac{\cos \phi'}{1 - \sin \phi'} \qquad (10.27)$$

EXAMPLE PROBLEM 10.1

General Remarks

This example problem illustrates the manner in which strength quantities are computed from test data.

Problem Statement

A series of drained triaxial compression tests on a normally consolidated clay indicated that $\phi' = 28°$ and $c' = 0$ for this soil. Determine the deviator stress at failure for a sample that was consolidated to 49 psi prior to shearing.

Solution

In a triaxial compression test the minor principal stress is maintained at the consolidation stress. Thus, $\sigma'_3 = 49$ psi at failure.

As discussed in Section 10.6.1, for the case of $c' = 0$, the major and minor principal stresses at failure are related according to Eq. (10.15), i.e.,

$$\sigma'_{1_f} = \sigma'_{3_f} \tan^2\left(45° + \frac{\phi'}{2}\right) \tag{10.1.1}$$

Substituting the known values gives

$$\sigma'_{1_f} = (49 \text{ psi})\tan^2\left(45° + \frac{28°}{2}\right) = 135.7 \text{ psi} \tag{10.1.2}$$

The deviator stress at failure is thus

$$\sigma'_{1_f} - \sigma'_{3_f} = 135.7 - 49.0 = \mathbf{86.7 \ psi} \tag{10.1.3}$$

EXAMPLE PROBLEM 10.2

General Remarks

This example problem illustrates the manner in which strength quantities are computed from test data.

Problem Statement

A sample of sand subjected to axisymmetric triaxial compression failed when the major and minor principal stresses were 11,600 and 3300 psf, respectively. Determine the angle of internal friction and the normal and shear stresses on the failure plane.

Solution

Assuming $c' = 0$ for the sand, then Eq. (10.13) gives

$$\sin \phi' = \frac{\sigma'_{1_f} - \sigma'_{3_f}}{\sigma'_{1_f} + \sigma'_{3_f}} = \frac{11,600 - 3300}{11,600 + 3300} = 0.557 \tag{10.2.1}$$

Thus,

$$\phi' = \sin^{-1}(0.557) = \mathbf{33.9°} \tag{10.2.2}$$

From Section 10.6.1, the normal stress acting on the failure plane is given by Eq. (10.22), i.e.,

$$\sigma'_{ff} = C - R \sin \phi' = \frac{1}{2}\left(\sigma'_{1_f} + \sigma'_{3_f}\right) - \frac{1}{2}\left(\sigma'_{1_f} - \sigma'_{3_f}\right)\sin \phi'$$

$$= \frac{1}{2}(11,600 + 3300) - \frac{1}{2}(11,600 - 3300)(0.557) = \mathbf{5138 \ psf} \tag{10.2.3}$$

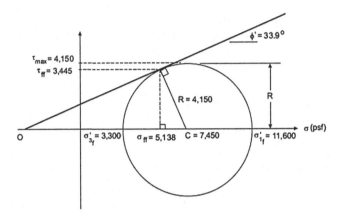

FIGURE EX. 10.2 Mohr's circle associated with the problem (not to scale).

Finally, the shear stress acting on the failure plane is given by Eq. (10.24), where

$$R = \sqrt{\left[\frac{1}{2}(\sigma_x - \sigma_y)\right]^2 + (\tau_{xy})^2} = \sqrt{\left[\frac{1}{2}(11,600 - 3300)\right]^2 + (0.0)^2} = 4150 \text{ psf}$$

$$(10.2.4)$$

Thus,

$$\tau_{ff} = R \cos \phi' = (4150 \text{ psf}) \cos(33.9°) = \textbf{3445 psf} \qquad (10.2.5)$$

Figure Ex. 10.2 shows the Mohr's circle associated with this problem. As evident from Figure Ex. 10.2, the shear stress acting on the failure plane (τ_{ff}) is less than the maximum in-plane shear stress $\tau_{max}(=R)$.

EXAMPLE PROBLEM 10.3

General Remarks

This example problem illustrates the manner in which strength quantities are computed from test data.

Problem Statement

A normally consolidated clay sample has an undrained shear strength equal to 1.00 kg/cm². In the laboratory the clay is found to have an angle of internal friction equal to 30 degrees and a cohesion intercept of zero. If a sample of this clay fails under undrained conditions, determine the effective principal stresses at failure.

Solution

The known undrained shear strength of the sample is written as follows:

$$\frac{1}{2}\left(\sigma'_{1_f} - \sigma'_{3_f}\right) = 1.00 \text{ kg/cm}^2 \qquad (10.3.1)$$

Since the cohesion intercept is zero, the principal stresses at failure are related by Eq. (10.11), i.e.,

$$\sigma'_{1_f} = \sigma'_{3_f}\left(\frac{1 + \sin \phi'}{1 - \sin \phi'}\right) = \sigma'_{3_f}\left(\frac{1 + \sin 30°}{1 - \sin 30°}\right) = 3.0\sigma'_{3_f} \qquad (10.3.2)$$

Substituting for the major principal stress, the undrained shear strength is rewritten as

$$\sigma'_{1_f} - \sigma'_{3_f} = [3.0 - 1]\sigma'_{3_f} = 2\left(1.00 \text{ kg/cm}^2\right) \Rightarrow \sigma'_{3_f} = \textbf{1.00 kg/cm}^2 \quad (10.3.3)$$

The major principal stress is thus

$$\sigma'_{1_f} = 2.00 + \sigma'_{3_f} = \textbf{3.00 kg/cm}^2 \qquad (10.3.4)$$

EXAMPLE PROBLEM 10.4

General Remarks

This example problem illustrates the manner in which strength quantities are computed from test data.

Problem Statement

The major and minor principal stresses measured in a soil element are equal to 200 and 100 kPa, respectively. The soil is a dry sand with an internal friction angle of 32 degrees and a cohesion intercept of zero. When the soil will be loaded, additional compressive stress increments of $\Delta\sigma$ and $\Delta\sigma/4$ will be superimposed on σ_1 and σ_3, respectively.

Determine the maximum possible value of $\Delta\sigma$ such that the shear strength will not be exceeded in the soil element.

Solution

Since $c' = 0$ for the sand, then at failure Eq. (10.11) gives

$$\sigma'_{1_f} = \sigma'_{3_f}\left(\frac{1 + \sin \phi'}{1 - \sin \phi'}\right) = \sigma'_{3_f}\left(\frac{1 + \sin 32°}{1 - \sin 32°}\right) = 3.255 \, \sigma'_{3_f} \qquad (10.4.1)$$

For the soil element in question,

$$\sigma'_{1_f} = (200 + \Delta\sigma); \quad \sigma'_{3_f} = \left(100 + \frac{\Delta\sigma}{4}\right) \qquad (10.4.2)$$

Thus,

$$(200 + \Delta\sigma) = 3.255\left(100 + \frac{\Delta\sigma}{4}\right) \qquad (10.4.3)$$

Solving for the stress increment gives

$$\left(1 - \frac{3.255}{4}\right)\Delta\sigma = (3.255)(100) - 200 \Rightarrow \Delta\sigma = \textbf{673.8 kPa} \qquad (10.4.4)$$

EXAMPLE PROBLEM 10.5

General Remarks

This example problem illustrates the manner in which strength quantities are computed for field data.

Problem Statement

The groundwater table in the deposit of fine sand shown in Figure Ex. 10.5 is located at a depth of 15 ft. The sand is, however, saturated beginning at a depth of 8 ft. The sand has a uniform dry unit weight of 100 lb/ft³, a specific gravity of solids (G_s) of 2.70, and a moisture content (w) of 12% for the depth range between 0 and 8 ft.

If the angle of internal friction for the sand is 31 degrees, what is the shear strength at a depth of 10 ft?

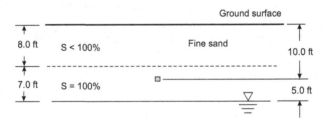

FIGURE EX. 10.5   Deposit of fine sand with capillary rise.

Solution

The requisite unit weights are first computed. In the upper 8 ft of the sand deposit,

$$\gamma = \gamma_d(1 + w) = (100 \text{ lb/ft}^3)(1 + 0.12) = 112.0 \text{ lb/ft}^3 \qquad (10.5.1)$$

The void ratio is computed using Case 1.4 of Chapter 1, i.e.,

$$\gamma_d = \frac{G_s \gamma_w}{1 + e} \Rightarrow e = \frac{G_s \gamma_w}{\gamma_d} - 1 \qquad (10.5.2)$$

Assuming that the void ratio is the same throughout the upper 10 ft of the sand layer gives

$$e = \frac{G_s \gamma_w}{\gamma_d} - 1 = \frac{(2.70)(62.4 \text{ lb/ft}^3)}{100 \text{ lb/ft}^3} - 1 = 0.685 \qquad (10.5.3)$$

The saturated unit weight is then computed using the expression developed in Case 1.8 of Chapter 1, i.e.,

$$\gamma_{sat} = \frac{\gamma_w(G_s + e)}{1 + e} = \frac{(62.4 \text{ lb/ft}^3)(2.70 + 0.685)}{1 + 0.685} = 125.4 \text{ lb/ft}^3 \qquad (10.5.4)$$

The vertical total stress at a depth of 10 ft is thus

$$\sigma_v = (112.0 \text{ lb/ft}^3)(8 \text{ ft}) + (125.4 \text{ lb/ft}^3)(2 \text{ ft}) = 1146.8 \text{ lb/ft}^2 \qquad (10.5.5)$$

The pore pressure at a depth of 10 ft is due to capillary rise in the 5 ft above the groundwater table, so

$$u = -(62.4 \text{ lb/ft}^3)(5 \text{ ft}) = -312.0 \text{ lb/ft}^2 \qquad (10.5.6)$$

Finally, the vertical effective stress at a depth of 10 ft is

$$\sigma_v' = \sigma_v - u = 1146.8 - (-312.0) = 1458.8 \text{ lb/ft}^2 \qquad (10.5.7)$$

The shear strength at a depth of 10 ft is thus

$$\tau = \sigma_v' \tan \phi' = (1458.8 \text{ lb/ft}^2)\tan 31° = \mathbf{876.5 \text{ lb/ft}^2} \qquad (10.5.8)$$

EXAMPLE PROBLEM 10.6

General Remarks

This example problem illustrates the manner in which strength quantities are computed from data obtained from some standard laboratory experiments.

Problem Statement

A sand sample is tested in a direct shear device. The vertical normal stress on the sample is 300 kPa. The horizontal shear stress at failure is equal to 210 kPa.

a) Compute the angle of internal friction for the sand and the magnitude and direction of the principal stresses at failure.
b) During a consolidated–drained triaxial test, a sample of the same sand failed at a principal stress difference of 130 kPa. Determine the magnitude of the initial confining stress, which remains constant in the test, and of the major principal stress at failure.
c) The sand considered earlier is to be used as backfill material behind a retaining wall. The stress state behind the wall consists of a vertical normal stress of 200 kPa and a horizontal normal stress of 80 kPa. Is this condition safe from failure?
d) The horizontal normal stress behind the wall mentioned in part (c) reduces as a result of movement of the wall. The vertical normal stress remains unchanged. Compute the value of the horizontal normal stress that will cause failure.

Solution

a) For sands the cohesion intercept (c) is zero. Figure Ex. 10.6A shows the Mohr's circle associated with the problem. Since $c = 0$, the value of the friction angle is computed as follows:

$$\tan \phi' = \frac{\tau_{ff}}{\sigma_{ff}} = \frac{210}{300} = 0.700 \Rightarrow \phi' = \textbf{35.0}° \qquad (10.6.1)$$

The radius of the Mohr's circle is computed from Eq. (10.24), i.e.,

$$\tau_{ff} = R \cos \phi' \Rightarrow R = \frac{\tau_{ff}}{\cos \phi'} = \frac{210 \text{ kPa}}{\cos 35°} = 256.4 \text{ kPa} \qquad (10.6.2)$$

The center of the Mohr's circle is next computed from Eq. (10.22), i.e.,

$$\sigma'_{ff} = C - R \sin \phi' \Rightarrow C = \sigma'_{ff} + R \sin \phi' = (300 \text{ kPa})$$
$$+ (256.4 \text{ kPa})\sin 35° = 447.0 \text{ kPa} \qquad (10.6.3)$$

The principal stresses are then computed using Eq. (4.39), i.e.,

$$\sigma_1 = C + R = 447.0 + 256.4 = \textbf{703.4 kPa} \qquad (10.6.4)$$

$$\sigma_2 = C - R = 447.0 - 256.4 = \textbf{190.6 kPa} \qquad (10.6.5)$$

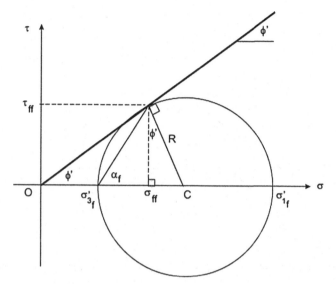

FIGURE EX. 10.6A   General Mohr's circle associated with part (a).

The angle between the stress point (300.0, 210.0) and the minor principal stress direction is computed as follows:

$$\tan 2\theta_{p_2} = \frac{210.0}{447.0 - 300.0} = 1.429 \Rightarrow 2\theta_{p_2} = 55.0° \Rightarrow \theta_{p_2} = 27.5° \quad (10.6.6)$$

The angle between the stress point (300.0, 210.0) and the major principal stress direction is thus

$$2\theta_{p_1} = 180° - 2\theta_{p_2} = 180° - 55.0° \Rightarrow \theta_{p_1} = 62.5° \quad (10.6.7)$$

It is timely to note that the magnitude of the maximum in-plane shear stress is given by Eq. (4.40)

$$|\tau_{max}| = R = 256.4 \text{ kPa} \quad (10.6.8)$$

which is greater than τ_{ff}. For completeness, the angle that the failure plane makes with the horizontal is computed using Eq. (10.21), i.e.,

$$\alpha_f = 45° + \frac{\phi'}{2} = 45° + \frac{35.0}{2} = 62.5° \quad (10.6.9)$$

This value could likewise have been computed using Eq. (10.27)

$$\tan \alpha_f = \frac{\cos \phi'}{1 - \sin \phi'} = \frac{\cos 35°}{1 - \sin 35°} = 62.5° \quad (10.6.10)$$

Figure Ex. 10.6B shows the detailed Mohr's circle associated with part (a) of this problem.

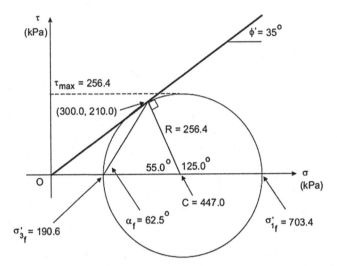

FIGURE EX. 10.6B Detailed Mohr's circle associated with part (a) (not to scale).

The same results could likewise be obtained using the pole method. Figure Ex. 10.6C shows the Mohr's circle associated with such a solution.

First the point (300.0, 210.0) is located, to scale, in the σ-τ plane. Since the cohesion intercept is zero, the Mohr–Coulomb failure line is passed through this point and the origin. The angle that this line makes with the σ-axis, which represents the friction angle, is then measured.

A line perpendicular to the failure line at the point (300.0, 210.0) is next drawn. Its intersection with the σ-axis locates the center (C) of the

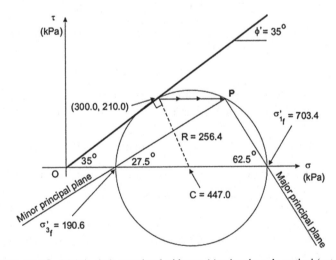

FIGURE EX. 10.6C Mohr's circle associated with part (a) using the pole method (not to scale).

circle (447.0, 0.0). The radius of the circle (R) is next measured to be 256.4 kPa.

The Mohr's circle is next drawn. Its intersection with the σ-axis gives the major and minor principal stresses (703.4 and 190.6 kPa, respectively).

Since the original failure stress state (300.0, 210.0) was determined from a direct shear test, these stresses act on a *horizontal* plane of the soil sample. A horizontal line is thus drawn from the point (300.0, 210.0) until it intersects the circle at point P. This point is the *pole* of the Mohr's circle of stress. The major and minor principal stresses act normal to the planes found by drawing lines from point P through the major and minor principal stress values.

b) From the given information,

$$\sigma'_{1_f} - \sigma'_{3_f} = 130 \text{ kPa} \tag{10.6.11}$$

A second equation is necessary to determine the magnitude of the initial confining stress. Using Eq. (10.13) gives

$$\sin \phi' = \frac{\sigma'_{1_f} - \sigma'_{3_f}}{\sigma'_{1_f} + \sigma'_{3_f}} = \frac{130 \text{ kPa}}{\sigma'_{1_f} + \sigma'_{3_f}} = \sin 35° \Rightarrow \sigma'_{1_f} + \sigma'_{3_f} = \frac{130 \text{ kPa}}{\sin 35°} \tag{10.6.12}$$

Subtracting Eq. (10.6.9) from Eq. (10.6.10) gives

$$2\sigma'_{3_f} = \frac{130 \text{ kPa}}{\sin 35°} - 130 \text{ kPa} = 96.6 \text{ kPa} \Rightarrow \sigma'_{3_f} = \textbf{48.3 kPa} \tag{10.6.13}$$

For completeness, from Eq. (10.6.9),

$$\sigma'_{1_f} = 130 \text{ kPa} + \sigma'_{3_f} = 130 + 48.3 = \textbf{178.3 kPa} \tag{10.6.14}$$

Figure Ex. 10.6D shows the detailed Mohr's circle associated with part (b) of this problem. If plotted to scale, the Mohr's circle could have been used directly to solve the problem. This solution involves drawing the circle so it just touches the failure line, inclined at 35 degrees. The intersection of the circle with the σ-axis gives the two principal stresses.

c) In solving this part of the problem, it is important to note that the vertical and horizontal normal stresses are principal stresses (i.e., the shear stress is zero). Thus,

$$\sigma_v \equiv \sigma_1 = 200 \text{ kPa}; \quad \sigma_h \equiv \sigma_3 = 80 \text{ kPa} \tag{10.6.15}$$

The next step in the solution is to construct the Mohr's circle associated with the given stress state. The σ-coordinate of the center of the circle is computed using Eq. (4.32), i.e.,

$$C = \frac{1}{2}(\sigma_1 + \sigma_3) = \frac{1}{2}(200 + 80) = 140.0 \text{ kPa} \tag{10.6.16}$$

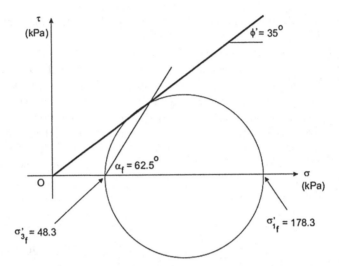

FIGURE EX. 10.6D Detailed Mohr's circle associated with part (b) (not to scale).

The radius of the Mohr's circle is next computed using Eq. (4.32), i.e.,

$$R = \sqrt{\left[\frac{(\sigma_1 - \sigma_3)}{2}\right]^2 + (\tau_{xy})^2} = \sqrt{\left[\frac{(200 - 80)}{2}\right]^2 + (\tau_{xy})^2} = 60.0 \text{ kPa}$$

(10.6.17)

Using the aforesaid values of C and R, the normal and shear stresses on the failure plane are next computed using Eqs. (10.22) and (10.24). From the former equation,

$$\sigma_f' = C - R \sin \phi' = 140.0 - (60.0)\sin 35° = 105.6 \text{ kPa}$$ (10.6.18)

where only a single "f" is used as the subscript because the normal stress computed earlier is not necessarily associated with a failure state. From Eq. (10.24),

$$\tau_f = R \cos \phi' = (60.0)\cos 35° = 49.2 \text{ kPa}$$ (10.6.19)

For the same value of normal stress, the failure would occur at a shear stress of

$$\tau_{ff} = \sigma_f \tan \phi' = (105.6 \text{ kPa})\tan 35° = 73.9 \text{ kPa}$$ (10.6.20)

Since $49.2 < 73.9$ kPa, the soil behind the wall will *not* be at a failure state. Figure Ex. 10.6E shows the Mohr's circle associated with this part of the problem.

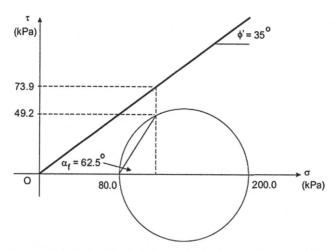

FIGURE EX. 10.6E Mohr's circle associated with part (c) (not to scale).

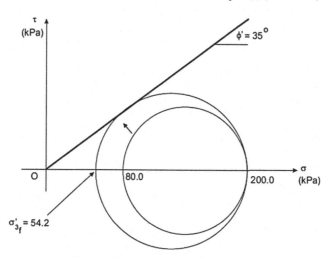

FIGURE EX. 10.6F Detailed Mohr's circle associated with part (d) (not to scale).

d) If the vertical normal stress remains unchanged, the horizontal normal stress behind the wall is computed using the second of Eq. (10.16); i.e.,

$$\frac{\sigma'_{3_f}}{\sigma'_{1_f}} = \tan^2\left(\frac{\pi}{4} - \frac{\phi'}{2}\right) \Rightarrow \sigma'_{3_f} = \sigma'_{1_f}\tan^2\left(\frac{\pi}{4} - \frac{\phi'}{2}\right)$$

$$= (200.0 \text{ kPa})\tan^2\left(45° - \frac{35°}{2}\right) = \mathbf{54.2 \text{ kPa}}$$

(10.6.21)

Figure Ex. 10.6F shows the Mohr's circle associated with this part of the problem.

Index

'*Note*: Page numbers followed by "f" indicate figures and "t" indicate tables.'

Printed in the United States
By Bookmasters